Electron and Ion
Spectroscopy of Solids

NATO ADVANCED STUDY INSTITUTES SERIES

A series of edited volumes comprising multifaceted studies of contemporary scientific issues by some of the best scientific minds in the world, assembled in cooperation with NATO Scientific Affairs Division.

Series B: Physics

RECENT VOLUMES IN THIS SERIES

The series is published by an international board of publishers in conjunction with NATO Scientific Affairs Division

A	Life Sciences	Plenum Publishing Corporation
B	Physics	New York and London
C	Mathematical and Physical Sciences	D. Reidel Publishing Company Dordrecht and Boston
D	Behavioral and Social Sciences	Sijthoff International Publishing Company Leiden
E	Applied Sciences	Noordhoff International Publishing Leiden

Electron and Ion Spectroscopy of Solids

Edited by
L. Fiermans, J. Vennik, and W. Dekeyser
Laboratory for Crystallography and the Study of Solids
State University Ghent
Ghent, Belgium

PLENUM PRESS • NEW YORK AND LONDON
Published in cooperation with NATO Scientific Affairs Division

Library of Congress Cataloging in Publication Data

Main entry under title:
Electron and ion spectroscopy of solids.

   (NATO advanced study institutes series: Series B, Physics; v. 32)
   "Based on lectures presented at the NATO Advanced Study Institute held in Ghent,
Belgium, August 29–September 9, 1977."
   Includes bibliographies and indexes.
   1. Electron spectroscopy. 2. Ions–Spectra. 3. Solids–Spectra. I. Fiermans, Lucien.
II. Vennik, J. III. Dekeyser, Willy. IV. NATO Advanced Study Institute, Ghent,
Belgium, 1977. V. Series.
QC454.E4E4   539.7'2112   78-6171
ISBN 978-1-4684-2819-3   ISBN 978-1-4684-2817-9 (eBook)
DOI 10.1007/978-1-4684-2817-9

Based on the lectures presented at the NATO Advanced Study Institute held in
Ghent, Belgium, August 29–September 9, 1977

©1978 Plenum Press, New York
Softcover reprint of the hardcover 1st edition 1978
A Division of Plenum Publishing Corporation
227 West 17th Street, New York, N.Y. 10011

Preface

Surface physics and chemistry have in recent years become one of the most active fields in solid state research. A number of techniques have been developed, and both the experimental aspect and the correlated theory are evolving at an extremely fast rate. Electron and ion spectroscopy are of major importance in this development.

In this volume, which contains edited and extended versions of eight sets of lectures given at the NATO Advanced Study Institute held at Ghent, Belgium, from August 29 to September 9, 1977, a review of the state of the art in these fields is given from both an experimental and a theoretical point of view. Electron emission techniques such as UPS (ultraviolet photoemission spectroscopy), XPS (x-ray photoemission spectroscopy), and AES (Auger electron spectroscopy) constitute the major part of this volume, reflecting the fact that they continue to be the most widely applied surface techniques.

Recent developments in the application of synchrotron radiation to angle-resolved photoelectron spectroscopy are extensively covered, from an experimental point of view by Prof. W. E. Spicer (Stanford University, U.S.A.) and from a theoretical point of view by Dr. A. Liebsch (Kernforschungsanlage Jülich, Germany). Emphasis is put on the study of energy bands in layered structures, and on chemisorption on well-defined surfaces. Chemisorption and catalysis on metals is treated in detail by Prof. G. Ertl (Universität München, Germany). This chapter contains a review of the application of the different surface techniques to specific surface systems.

Fundamental problems involved in the interpretation of XPS and AES spectra are discussed in chapters by Dr. G. K. Wertheim (Bell Laboratories, U.S.A.) and Dr. T. E. Gallon (University of York, U.K.), respectively. A general introduction to the theory of photoemission is given in a chapter by Prof. G. D. Mahan (Indiana University, U.S.A.). The chapter by Dr. H. D. Hagstrum (Bell Laboratories, U.S.A.) in some respects can be considered to deal with the transi-

tion from electron- and photon-stimulated electron emission techniques to techniques where ions are used as primary beams. In this
chapter Dr. Hagstrum comprehensively reviews the results obtained
with INS (ion-neutralization spectroscopy) in the study of chemisorption. A comparison with UPS is made. The last chapter on SIMS
(secondary ion mass spectroscopy) by Dr. H. W. Werner (Philips Research Laboratories, The Netherlands) is an extensive review of its
application to surface studies.

The editors are grateful to the lecturers of this Institute
for providing extended lecture notes. In most cases this involved
a considerable writing task in order to cover the subject in a
comprehensive way.

The support of the NATO Science Committee is gratefully acknowledged. Our gratitude also goes to the authors and editors of
books and periodicals who granted permission to reproduce figures,
diagrams, or other material.

Finally we also wish to thank Plenum Press for providing the
necessary support for the publication of these Proceedings.

                                                L. Fiermans
                                                J. Vennik
                                                W. Dekeyser

Ghent, December 1977

Contents

Studies of Adsorbate Electronic Structure Using
         Ion Neutralization and Photo-
         emission Spectroscopies   273
   H. D. Hagstrum

Introduction to Secondary Ion Mass Spectrometry (SIMS) . .   324
   H. W. Werner

THEORY OF PHOTOEMISSION

G.D. Mahan

Physics Department
Indiana University
Bloomington, Indiana 47401

I. INTRODUCTION

These lectures are intended to be an introduction to the
theory of photoemission. They are not intended to be a comprehensive review. Whatever inclinations we had to write a review soon
dissolved when we examined the literature. There are many recent
reviews--a partial list is given with the references. The abundance of good reviews relieved us from the desire to write one.
Instead, we have picked some aspects of the subject which are
interesting, and discuss them in detail. These divide into two
catagories. The first is the one electron theory of photoemission,
as applied to solids. The second is the many body theory as
applied to photoemission in metals. There are other large areas
which have been omitted for lack of space. One is photoemission
from surfaces. This includes ideal surfaces, and also adsorbates.
These are interesting and active areas.

Sec. II is on the one electron theory of photoemission.
The outgoing wave theory is used to derive the three step model.
Sec. III discusses the angular dependence of photoemission, as
applied to simple metals. We also present a review of the experiments and theories on silver, which has been much studied. Sec.
IV discusses plasmon losses, with the emphasis on bulk plasmons.
The distinction between intrinsic and extrinsic plasmon excitation
is explained, and the experiments are reviewed. Again, lack of
space prevents a detailed theory of surface plasmon loss. Sec.
V is on many body processes in XPS: phonons, Auger decay, and
electron-electron effects. We review the recent discussion over
interference between phonons and Auger decay.

II.   ONE ELECTRON THEORY

A. History

The one electron theory of photoemission has usually been divided into two catagories: surface and volume properties. The research of the past few years has shown that this distinction is less clearly delineated than once was thought.[1] Most electrons come from close to the surface, or from regions which are influenced by surface effects. For example, the optical photons which are causing the photoemission may be represented by a classical electric field E(z). Feibelman[2,3] and others[4,5] have shown that E(z) has friedel oscillations near the surface, which penetrate far into the surface. Thus photoelectrons which emanate from several atomic layers inside of the surface may not experience the bulk value of E(z). It is, in fact, remarkable that photoemission measures any bulk property at all.

We shall concentrate on a discussion of the volume contribution to photoemission. The surface effect is equally important, but is more difficult to discuss well, and will be omitted for lack of time.

The volume effect in photoemission has been invariably treated as a three step process. These three steps are:
1. An electron absorbs a photon, and thereby undergoes an optical transition to an excited state.
2. An electron, in this excited state, is transported to the surface. The electron is assigned a mean free path L, and only those electrons in the depth L from the surface arrive at the surface. Berglund and Spicer[6] refined this concept by introducing the escape cone for each electron.
3. The electron is transmitted through the surface with a probability T. This quantity is energy and directional dependent.

The three step model appears in the earliest theories of photoemission, and has been a constant theme since then. For example, Fan (1945)[7] wrote that the external current is J = eLNT, where N is the absorption probability, and the other quantities are defined above. This has the three factors, one for each step. A similar theory was used by Apker, Taft, and Dickey (1948).[8] Kane, Gobeli, and Allen (1962)[9,10] used the theory in the form which is popular today--in wave vector space. Thus the absorption, mean free path, and surface transmission are computed for each state k, and then one averages over the Brillouin zone of available states. Berglund and Spicer (1964)[6] developed the three step model in energy space. This form is much easier, and was widely

used to compare with experiment. Koyama and Smith (1970) [11]
discussed again the three step model in k-space, and used it
successfully to interpret a number of experiments on simple metals.
Simultaneously, Mahan (1970) [12,13] derived the three step formula
from outgoing wave theory. This derivation obviously came well
after numerous applications. Subsequently, Schaich and Ashcroft
(1971) [1] derived the same result from their quadratic response
formulism. This result has now been rederived by a variety of
techniques, [14,15] but we shall use the outgoing wave formulism
here, since it seems easiest. The outgoing wave formulism was
first applied to photoemission theory by Adawi, [16] who used it
to discuss the surface effect.

B.   Atomic Photoionization

Our theoretical discussion will begin by treating photoionization of an atom. This is chosen as a simple system, for which
it is easy to visualize the physics, and one can concentrate on the
mathematics.

We begin the discussion of the outgoing wave formulism. Let
the system be described by a Hamiltonian

$$H = H_0 + H'$$

The Hamiltonian H_0 describes the atom in the absence of an external radiation field, and it has eigenstates ψ_n and eigenvalues E_n

$$H_0 \psi_n = E_n \psi_n \tag{1}$$

The Hamiltonian

$$H' = \frac{1}{2}\frac{e}{m} (\underset{\sim}{p}\cdot\underset{\sim}{A} + \underset{\sim}{A}\cdot\underset{\sim}{p})$$

is from the external radiation field which causes the photoionization. We assume that it has a single frequency ω. This external
radiation, and its interaction H', change the eigenstates of the
system. This may be described by first order perturbation theory

$$\Psi_n(\underset{\sim}{r}) = \psi_n(\underset{\sim}{r}) + \sum_\ell \frac{\psi_\ell(\underset{\sim}{r}) < \ell |H'|n>}{E_\ell - E_n - \omega}$$

The second term may be written in terms of the Green's function,
which is defined as

$$G(\underset{\sim}{r},\underset{\sim}{r}';E) = \sum_\ell \frac{\psi_\ell(\underset{\sim}{r})\psi_\ell^*(\underset{\sim}{r}')}{E_\ell - E} \tag{2}$$

Thus we can write

$$\Psi_n(r) = \psi_n(r) + \int d^3r' \, G(\underset{\sim}{r},\underset{\sim}{r}';E_n+\omega)H'\psi_n(\underset{\sim}{r}')$$

The wave function $\Psi_n(r)$ can be visualized in terms of scattering theory. The process is shown in Fig. 1. This describes an inelastic scattering process whereby a photon goes into the atom, and an electron comes out. We can think of the state $\underline{n}$ as the initial state of the electron in the atom, so take $E_n < 0$ and $E_n + \omega > 0$. Thus the Greens function is evaluated in the outgoing state. Since the electron is leaving the atom, it has a finite amplitude far from the atom. So evaluate the wave function asymptotically

$$\lim_{r \to \infty} \Psi_n(r) = \lim_{r \to \infty} \int d^3r' \, G(\underset{\sim}{r},\underset{\sim}{r}';E_n+\omega)H'\psi_n(\underset{\sim}{r}')$$

The first term $\psi_n(r)$ is absent since it is a bound state, and vanishes far from the atom. From scattering theory it can be shown that the asymptotic form for the Greens function is

$$\lim_{r \to \infty} G(\underset{\sim}{r},\underset{\sim}{r}',E_n+\omega) \to \frac{e^{ipr}}{r} f(\underset{\sim}{p},\underset{\sim}{r}') \tag{3}$$

where the wave vector p is

$$p^2 = 2m(E_n + \omega)$$

The current dI in a small area $r^2 d\Omega$ of solid angle $d\Omega$ is given by

$$dI = \frac{e\hbar}{m} r^2 d\Omega \, Im[\Psi_n(r)^* \nabla \Psi_n(r)]$$

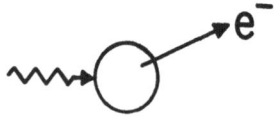

Fig. 1

Thus the current per unit solid angle is [13]

$$\frac{dI}{d\Omega} = \frac{e\hbar p}{m} \sum_{n} \left| \int d^3r' \ f(\underset{\sim}{k},\underset{\sim}{r}') \ H' \psi_n(\underset{\sim}{r}') \right|^2$$

where we have now summed over all possible initial states $\underline{n}$. Of course, the total current is obtained by integrating over solid angle

$$I = \int d\Omega \left(\frac{dI}{d\Omega} \right)$$

Incidently, angular measurements are indeed made for the atomic photoionization measurements. If polarized light is used, then the formula for the angular dependence for atoms is [17]

$$\frac{dI}{d\Omega} = \frac{I_o}{4\pi} \ [1 + \beta P_2(\theta)]$$

where θ is the angle between $\underset{\sim}{A}$ and the outgoing electron direction $\underset{\sim}{k}$. The constant β is frequency dependent. It represents an interference term between outgoing waves in different angular momentum states. [17]

The photoionization intensity is determined by the amplitude $f(\underset{\sim}{p},\underset{\sim}{r}')$ which comes from the asymptotic limit of the Greens function. In order to understand this function, let us first examine it in a simple case. This is for plane waves--the case of no interaction. Then the Greens function is

$$G(\underset{\sim}{r},\underset{\sim}{r}',p^2/2m) = \frac{2m}{\hbar^2} \int \frac{d^3k}{(2\pi)^3} \ \frac{e^{i\underset{\sim}{k}(\underset{\sim}{r}-\underset{\sim}{r}')}}{p^2-k^2+i\delta}$$

The $i\delta$ factor in the denominator is inserted to give the outgoing wave boundary conditions, which are desired. This integral is standard,

$$G(\underset{\sim}{r},\underset{\sim}{r}',p^2/2m) = -\frac{2m}{\hbar^2} \frac{1}{4\pi R} \ e^{ipR}$$

$$R = |\underset{\sim}{r}-\underset{\sim}{r}'|$$

Now we take the limit of $r \to \infty$.

$$\lim_{r\to\infty} R = \lim_{r\to\infty} [r^2-2rr'\cos\theta+r'^2]^{\frac{1}{2}} = r - r'\cos\theta + 0(r'^2/r)$$

The factor $1/R$ may simply be replaced by $1/r$ since the correction factors may be neglected. However, in the exponent we must be more careful, since the factor $pr \gg pr'$ but the latter factor still gives an important phase factor in the integrand. Thus the asymptotic form is

$$\lim_{r \to \infty} G(r,r',p^2/2m) = - \frac{m}{2\pi\hbar^2} \frac{1}{r} e^{ipr} e^{-ipr'\cos\theta}$$

so that

$$f_n(p,r) = - \frac{m}{2\pi\hbar^2} M_n$$

$$M_n = \int d^3r' \ e^{-ipr'\cos\theta} H' \psi_n(r')$$

This produces a differential cross section of the form

$$\frac{dI}{d\Omega} = \frac{e\hbar p}{m} \left(\frac{m}{2\pi\hbar^2}\right)^2 \sum_n |M_n|^2 \tag{4}$$

where M_n is the matrix element defined above. Of course, we have only defined it for the case of plane waves. The term Born Approximation is applied to the procedure of using plane waves in evaluating M_n. We caution the reader that this is a very poor approximation, which should be avoided since it gives wildly poor results. We only describe this approximation here in order to provide some simple theoretical examples. The $\cos\theta$ factor in M_n comes from the angle between r and r'. The direction r is that of the outgoing electron, with wave vector p. Let us define the direction of p to coincide with r, so that $p = p\hat{r}$, and $pr'\cos\theta = p \cdot r'$.

$$M_n = \int d^3r' \ e^{-ip \cdot r'} H' \psi_n(r')$$

We note that our cross section has the form of Fermi's Golden Rule. This Rule gives the rate for any transition as

$$w = \frac{2\pi}{\hbar} \int \frac{d^3p}{(2\pi)^3} |N|^2 \delta(E_n + \omega - \frac{p^2}{2m})$$

Doing the integral gives a result similar to the above result (4) for our photoemission current. In the Golden Rule, the matrix element N is taken between the initial and final electronic states.

$$N = \int d^3r' \ \psi_f^* H' \psi_n$$

This formula has a basic and important difference from our result. In the photoemission experiment, the final electronic state represents an outgoing electronic state, so that one thinks of an outgoing electronic wave function for ψ_f. However, our derivation shows that the matrix element M_n does not contain an outgoing wave, but instead a plane wave. These are different. For example, in the $\ell = 0$ angular momentum state, a plane wave has the form

$$j_0(kr) = \sin(kr)/(kr)$$

while the outgoing wave has the form

$$h_0(kr) = -ie^{ikr}/(kr)$$

Thus the formula for photoemission does indeed have the form of a Golden Rule, except with the important difference that the final state wavefunction in the matrix element is not taken to be an outgoing wave, in spite of the fact that the experiment does measure outgoing electrons. This is an important point, the reason we went through the above analysis.

So far we have been talking about plane waves. What about the actual case of interest, which must include the actual wave functions in the atom. In this case one can show that the exact matrix element is

$$M_n = \int d^3r' \psi_p^*(r') \, H' \psi_n(r') \tag{5}$$

The wavefunction ψ_p is for an electron of positive energy $p^2/2m$ in the vicinity of the atom. It is a solution to equation (1). However, the boundary conditions are important. We expand it in angular functions

$$\psi_p(r) = \sum_\ell (2\ell+1) \, i^\ell \, P_\ell(\hat{p}\cdot\hat{r}) \, R_\ell(pr)$$

The boundary conditions dictate that the radial function asympotically go to

$$\lim_{pr>>\infty} R_\ell(pr) \to \frac{1}{pr} \sin[pr - \frac{\ell\pi}{2} + \delta_\ell]$$

which is the standing wave form, and not the outgoing wave form.

Thus our Greens function analysis has shown us that photoemission can be considered a process which is described by the Golden Rule, but with a careful choice of the final state wavefunction. The same will be true in solids.

It is important to keep account of the dimensional units. The
interaction H' is ergs, and the bound state wavefunction ψ_n is
$cm^{-3/2}$ since we assume it is normalized

$$\int d^3r \, \psi_n^2 = 1$$

The final wave function is dimensionless, so M_n has the units of
erg $cm^{3/2}$. Now it is easy in (4) to show that the current has
the units of esu/sec.

There is a problem about relaxation. Although this has recently been recognized in Solid State Physics, it has been appreciated
much longer in Atomic Physics. The question is which potential
to use to calculate the final state wavefunction ψ_p. We write
H as a kinetic energy term plus a potential energy term V. One
form of V, call it V_G, is for the atom in the ground state.
However, after removing the one electron, all of the atomic
orbitals relax to a new configuration, and hence define an
atomic potential for the excited state V_X. These two potentials
are different; e.g., they describe atoms which differ by one
electron. Which one should be used to compute ψ_p? Both are
used, and they have become so common that names have been
attached to these methods. The Sudden Approximation is using V_G.
Here the rational is that the electron leaves the atom suddenly,
before the atom has an opportunity to adjust to the new configuration. This approximation is only valid, if at all, for electrons
of high kinetic energy. The other method is using V_X, and this
is the Adiabatic Approximation. Here the explanation is that the
electron is removed slowly from the atom, so that the atom does
have time to relax. This obviously applies to electrons of
low kinetic energy. One should not think that these are the
only approximations employed by theorists. The imagination is
much more fertile than that. For example, some theorists use
fractional charges in states,which is an intermediate case. [18]
However, there has been recent work in which these questions
have been resolved correctly, with the use of diagrammatic
techniques applied to higher order correlation functions.[19] Thus
great progress has been made in atomic calculations, and one
can hope for similar improvements in Solid State calculation.

C. Photoemission From a Solid

Fig. 2 depicts photoemission from a solid. Again it is an
inelastic scattering process, with a photon in and an electron
out. In fact, we can also apply the outgoing wave ideas from
the atomic case. The current is still given by (4), and the
matrix element by (5). However, the wavefunction ψ_k^* has a
special interpretation. It now represents an ingoing wave!

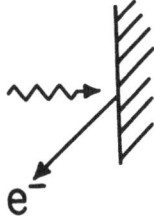

Fig. 2

Again we illustrate the point by considering a simple example. Consider the one dimensional Greens function for a finite step potential as shown in Fig. 3. This Greens function is defined in the usual way as

$$G(z,z'; k^2/2m) = \frac{2m}{\hbar^2} \int \frac{dk'}{2\pi} \frac{\phi_{k'}(z)\phi_{k'}(z')}{k^2-k'^2+i\delta}$$

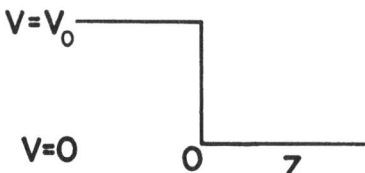

Fig. 3 Step potential at a metal surface.

where the wave functions can be given explicitly. These integrals
can be done to obtain the exact Greens function. Take $z < 0$
as outside of the solid, and $z > 0$ as inside. Then the Greens
function we want is for one variable, say z, to be outside and
outgoing. This Greens function is $(z < z')$

$$G(z,z', \hbar^2 k^2/2m) = \frac{im}{p} e^{-ipz} \{ e^{ipz'} - e^{-ipz'} \frac{(k-p)^2}{k_o^2} \} \quad z' < 0$$
$$z << 0$$

$$= \frac{mi}{p} e^{-ipz} \{ \frac{2}{p+k} e^{ikz'} \} \qquad\qquad z' > 0$$

where $p = [2m(E-V_o)]^{\frac{1}{2}}$ is the wave vector for $z < 0$, and $k = (2mE)^{\frac{1}{2}}$
for $z > 0$. There is a simple way to represent this answer.
We can write it as

$$G(z,z', \hbar^2 k^2/2m) = \frac{im}{p} e^{-ipz} \phi^>(p,z')$$

$$\phi^>(p,z) = e^{ipz} + R e^{-ipz} \qquad z < 0$$

$$= T e^{ikz} \qquad\qquad z > 0$$

where $\phi^>$ is a wavefunction for a wave incident on the solid from
the left. Thus to the left of $z = 0$ it is composed of an incident
$\exp(+ipz)$ and a reflected $R\exp(-ipz)$ part. To the right of the
step is a transmitted wave $T\exp(ikz)$. The matrix element for
photoemission from this one dimensional solid would be

$$M = \int dz \; \phi^>(p,z) H' \phi_i(z)$$

Here the final state is represented by the ingoing wave $\phi^>$, and
this is not complex conjugated! One can show that the above
form is quite general, and applies to any one dimensional case.
The matrix element for photoemission is given in terms of the
wave of unit amplitude incident towards the solid from the outside. Outside, far from the surface, it will have an incident
and reflected component, while inside far from the surface it
will have a transmitted component. In the surface region it
will have a complex form which is obtained by solving Schrodingers
equation for the appropriate potential function, and for the
boundary conditions that were just enumerated.

The use of an incoming wavefunction to calculate the rate
of an outgoing process is common in scattering theory. This
formulism is used in many branches of physics. Photoemission
theorists practice one-up-manship by seeing who can list the
earliest reference to this technique. [20]

What about three dimensions? The answer has the same form [13]

$$M = \int d^3r \; \phi^>(\vec{p},\vec{r}) H' \; \phi_i(r) \tag{6}$$

where the incoming wave $\phi^>$ has unit amplitude, and approaches
the surface from the direction $\vec{p}$. Again one will get reflected
waves outside, and transmitted waves inside. There will usually
be many reflected and transmitted waves for a single ingoing wave.
In reflection this is experimentally verified in LEED experiments,
wherein a single electron beam shot at the surface produces many
spots in reflection. This is thought to happen in transmission
as well, for the same reason, because an electron wavefunction
is a Block wave with many wavevector components. This analogy
to LEED was first pointed out by Mahan. [13]

D. Three Step Model

The three step model was described earlier in Sec. II-A.
The three steps are: (1) photon absorption, (2) transport of the
excited electron to the surface, and (3) transmission through the
surface. Now we would like to show how eqn. (4) and (6) can be
used to describe this model.

It is assumed that the incoming wave $\psi^>$ is transmitted through
the surface, and produces internal waves in the solid with wavevector k_{zi} with transmission amplitude T_{oi} for going from outside
to inside.

$$\psi^> = \sum_j T_{oi}(k_{zj}) \; e^{ik_{zj}z} \; e^{i\vec{k}_{\parallel} \cdot \vec{R}}$$

Thus the photoemission intensity is

$$\frac{dI}{d\Omega} = \frac{ep}{m} \left(\frac{m}{2\pi\hbar^2}\right)^2 \sum_{j,n} |T_{oi}|^2 \left| \int e^{ik_j \cdot r'} H' \psi_n(r') \right|^2$$

where the summation runs over initial states, which must include
a summation over initial atoms. This summation yields a volume
LA from which photoemission is effective, where A is the area
and L is the electron escape depth. The depth could also be
limited by the penetration distance of the x-rays or ultraviolet
light, but this is usually less important than the electron escape
depth.

$$\frac{dI}{d\Omega} = \frac{ep}{m} \left(\frac{m}{2\pi\hbar^2}\right)^2 LA \sum_j |T_{oi}|^2 |M_n(k_j)|^2$$

The transmission factor T_{oi} is for electrons going from outside
to inside a solid, whereas in the experiment, they are going the
other way. Thus we really wish to express the answer in terms of
the transmission factor T_{io} for an electron going from inside to
the outside. This is simple, since Fredkin and Wannier [21] showed,
using time-reversal arguments--that there is a simple relationship
between these quantities

$$k_z^2 \, T_{oi}^2 = T_{io}^2 \, p_z^2$$

III.   ANGULAR DEPENDENCE IN SIMPLE METALS

A.   Interband Transitions

Mahan predicted that the photoemission from simple metals
would be dramatically angular dependent. [12,13] This has proved
to be the case. [22-26] In fact, most metals show angular dependence.
This angular dependence is a simple consequence of the energy
band structure of the solid. The first step in the three step
model of photoemission is a simple optical absorption process,
and we need to examine how this depends upon the energy bands of
the solid.

The optical absorption may be expressed as the imaginary part
of the dielectric function, $\varepsilon_2(\omega)$. In a simple metal such as
sodium, this has the dependence upon frequency illustrated in
Fig. 4. There is a low frequency region called Drude, which
is caused by electrons absorbing light while interacting with
phonons. At a threshold frequency ω_T there is a rise in the
absorption caused by interband transitions. It is this latter
step which is important in causing the angular dependence in
photoemission.

A free electron cannot absorb light. This is a simple
momentum mismatch. The photon, at the speed of light, cannot
provide the electron with enough momentum to go with the energy
it acquires. That is, if we assume an electron of momentum
k absorbs a photon of momentum q and frequency cq, the energy
conservation is

$$\frac{\hbar^2 k^2}{2m} + \hbar c q \overset{?}{=} \frac{\hbar^2}{2m} (\vec{k}+\vec{q})^2 = \frac{\hbar^2}{2m} (k^2 + 2kq \cos\theta + q^2)$$

If we divide all terms by $\hbar q$, we get that

$$v \cos\theta \overset{?}{=} c - \hbar q/2m$$

where v is the electron velocity. This is obviously not equal

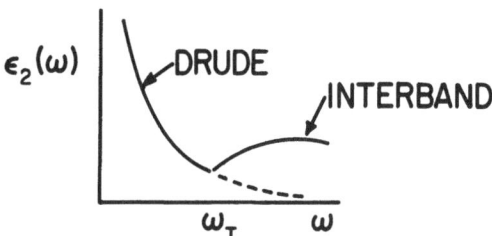

Fig. 4 Imaginary part of the dielectric function of sodium. A Drude tail at low frequencies, and an interband term beginning at ω_T.

to c, the speed of light, and the other term $\hbar q/2m$ is too small to make up the difference, so the above equation cannot be satisfied.

In a solid, we think of the photon wave vector q as being sufficiently small that we can neglect it for most experiments. The photon then provides the energy to the excited electron, and the wave vector comes from some other source. In Drude absorption, this is phonons. The phonons are a perfect complement to photons, since phonons have large wavevector and negligible energy, at least for most purposes. Thus the Drude term is described by the classical formula

$$\varepsilon(\omega) = \varepsilon_1(\omega) + i\varepsilon_2(\omega) = 1 - \frac{\omega_p^2}{\omega(\omega+i/\tau)}$$

where τ is a relaxation time which is nearly frequency independent. [28] The Drude absorption process may be represented by Fig. 5a. An electron (solid line) enters, absorbs a photon (wavy line) and emits the phonon (dashed line). The phonon may be either absorbed or emitted, and either before or after the photon.

The interband transition is a similar process, as shown in Fig. 5b. Here the electron gains momentum by scattering from the crystalline potential, which comes in discrete units called

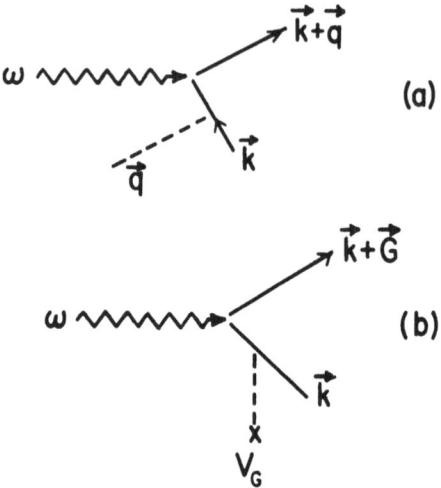

Fig. 5   (a)   The Drude absorption, wherein a photon and phonon
         scatter an electron.   (b)   The interband absorption,
         where the electron scatters from the crystal potential
         V_G while absorbing the photon.

reciprocal lattice vectors $\underset{\sim}{G}$.   The crystalline potential $V(r)$
in a solid can be expanded as a fourier series in these lattice
vectors

$$V(r) = \sum_{\underset{\sim}{G}} V_{\vec{G}} \, e^{i\vec{G}\cdot\vec{r}} \tag{7}$$

Thus interband transitions occur because the electron scatters
from the crystalline potential while absorbing the photon.

The more traditional picture of interband absorption is shown
in Fig. 6.   An electron starts in a state $\underset{\sim}{k}$ and has a vertical
transition shown by the arrow.   However, this is the picture in
the reduced zone scheme.   In an extended zone scheme, the transition really changes the wavevector by $\underset{\sim}{G}$, so that the transition
is really the diagonal arrow from $\underset{\sim}{k}$ to $\underset{\sim}{k}+\underset{\sim}{G}$.   Of course, these

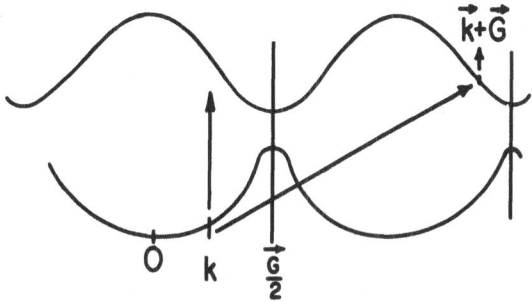

Fig. 6   Band picture of optical absorption: the transition is
         vertical in a reduced zone picture, but diagonal in an
         extended zone picture.

are the same in the reduced zone scheme. However, in photoemission
we will show that it is useful to think in an extended zone scheme.

In discussing the interband transitions, the photon wavevector q will be neglected compared with the large lattice vector
$\underset{\sim}{G}$. Energy conservation requires that

$$\frac{k^2}{2m} + \omega = \frac{1}{2m} \, (\underset{\sim}{k} + \underset{\sim}{G})^2$$

or

$$\omega = \frac{kG}{m} \, \cos\theta + E_G$$

$$E_G = G^2/2m$$

We may write this as

$$-1 \leq \frac{\omega - E_G}{kG/m} = \cos\theta \leq 1$$

where we have inserted the limits on the value of $\cos\theta$. If we
use these limits to restrict allowed values of ω, this gives

$$\omega_1 \equiv E_G - \frac{kG}{m} \leq \omega \leq E_G + \frac{kG}{m} \equiv \omega_2 \tag{8}$$

The Brillouin zone boundary is $G/2$. In metals such as the alkalis, the Fermi surface of diameter k_F does not touch the zone face, so that $k_F < G/2$. Then the threshold condition is the left hand inequality

$$\omega > \omega_T = E_G - \frac{k_F G}{m} > 0$$

Of course, this is changed by band gaps, or if the effective mass differs from unity.

The intensity of the interband absorption will be calculated next. The resulting formula is due to Wilson and Butcher. The crystalline potential is put in by perturbation theory. Thus the wavefunction in (7) causes a first order change in the initial and final wave functions of

$$\psi_k^{(i)} = e^{i\underset{\sim}{k} \cdot \underset{\sim}{r}} + \sum_{G'} \frac{V_{G'} \, e^{i\underset{\sim}{r} \cdot (\underset{\sim}{k} + \underset{\sim}{G'})}}{\varepsilon_{\underset{\sim}{k}} - \varepsilon_{\underset{\sim}{k} + \underset{\sim}{G'}}}$$

$$\psi_{k+G}^{(f)} = e^{i\underset{\sim}{r} \cdot (\underset{\sim}{k} + \underset{\sim}{G})} + \sum_{G''} \frac{V_{G''} \, e^{i\underset{\sim}{r} \cdot (\underset{\sim}{k} + \underset{\sim}{G} + \underset{\sim}{G''})}}{\varepsilon_{k+G} - \varepsilon_{k+G+G''}}$$

In the initial wave function $\psi_k^{(i)}$, the only term we really want is that with $G' = G$. In the final wave function, we want $G'' = -G$. Then when we calculate the matrix elements over the volume V_o of the unit cell

$$\hat{\varepsilon} \cdot P_{fi} = \frac{\hbar}{i} \int \frac{d^3 r}{V_o} \, \psi_{k+G}^{(f)} \, \hat{\varepsilon} \cdot \nabla \psi_k^{(i)} = \frac{\hbar}{V_o} \int d^3 r \, \{ e^{-i\underset{\sim}{G} \cdot \underset{\sim}{k}} \, \varepsilon \cdot \underset{\sim}{k} +$$

$$\sum_{G'} \frac{\hat{\varepsilon} \cdot (k + G') V_{G'} \, e^{i\underset{\sim}{r} \cdot (G+G')}}{\varepsilon_k - \varepsilon_{k+G'}} + \sum_{G''} \frac{\hat{\varepsilon} \cdot k \, e^{-i\underset{\sim}{r} \cdot (G+G')}}{\varepsilon_{k+G} - \varepsilon_{k+G+G'}} V_{G''} \}$$

we only keep terms to first order in V_G. Thus we are assuming that V_G is small and we are using it as the expansion parameter. The next terms in this series, of order $V_G V_{G'}$ have been examined, and change the result by about 20%. Here we shall just keep the first order terms in V_G. These give

$$\hat{\epsilon} \cdot \underset{\sim}{P}_{fi} = \frac{\hat{\epsilon} \cdot \underline{G} \ V_G}{\hbar \omega} \tag{9}$$

The formula for the optical absorption rate is

$$\epsilon_2(\omega) = \frac{8\pi^2 e^2}{m^2 \omega^2} \int\limits_{k < k_F} \frac{d^3 k}{(2\pi)^3} \ |\hat{\epsilon} \cdot \underset{\sim}{P}_{fi}|^2 \ \delta(\epsilon_k + \hbar \omega - \epsilon_{k+G})$$

We have included a factor of two for spin degeneracy, since we are assuming that spin up and down contribute equally. Since our matrix element (9) is independent of $\underset{\sim}{k}$, the integral reduces to

$$\epsilon_2(\omega) = \frac{8\pi^2 e^2}{m^2 \omega^2} \ \underset{\underset{\sim}{G}}{\Sigma} \ \left(\frac{V_G}{\hbar \omega}\right)^2 (\hat{\epsilon} \cdot \underset{\sim}{G})^2 \ \frac{1}{(2\pi)^2} \int\limits_0^{k_F} k^2 dk \int\limits_{-1}^{1} d(\cos\theta)$$

$$\delta(\hbar\omega - E_G - \frac{kG}{m} \cos\theta)$$

The $\cos\theta$ integral only has a finite value if the frequency ω is in the range prescribed earlier (8).

$$\int\limits_{-1}^{1} d(\cos\theta) \ \delta(\omega - E_G - \frac{kG}{m} \cos\theta) = \frac{m}{kG} \ \text{if} \ -1 \leq (\omega - E_G) \frac{m}{kG} \leq 1$$

The inequality is used to limit the k integral

$$\frac{m}{G} \int\limits^{k_F} k dk \quad = \frac{m}{2G} \ \{ k_F^2 - \frac{m^2}{G^2} (\omega - E_G)^2 \}$$

$$\frac{m}{G} \ |\omega - E_G|$$

$$= \frac{m^3}{2G^3} \ (\omega - \omega_1)(\omega_2 - \omega)$$

$$\omega_{1,2} = E_G \mp \frac{k_F G}{m}$$

The frequency ω_1 is the same as the threshold frequency ω_T. The Wilson-Butcher equation for the interband optical absorption is

$$\epsilon_2(\omega) = \frac{me^2}{(\hbar\omega)^4} \ \underset{G}{\Sigma} \ \frac{(\hat{\epsilon} \cdot \underset{\sim}{G})^2 V_G^2}{G^3} \ (\omega - \omega_1)(\omega_2 - \omega)$$

absorption only exists in the frequency region $\omega_2 > \omega > \omega_1$. The
summation over G is taken for all values of G which have $\omega_2 > \omega > \omega_1$.
More values of G must be included at high frequency. This theory
is only valid for s-p like bands in free electron metals. The
d- or f- bands are too localized, and do not have free electron
like behavior.

B.   Internal Photoemission: Direction

In which directions are the electrons going? This determines
the angular dependence. We shall see that the electrons in the
final state are going in cones, where the cones are centered
about the direction $\underset{\sim}{G}$.

It is convenient to change our notation. Since our attention
is on the electrons in the final state, call this wave vector k.
Thus the initial state wave vector is $k - \underset{\sim}{G}$. We are using an
extended zone scheme in which $\underset{\sim}{k}$ is now a large wave vector, and
is outside of the first Brillouin zone or Fermi sea, while $\underset{\sim}{k}-\underset{\sim}{G}$
is inside both. Now energy conservation is

$$\frac{(k-G)^2}{2m} + \omega = \frac{k^2}{2m}$$

or

$$\omega + E_G = \frac{kG}{m} \cos\theta$$

If we solve for k, the final state energy is

$$E_f = \frac{k^2}{2m} = \frac{(\omega+E_G)^2}{4E_G \cos^2\theta} \equiv \mathcal{N}\cos^2\theta \tag{10}$$

The energy has been expressed as a constant factor $(\omega+E_G)^2/4E_G$
divided by $\cos^2\theta$. The angle θ is between the directions k and $\underset{\sim}{G}$,
as in Fig. 7. Thus all electrons which have the same angle with
respect to G, have the same energy. Electrons of the same energy
have a conical distribution about G. Of course, any metal has
many G's in different directions, and different magnitudes. Each
direction $\underset{\sim}{G}$, for which optical excitation is permitted by $\omega_2 > \omega > \omega_1$,
will have a cone of electrons about it. [12,13]

The maximum cone angle can be determined by a simple argument.
The maximum initial energy for an electron is the Fermi energy
E_F, so the maximum final energy is

$$E_f = E_i + \omega \leq \omega + E_F$$

If we combine this inequality with (10) we obtain

$$\frac{(\omega + E_G)^2}{4E_G(\omega + E_F)} \leq \cos^2\theta \leq 1$$

We have added on the right the obvious inequality that the $\cos\theta$ has a magnitude less than unity. If the above inequality is treated as a restriction on ω, one obtains

$$\omega_2 \geq \omega \geq \omega_1$$

Of course, this is the same restriction we obtained above for the optical absorption. We now see that our earlier restriction was simply that the cone angle be real and not complex.

The physics of this angular dependence is very appealing. If we measure in any direction (θ, ψ) inside of the solid, we will find optically excited electrons of only one energy. This presumes that cones of electrons from different G's do not over-lap in direction, which is satisfied at lower optical frequencies. Our calculations have assumed that we have a free electron metal. Actual metals have energy bands which are different from ideal because of the crystalline potential. However, the excited electron must be in a state $\underset{\sim}{k}_f$ which has a definite energy and direction. If one could measure the energy and angular dependence of the photoemitted electron, then the excited state $\underset{\sim}{k}_f$ is entirely determined. However, in the reduced zone scheme, the initial state has the same wavevector and an energy which is lower by ω. Thus the energy bands, which are energy vs. direction plots, are uniquely determined for both the initial and final state. Thus the angular dependence of photoemission has the promise of a detailed mapping of the energy band structure of the solid.

There are significant impediments to this procedure. One is that the direction of $\underset{\sim}{k}$ is ill defined by both experimental resolution and finite mean-free-path effects. Second, the elec-tron energy is ill defined because the electron can lose energy by exciting both bulk and surface plasmons, as well as phonons and electron-hole pairs. Third, the angular measurements must be made outside of the surface, and not inside. The electron changes its direction upon leaving the solid. A very rough surface will diffusely scatter the electrons, and all angular information will be lost. These difficulties are not insur-mountable, since angular measurements do show anisotrophies. But care must be used in interpreting data.

C.   Internal Photoemission: Intensity

The intensity of photoemission is the actual number of electrons going in any direction at any energy. It must be proportional to the flux F of photons in the solid. This photon flux is

$$F = \frac{c}{n} A_q \left(\frac{N_q}{V} \right)$$

where n is the refractive index, and N_q is the number density of photons in the volume V=AL. The internal flux of photons is not the same as the external flux impinging on the surface of the solid. One must apply classical matching procedures at the surface to find the photon fraction which gets reflected and transmitted. The interaction term

$$H' = \frac{e}{m} A_q \,\hat{\varepsilon} \cdot \underset{\sim}{p}$$

contains the vector potential, whose square is proportional to the photon density

$$A_q^{\,2} = \frac{2\pi\hbar c^2}{n^2 \omega} \frac{N_q}{V}$$

The internal photoemission is computed with eqn. (4). Since we want internal photoemission, then the final state in (6) is just a plane wave $\psi^> \equiv \exp(-ik_f \cdot \underset{\sim}{r})$. The matrix element is evaluated as in (9), except that wave vector conservation is explicitly retained

$$M = \frac{e}{mc} A_q \left(\frac{\hat{\varepsilon} \cdot \underset{\sim}{G} V_G}{\hbar\omega} \right) V^{\frac{1}{2}} D$$

$$D = \delta_{\underset{\sim}{k_i} + \underset{\sim}{G} - \underset{\sim}{k_f}}$$

The $V^{\frac{1}{2}}$ arises from our funny convention that the initial state

$$\Psi_i = \frac{1}{\sqrt{V}} e^{i k_i \cdot \underset{\sim}{r}} \left[1 + \sum_{G'} \frac{V_{G'} e^{iG' \cdot r}}{\varepsilon_k - \varepsilon_{k+G'}} \right]$$

is normalized but the final state is not. From equation (4) we get

$$\frac{dI}{d\Omega} = \frac{2e\hbar}{m} \left(\frac{m}{2\pi\hbar^2} \right)^2 \left(\frac{eA_q}{mc} \right)^2 V\Sigma \frac{(\hat{\varepsilon} \cdot \underset{\sim}{G})^2}{\omega^2} V_G^2 H$$

$$H = \sum_{k_i} \delta_{\underset{\sim}{k_i}+G-k_f} \; k_f = \int \frac{d^3k_i}{(2\pi)^3} \; \delta^3(\underset{\sim}{k_i} + \underset{\sim}{G} - \underset{\sim}{k_f}) \; k_f$$

The factor H is set-off above because its evaluation is difficult. At first it appears that the three dimensional delta function eliminates the three dimensional delta function, so that $H = k_f$. The factor k_f was called p in eqn. (4). However, the evaluation of H is more subtle than this, since k_i is a function of k_f because of energy conservation

$$k_f = \sqrt{k_i^2 + 2m\omega}$$

The integral must be done carefully. Call $\underset{\sim}{k_f}$ the "z" direction. If $\underset{\sim}{k_i}$ and $\underset{\sim}{k_f}$ make an angle θ', then

$$k_{iz} = k_i \cos\theta'$$
$$k_{ix} = k_i \sin\theta' \cos\psi'$$
$$k_{iy} = k_i \sin\theta' \sin\psi'$$

In this coordinate system the integral becomes

$$H = \int_0^{k_f} dk_i \; k_i^2 [k_i^2 + 2m\omega]^{\frac{1}{2}} \int_0^{\pi} d\theta' \sin\theta' \int_0^{2\pi} d\psi' \; \delta(G_z + k_i \cos\theta'$$
$$- \sqrt{k_i^2 + 2m\omega}) \; \delta(G_\perp + k_i \sin\theta' \cos\psi') \; \delta(k_i \sin\theta' \sin\psi')$$

The three integrals give, in turn, the result

$$\int_0^{2\pi} d\psi' \; \delta(k_i \sin\theta' \sin\psi') = \frac{1}{k_i \sin\theta'}$$

$$\int_0^{\pi} d\theta' \; \frac{\sin\theta'}{k_i \sin\theta'} \; \delta(G_\perp - k_i \sin\theta') = \frac{1}{k_i \sqrt{k_i^2 - G_\perp^2}}$$

$$H = \int_0^{k_f} k_i \, dk_i \left(\frac{k_i^2 + 2m\omega}{k_i^2 - G_\perp^2} \right)^{\frac{1}{2}} \delta(G_z + \sqrt{k_i^2 - G_\perp^2} - \sqrt{k_i^2 + 2m\omega}) = \frac{k_f^2}{G_z}$$

$$H = \frac{m^2(\omega + E_G)^2}{G_z^3}$$

where we have used eqn. (10). Thus we obtain the final result
for the intensity of internal photoemission [13]

$$\frac{dI}{d\Omega} = \frac{e\alpha FLm}{2\pi n(\hbar\omega)^3} \sum_G V_G^2 \frac{(\hat{\epsilon}\cdot\underset{\sim}{G})^2}{(\underset{\sim}{G}\cdot\hat{k}_f)^3} (\omega + E_G)^2 \tag{11}$$

where $\alpha = e^2/\hbar c$. This intensity is for electrons in the cone. It
gives the number per unit solid angle which are going in any
direction. The energy of these electrons is given by (10). Thus
the intensity at each energy per unit solid angle is obtained by
adding a delta function for energy (11)

$$\frac{d^2I}{d\Omega dE} = (\frac{dI}{d\Omega}) \delta(E - \Lambda/\cos^2\theta)$$

$$\Lambda = (\omega + E_G)^2/4E_G$$

The energy is listed as a function of the independent variable,
which is the angle θ. Another interesting quantity is the intensity per unit energy, which is obtained by integrating over all
solid angle

$$\frac{dI}{dE} = \int d\Omega \frac{d^2I}{d\Omega dE} = 2\pi \int_{-1}^{1} d(\cos\theta) \delta(E - \Lambda/\cos^2\theta)(\frac{dI}{d\Omega})$$

Of course, this integral is easy to do, since the energy delta
function takes out the $\cos\theta$ integration, and gives [11,13]

$$\frac{dI}{dE} = \frac{e\alpha FL}{(\hbar\omega)^3} \sum_G \frac{(\hat{\epsilon}\cdot\underset{\sim}{G})^2}{G} V_G^2 \qquad\qquad E_F + \omega \geq E \geq \Lambda$$

$$= 0 \qquad\qquad\qquad\qquad\qquad \text{otherwise} \tag{12}$$

The intensity as a function of energy is a constant. It has the
distribution shown in Fig. 8, with a rectangular distribution.
This result was first derived simultaneously by Koyama and Smith,[11]
and by Mahan.[13] It has not been possible to measure this rectangular distribution inside of a solid. However, external photoemission results, which will be summarized below, do indicate that
this rectangular shape is found.

The result (11) also contains features which are expected
from the optical absorption. The intensity depends upon the
potential V_G as well as on the polarization of the light $(\hat{\epsilon}\cdot\underset{\sim}{G})^2$.

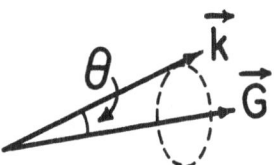

Fig. 7   The optically excited electron has a wavevector k which
         makes an angle θ with respect to G.

D.   Experiments on Ag

It is easiest for the theorist to discuss experiments on
free electron metals such as sodium, magnesium, and aluminum.
These are more difficult for the experimentalist. It is hard to
get smooth surfaces on single crystal faces, which are needed
for good angular measurements. The angular dependence of photoemission was first tested on metallic silver.

At first Ag does not appear to be a free electron metal,
since many of its optical properties are dominated by d-bands.
However, the d-bands in Ag are more tightly bound than in Cu
or Au--which gives it its "silver"color. The energy band
structure near the L-point, or the (111) direction, is shown

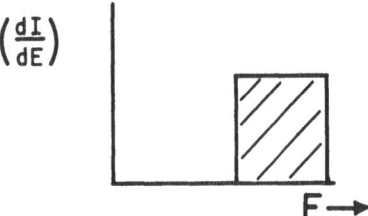

Fig. 8   The internal energy distribution of optically excited
         electrons in the cone model.

in Fig. 9 near the Fermi energy. The low energy optical excitations
only involve s-p bands, and are the type shown by the arrow. The
Fermi energy is in the band gap--since the Fermi surface is
connected.

The energy gap at the zone edge 4.2 eV is quite large. [23]
These energy gaps are usually put in with a two band model. [11,13]
This model includes the band curvature near the zone face. A
photoemission calculation with the two band model shows that
little of our prior results are changed. One still gets cones
centered about $\underset{\sim}{G}$. [13]

The first experimental verification of these ideas was by
Koyama and Smith. [11] They did not discuss the angular dependence,
but instead the energy distribution curves dI/dE. They observed
that the results of Berglund and Spicer, shown here in Fig. 10,
do have the step distribution expected in eqn. (12), and illustrated in Fig. 8. These distributions have since been remeasured
by other experimentalists, and improved background subtraction
makes them look very square. Koyama and Smith also showed that
the energy limits of the box also followed the model. This
behavior is shown in Fig. 11, where the black and hollow circles
correspond to their counterpart in Fig. 10. The solid line is

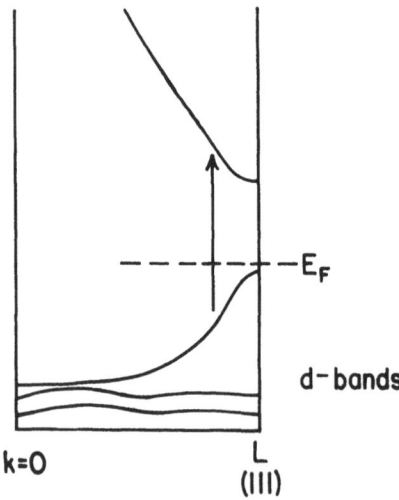

Fig. 9   Schematic picture of energy bands of Ag near the L-point.
         The transition shown by the vertical arrow is free elec-
         tron-like. The Ag band structure has been done by
         N.E. Christensen, phys. stat. solidi (b) 54, 551 (1972).

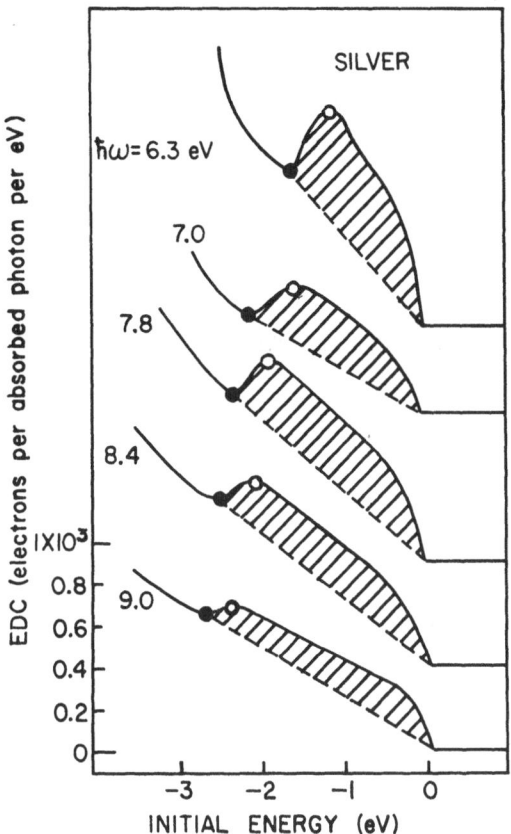

Fig. 10 Experimental EDC in Ag showing the box-like distributions
         as predicted by the cone model. From Berglund and Spicer,
         quoted in R.Y. Koyama and N.V. Smith, Phys. Rev. <u>B2</u>,
         3049 (1970).

the theory, which is the two band model version of our earlier
result in (12)

$$E_F > E_i > \frac{1}{4E_G} \; [(\omega + E_G)^2 - 4V_G^2 \,] - \omega$$

except here they plot initial energy. These crystals are oriented so that the (111) reciprocal lattice vector is normal to the
surface plane. Then the surface transmission factor T is the
same for all electrons emerging in the cone, and the rectangular

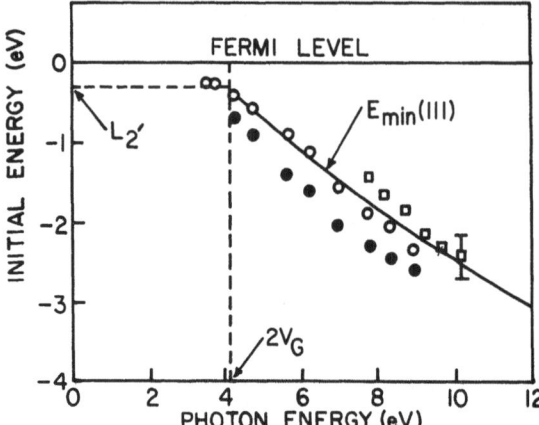

Fig. 11   Initial energy threshold variation with photon frequency.
          Solid line is theory, points are data of Berglund and
          Spicer. Quoted in R.Y. Koyama and N.V. Smith, Phys.
          Rev. B 2, 3049 (1970).

energy distribution curve stays that shape even outside of the
crystal. [11,13]   In other geometries, where the $\underset{\sim}{G}$ points in some
arbitrary direction with respect to the surface, the surface
transmission factor imparts a sawtooth shape to the EDC (energy
distribution curve).

The same experimental geometry was used in the experiment of
Gustafsson, Nilsson and Wallden, [22] who did the first angular
measurements. Their experimental configuration, and results, are
shown in Fig. 12. They collected the electrons in three concentric rings. Each ring should record electrons of a different
energy, and this appears to be the case. This seemed to verify
the cone model.

There appeared two difficulties to this success for the cone
model. In the experimental geometry of Fig. 12, one sees that
the light vector is perpendicular to G, so that $\hat{\epsilon} \cdot G = 0$. Thus
there should be no matrix element for optical excitation, and no
photoemission. This difficulty was resolved by Schaich. [29]
He calculated matrix elements with more OPW's than just the
two band model, and found that optical excitation and photoemission
was possible even if $\hat{\epsilon} \cdot G = 0$. This result has been confirmed
by other calculations. [23-30]

The other difficulty is that the experiment was repeated on

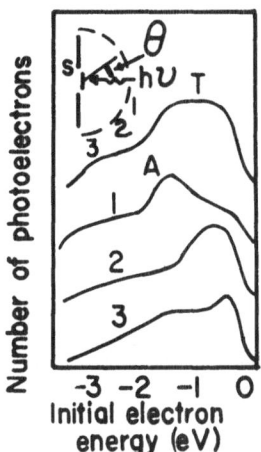

Fig. 12   First angular measurements in photoemission, on Ag.
          Electrons collected on different concentric rings
          were measured at different energies. From T. Gustafsson,
          P.O. Nilsson, and L. Wallden, Phys. Lett. 37A, 121 (1971).

the (100) face of Ag with a similar result. [31] Since the (100)
lattice vector does not make cones, then this result violates the
cone model--the (111) lattice vector should not make cones in the
(100) direction. This puzzle was resolved by the work of Gerhardt
and his colleagues. [23] They pointed out that a significant fraction of the electrons were diffusely scattering from the surface,
and destroying angular information. However, they were able to
subtract the diffusely scattered electrons by using polarized
light. Their experimental configuration, with results for copper,
are shown in Fig. 13. In their geometry, $\hat{\varepsilon} \cdot G = 0$, so the optical
excitation depends, according to Schaich, upon the component of
$\tilde{k}$ which is perpendicular to G.   The result of this is that for
$\hat{\varepsilon} \cdot G = 0$, the electrons come out in the direction of the light
polarization $\hat{\varepsilon}$. They reasoned that electrons coming out with
$k_\perp \hat{\varepsilon}$, called $N_\perp(E)$, are from surface scattering. Those with
$k_{||} \hat{\varepsilon}$, $N_{||}(E)$, contain some which are scattered, and some which
are not. They assume that the surface scattering is the same for
both directions, so those which are not surface scattered are

$$N = N_{||}(E) - N_\perp(E)$$

For copper, at 30°, the same number come out for $N_\perp(E)$ and
$N_{||}(E)$, showing that all are due to surface scattering. At 45°
and 60° there is a difference, and this difference are those
electrons which do not scatter. These results are in agreement
with band structure, which predict that only the second two cases

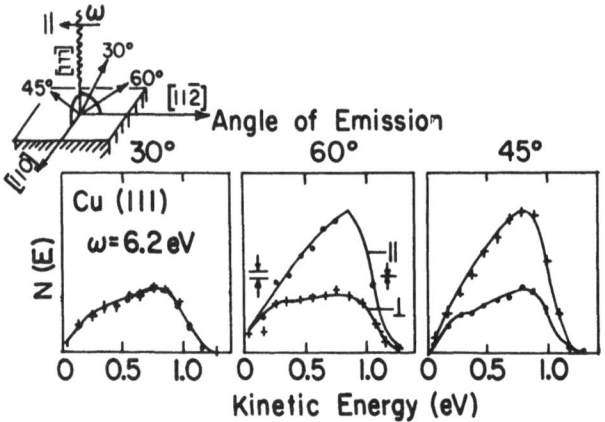

Fig. 13 Measurements with polarized light, showing that some
electrons are diffusely scattered at surface, and some
are not. From H. Becker, et al, Phys. Rev. B 12
2084 (1975).

should have unscattered electrons coming out. In these experiments they are measuring at a single frequency and a single angle,
so electrons at each angle should have one energy. The observed
width is due to experimental resolution in the angular aperture.
They report similar results for Ag, and confirm the three step
model. Those electrons which they observe leaving without scattering are just those which one would expect from a direct optical
transition, followed by the excited electron leaving the surface.

Angular measurements have now been made on many other metals
and semiconductors, and nearly all measurements show angular
dependence on single crystal samples. [22-27] These generally
agree with known band structure. There are, however, definite
angular dependence from surface photoemission, as well as that
from adsorbed atoms on the metal surface. Angular measurements
are now the standard way of doing photoemission experiments.

IV. PLASMON SATELLITES IN XPS

A. X-Ray Photoelectron Spectroscopy

Photoemission from core levels of atoms was developed by
chemists as a means of chemical analysis. [32] They call it ESCA
(electron spectroscopy chemical analysis), while physicists do
the same experiment and call it XPS. It is an excellent means of
studying not only core levels in solids, but also the many body
processes which accompany photoemission. The electron in the core
state has a fixed energy E_c. If it gains energy ω by optical
excitation, then its final energy should be $E_c + \omega$, which is also
fixed. Thus the photoemission measurement should determine that
all emitted electrons have the same energy. In fact, measurements show that many electrons lose energy somewhere. These
are many body processes, and thus XPS is an excellent way of
studying them. The dominant loss mechanism is caused by surface
and bulk plasmons.

Before we launch a discussion of the many body physics, it
is best to summarize the one-electron photoemission theory from
a localized core level. Again we shall concentrate on the internal photoemission in the solid, to avoid treating the refraction
problem of the surface. The refraction has been treated, and
is algebraically complicated without improving the physics. [13]

The initial wavefunction $\phi_n(\underset{\sim}{r}-\underset{\sim}{R}_j)$ is a localized orbital
centered at $\underset{\sim}{R}_j$. Using (4) we get that the angular intensity is [13]

$$\frac{dI}{d\Omega} = \frac{e\alpha FLV}{2\pi\hbar^2\omega n}\ n_0\ (\hat{\epsilon}\cdot\underset{\sim}{P}_{kn})^2$$

where the matrix element is

$$\underset{\sim}{P}_{kn} = \frac{\hbar}{i}\int d^3r\ \psi_k^>(r)\nabla\psi_n(r)$$

We have assumed that the final state inside the solid is $\psi_k^>$. It
is a poor approximation to assume that $\psi^>$ is a plane wave
$\exp(-i\underset{\sim}{k}\cdot\underset{\sim}{r})$ since the integral is mostly in the core region, where
this approximation is bad. The factor n_0 is the density of atoms
in the solid which are absorbing. Since all of the electrons exit
at the same energy in a one-electron picture, then the intensity
per unit energy per solid angle is

$$\frac{d^2I}{dEd\Omega} = \frac{dI}{d\Omega}\ \delta(E - \omega - E_c)$$

The actual energy distributions are not infinitely sharp, like a
delta function. Several factors combine to broaden the spectra:
Auger decay, phonons, and electron-electron interactions. The
Auger decay determines the lifetime of the hole, and imparts a
lorentzian broadening to the spectral shape. In many cases
this is the most significant broadening mechanism, but not always.

The XPS spectra for the aluminum 2s state is shown in Fig.
14. [33] The peak labeled P_0 is the central peak at an energy E_0.
The main series of satellite peaks occurs at energies $E_0 - n\hbar\omega_p$
where n has integer values. These peaks are labeled P_n. A
second series of peaks labeled S_n are due to surface plasmon
emission. They are at an energy $\hbar\omega_{sp}$ below the preceding satellite
P_{n-1}. These energies agree with the classical values

$$\hbar\omega_p = \hbar[4\pi\frac{n_0}{m}e^2]^{\frac{1}{2}}$$

$$\hbar\omega_{sp} = \frac{1}{\sqrt{2}}\hbar\omega_p$$

These plasmon satellites are an important feature of all XPS
spectra of metals.

B. Electron Mean Free Path

An energetic electron, traveling in a metal, can directly
excite bulk plasmons.[34-36] This process is responsible for many
of the plasmon satellite peaks shown in Fig. 14. An electron

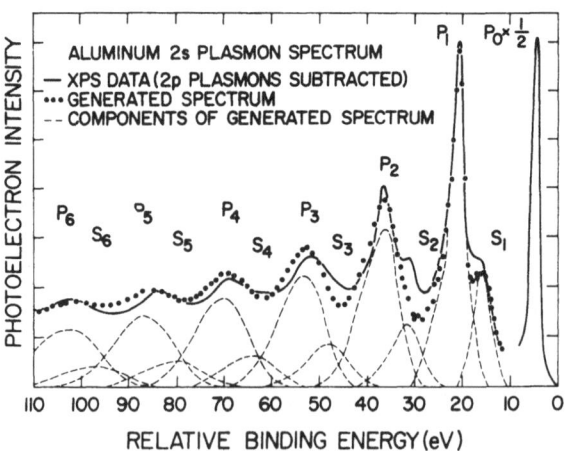

Fig. 14   Aluminum 2s XPS spectrum. From W.J. Pardee, et al,
Phy. Rev. B 11, 3614 (1975).

often leaves the solid with energy $E_0 - n\hbar\omega_p$ because it has excited n plasmons on its way out of the crystal. This process is
called <u>extrinsic</u> plasmon emission. This is to be distinguished
from <u>intrinsic</u> plasmon emission, which happens at the time of
optical excitation. There is great interest, both experimental
and theoretical, in distinquishing between these two processes.
Here we shall discuss the extrinsic process.

An electron traveling in a solid emits plasmons at an average
rate. If this rate is divided by the electron velocity, one
gets the inverse mean free path L which has been defined earlier.
Of course, L is the total mfp from all loss processes. But for
energetic particles the plasmon loss totally dominates, so we
can identify L with this quantity. Ferrell has shown that the
rate of scattering from $\underset{\sim}{k} \to \underset{\sim}{k+q}$ is

$$\frac{1}{L(k)} = -\frac{2}{\hbar v} \int \frac{d^3q}{(2\pi)^3} \left(\frac{4\pi e^2}{q^2}\right) \text{Im} \frac{1}{\varepsilon(q, \varepsilon_k - \varepsilon_{k+q})}$$

where $\varepsilon(q,\omega)$ is the longitudinal dielectric function of the
solid. We will now proceed by circular reasoning. It will be
assumed that the plasmon emission occurs at long wavelengths--at
small values of q. This will be shown to be correct afterwards.
Incidently, this fact is also amply confirmed by electron energy
loss experiments. [36]   So at small q

$$\lim_{q\to 0} \varepsilon(q,\omega) = 1 - \omega_p^2/\omega^2$$

$$\frac{1}{\varepsilon} = \frac{1}{1-\omega_p^2/\omega^2} = 1 + \frac{\omega_p}{2} \left[\frac{1}{\omega-\omega_p+i\delta} - \frac{1}{\omega+\omega_p+i\delta} \right]$$

so that

$$\text{Im} \frac{1}{\varepsilon} = -\frac{\pi}{2} \omega_p \delta(\omega-\omega_p).$$

This delta function is inserted into the integral for L(E). By
changing variables to $\underset{\sim}{k'} = \underset{\sim}{k+q}$, the integral becomes

$$\frac{1}{L} = \frac{e^2\omega_p}{\hbar v} \int k'^2 dk' \, \delta(\frac{k^2}{2m} - \omega_p - \frac{k'^2}{2m}) \int_{-1}^{1} d(\cos\theta) \frac{1}{k^2+k'^2-2kk'\cos\theta}$$

which is evaluated to give

$$\frac{1}{L} = \frac{\hbar\omega_p}{2a_B E} \ln\left| \frac{\sqrt{E} + \sqrt{E-\omega_p}}{\sqrt{E} - \sqrt{E-\omega_p}} \right|$$

For high energies $E >> \omega_p$ this can be approximated by

$$\frac{1}{L} = \frac{\hbar\omega_p}{2a_B E} \ln\left(\frac{4E}{\omega_p}\right)$$

which is similar to the old formula of Bethe for energy loss in solids. [37]

This mean free path is zero for $E < \omega_p$ since obviously plasmons cannot be created in this case. However, for $E > \omega_p$ it rapidly becomes very a short distance, and increases slowly with energy. Most metals have exactly this feature for $L(E)$. Many metals can be fit to a similar curve, which is shown in Fig. 15. [38] Thus long mfp are obtained either at low energy or at high energy, but not inbetween. There photoemission must occur near the surface, since otherwise the electrons lose much energy on the way out.

Next we will show that an energetic electron changes its direction very little when it emits a plasmon. The angular dependence of the above scattering is given by eliminating the angular integrals for the above rate calculation

$$\frac{dw}{d\Omega} = \frac{e^2\omega_p}{2\pi} \int_0^\infty k'^2 dk' \frac{\delta\left(\frac{k^2}{2m} - \omega_p - \frac{k'^2}{2m}\right)}{k^2 + k'^2 - 2kk'\cos\theta}$$

The delta function determines the value of k'. For high energies this is

$$k' = \sqrt{k^2 - 2m\omega_p} \simeq k - \omega_p/v$$

If we now use the small angle approximation on $\cos\theta = 1 - \theta^2/2$ we obtain the rate of scattering per unit solid angle to be

$$k^2 + k'^2 - 2kk'\cos\theta \simeq (k-k')^2 + kk'\theta^2 \simeq k^2[\theta_0^2 + \theta^2]$$

and [35]

$$\frac{dw}{d\Omega} = \frac{e^2\omega_p}{2\pi vh}\left(\frac{1}{\theta^2 + \theta_0^2}\right)$$

$$\theta_0 = \frac{\omega_p}{2E}$$

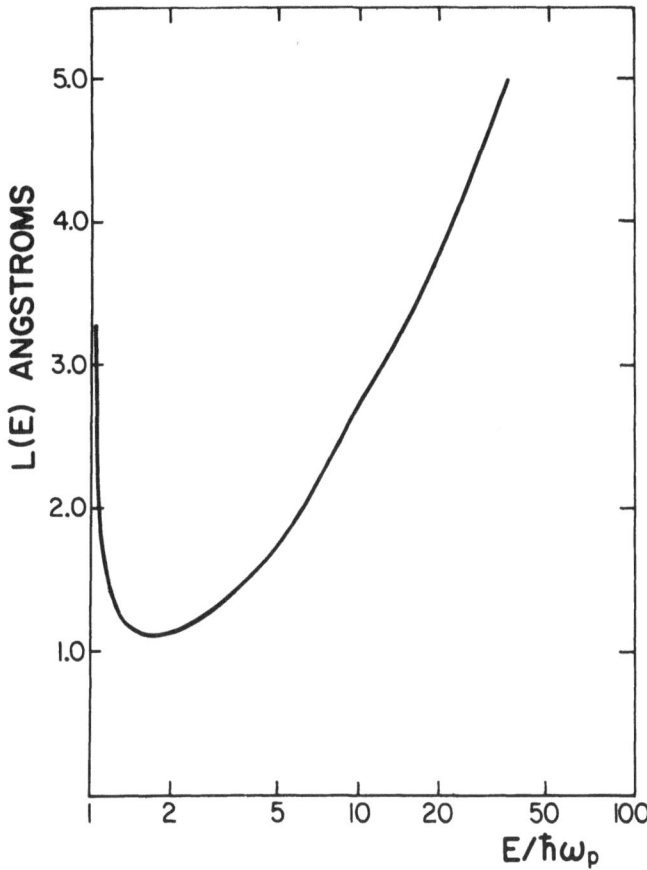

Fig. 15   Mean free path for electrons, from plasmon emission,
          vs energy

Thus the angular distribution is sharply peaked in the forward
direction.[36]   The width is small, since plasmon energies are
5-15eV in metals, while E is typically a kilovolt. Thus θ_0
is typically 1% of a radian or less. Thus an energetic electron
is scattered only a small angle when it emits a plasmon. This
conclusion does not apply to low energy electrons.

In an XPS experiment, an excited electron typically has a
kinetic energy of a kilovolt. After emitting one plasmon, neither
its energy nor direction are significantly changed. Thus the
probability of emitting a second one is identical to that of
a first. This leads to a poisson distribution, where the probability of emitting n plasmons is

$$P_n = e^{-Q} \frac{Q^n}{n!}$$

After going a distance Z in the solid, the factor Q is the probability of one plasmon emission

$$Q = Z/L$$

The exponential prefactor is included to normalize the distribution,
so that the probability of all events is

$$\sum_{n=0}^{\infty} P_n = 1$$

Electron energy loss experiments through metal foils has verified
the details of this theory.[36] Each scattering changes the direction a small amount, and has a lorentzian distribution. The
probability of making n plasmons is a poisson distribution, where
Q is indeed t/L, where t is the thickness of the foil.

C. Extrinsic vs. Intrinsic

Photoemission creates two excitations in the solid. The
excited electron, and the hole it leaves behind. The hole also
is coupled to plasmons. In XPS the hole is fixed in space a
distance Z from the surface. It interacts with plasmons also,
and causes plasmons to be created when it is created. However,
the answer we want is not just the summation of electron and hole
terms, since there are interference terms too.

There is a simple approximation which, once made, gives the
answer in a simple way. This approximation is to assume that at
time t = 0 the electron leaves the hole in a straight line
trajectory, and exits from the solid, Fig. 16. This action of
the electron and hole define a time varying electric field, from
which one can calculate the quantum response of the plasmon system.
This is actually a good approximation, since we have already
illustrated that the electron does not scatter much while emitting
plasmons. Mahan introduced this approximation in 1973,[39] and
recent work by Šunjić and his associates has confirmed the accuracy
of the simple method. [40-42]

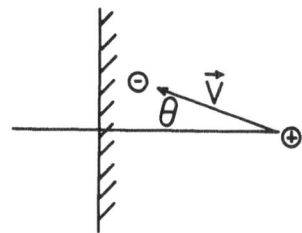

Fig. 16

If a_k are destruction operators for bulk plasmons, then the plasmon system can be represented by the Hamiltonian

$$H = \sum_k [\hbar\omega_p (a_k^+ a_k + 1/2) + \Gamma_k (a_k + a_k^*)]$$

$$\Gamma_k = (\frac{4\pi e^2 \omega_p}{Vk^2})^{\frac{1}{2}} [\sin k_z Z - e^{ik_{||} \cdot \rho_{||}(t)} \sin(k_z Z_e)]$$

The hole is fixed at Z, $\rho = 0$, while the electron is at $Z_e = Z - v_z t$, $\rho_{||} = v_{||}t$. We shall only discuss the case of normal emission, where $v_z = v, v_{||} = 0$. Once the electron has left the solid, when $t > Z/v$ the electron term is zero. This is because bulk plasmons only interact with charges when they are inside the solid, and the interaction ceases outside. This is the reason that the interaction vanishes at the surface as $\sin(k_z Z)$ for the electron and hole. The method of solution follows Lucas and his collaborators. [43] We solve the equation

$$H\Psi(t) = i\frac{d\Psi}{dt}$$

which has the solution

$$\Psi(t) = e^{iH_o t} e^{-\frac{1}{2}|I|^2} e^{-iI a^*+} e^{-iIa} \Psi(0)$$

$$I(t) = \int_0^t dt' e^{-\omega_p t'} \Gamma(t')$$

The state of n plasmons is

$$|n\rangle = \frac{(a^+)^n}{\sqrt{n!}} |0\rangle$$

and this is projected onto the time dependent wave function

$$\langle n|\Psi(t)\rangle = e^{-\frac{1}{2}|I|^2} \frac{(iI)^n}{\sqrt{n!}}$$

The probability of leaving the system with n-plasmons is

$$P_n = \lim_{t\to\infty} |\langle n|\Psi(t)\rangle|^2 = e^{-Q} \frac{Q^n}{n!}$$

where we now sum over all modes

$$Q = \frac{V}{4\pi^3} \int_{k_z>0} d^3k \, |I_k(t=\infty)|^2$$

The quantity $I_k(t=\infty)$ is easy to calculate

$$I_k(\omega) = \left(\frac{4\pi e^2 \omega_p}{k^2 V}\right)^{\frac{1}{2}} \frac{k_z v}{\omega_p^2 - (k_z v)^2} [e^{iz/Z_0} - \cos(k_z Z)$$

$$- ik_z Z_0 \sin(k_z Z)]$$

$$Z_0 = v/\omega_p$$

The next step in the calculation is to do the wave vector integrations. But these diverge unless some cut-off is used at a maximum value of k. This divergence arises from the hole self energy

$$M(Z) = \frac{V}{4\pi^2} \frac{(4\pi e^2 \omega_p)}{V\omega_p} \int_{k_z>0} \frac{d^3k}{k^2} \sin^2(k_z Z)$$

which diverges at large k. Physically the cut-off occurs because of the finite size of the Brillouin zone. We first evaluate the k_z integral since this converges, and then use the cut-off k_c in the other integrations. The hole self energy is

$$M(Z) = \frac{e^2}{2} \{ k_c - \frac{1}{2Z} (1 - e^{-2Zk_c}) \}$$

This hole self energy vanishes at $Z = 0$, but rises sharply as a function of Z if k_c is not small. The evaluation of all the terms in Q yields

$$Q = \frac{e^2}{2\hbar v}\{\ \delta - \frac{1}{2\lambda}\ (1-e^{-2\lambda\delta}) + \lambda\ln(1+\delta^2) - 3\ \tan^{-1}\delta + \frac{2\delta}{1+\delta^2}$$

$$+ \int_0^\delta dx[\frac{e^{-2\lambda x}}{1+x^2} + \frac{4x^2 e^{-\lambda x}}{(1+x^2)^2}\ (\cos\lambda + x\ \sin\lambda)\,]\}$$

(13)

$$\delta = k_c z_0 \qquad\qquad \lambda = z/z_0$$

$$z_0 = v/\omega_p$$

The first term is the hole self energy divided by ω_p. The second term is

$$\frac{e^2}{2\hbar v}\ \lambda\ln(1+\delta^2) = z/L'$$

$$1/L' = \frac{e^2 \cdot \omega_p}{E}\ \ln(1 + \frac{4E}{\omega_p}\ \frac{k_c^2}{2m\omega_p})$$

which is just the electron mean free path term. This is not the same mfp which we evaluated in the prior section, since the earlier one did not depend upon k_c. But the two are similar. The different result in the two derivations arises from the present condition that the electron cannot change its direction while emitting plasmons. The quantity in brackets in (13) is plotted in Fig. 17.

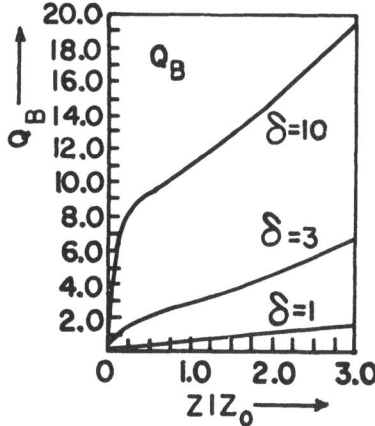

Fig. 17   The average number of plasmons made during an XPS transition from a depth Z.

For small values of δ, the hole term is negligible. But for
large values of δ, Q rises sharply at small Z because of the hole
self energy. The linear dependence of Q_o at high Z is caused by
the mfp term. Since typically $Z_0 \sim 4$-5 Å, or one atomic layer,
then we can approximate Q by the form

$$Q = \beta + Z/L$$

The first term, which is a constant, is called the intrinsic
contribution to Q. The second term is due to the mfp, and is
the extrinsic. We will adopt these as the conventional definitions
of intrinsic and extrinsic. [40] The intrinsic term is usually
viewed as arising from the optical excitation, and it should be
independent of Z. This concept is too simplistic, since our
derivation shows that there are interference terms between the
electron and hole. Related definitions of intrinsic and extrinsic
have been provided by Šunjić and Šokčević. [41]

We adopt the model that the probability of creating n plasmons by an electron from a depth Z is

$$P_n(Z) = e^{-(\beta + Z/L)} \frac{(\beta + Z/L)^n}{n!}$$

Of course, not all electrons come from the same depth. One must
average over the possible depths. This average is taken to be

$$P_n = \frac{1}{D} \int_0^\infty dZ e^{-Z/D} P_n(Z) \tag{14}$$

The factor $\exp(-Z/D)$ could represent the depth profile of emitted
electrons, and then D^{-1} is the absorption constant for x-rays.
However, this attenuation constant is small, so that other factors
are important. An electron in the kilovolt range could knock
another electron from a core level, and thereby lose lots of
energy. It would disappear from the spectrum, as measured in
Fig. 14. We only see those electrons which fail to do this, and
$\exp(-Z/D)$ can be viewed as the probability that they get to the
surface without such a major energy loss. Then D^{-1} is the mfp
for electron energy loss by core level excitation.

The integral in (14) cannot be evaluated in closed form.
There are two limits which are interesting. These are:

All Extrinsic: $Q = Z/L$ $(\beta=0)$

$$P_n = \frac{L}{D} \frac{1}{(1+L/D)^{n+1}}$$

All Intrinsic, $Q = \beta$ $(L = \infty)$

$$P_n = e^{-\beta} \frac{\beta^n}{n!}$$

The general result, with a mix of extrinsic and intrinsic, is [33]

$$P_n = e^{-\beta(\frac{L}{D})} \sum_{m=0}^{n} \frac{\beta^m}{m!} \frac{1}{(1+L/D)^{n-m+1}}$$

The two limiting cases, all extrinsic or all intrinsic, predict different behavior at large n. The intrinsic probability falls off as a poisson distribution. This has a rapid fall-off at large n because of the n! factor. The extrinsic case falls off much more slowly with increasing n. This is why plots like Fig. 14 are interesting at large values of n. The sequence of peak intensities with increasing n can distinquish between the extrinsic or intrinsic mechanisms. In general, if one has a large number of plasmon satellites, then the sequence of peak intensities with n should be sufficient to determine both quantities β and L/D.

This fitting procedure was done for the 2s cores of sodium, magnesium, and aluminum. [33] Both magnesium and aluminum could be well fit to a model which had all extrinsic--and no intrinsic. The sodium data could not be fit with any set of two values for intrinsic or extrinsic. The magnesium and aluminum fits showed that $L/D \sim 0.1$. The lack of intrinsic is a little surprising, since most theoretical treatments, from Fig. 17 onwards, have predicted that a finite fraction is intrinsic. [39-41] One should keep in mind that the data analysis must subtract off the background absorption, and this is sensitive at high satellite number. It is also the high satellites which are crucial in separating intrinsic from extrinsic.

There have been several new experiments reported to probe this phenomena. Flodstrom measured the strength of the valence band plasmon satellite in UPS spectra, and found it totally lacking. [44] This was a surprising result. Recently they have measured the XPS plasmon strength as a function of the kinetic energy of the outgoing electron. [45] Synchrotron radiation was used as a variable source of photon energy. They found that the plasmon satellite intensity dropped as the kinetic energy dropped, and the relation appears to go roughly as $E^{\frac{1}{2}}$. This has also been confirmed by Shirley's group. [46]

All of the above discussion pertains to bulk plasmon excitation. The surface plasmons are also excited in XPS. The surface plasmon excitation probability appears to be independent of final electron kinetic energy, in spite of theoretical predictions to the contrary. [45],[46] Shirley has also measured the angular dependence of the ratio of surface to bulk plasmon intensities, and that agrees better with the theory than the separate intensities.[46]

V.   COLLECTIVE EFFECTS

In this section we are concerned with the shape of the central peak in XPS spectra--the peak labeled P_0 in Fig. 14. Earlier it was mentioned that the spectral shape of this peak was influenced by phonons, Auger decays, and electron-electron interactions. It was once thought that one could compute each of these effects separately, and obtain the final spectra by convoluting them. This procedure, which is followed by experimentalists when interpreting data, was seriously challenged by Šunjić and Lucas. [47] This has started a lively debate. The subject appears complicated, and likely to take some time to sort out. Here we can only provide an introduction to this interesting subject. We begin by discussing each topic separately, before mixing them up--so to speak.

A.   Phonons

Since the core hole is a localized electronic state, the phonon coupling to it is easy to solve mathematically. The details were worked out twenty years ago for F-Centers in alkali halides. The model is quite simple, and has been applied to everything.[48-51] The Hamiltonian

$$H = \varepsilon_c d^+ d + \sum_q [\omega_q b_q^+ b_q + d^+ d \, M_q (b_q + b_q^*)]$$

describes how phonons (b_q) interact with hole states (d). The matrix element M_q provides the interaction between the phonons and core hole. The Hamiltonian can be solved by a transformation

$$H' = e^s H e^{-s} , \quad s = d^+ d \sum_q \frac{M_q}{\omega_q} (b_q^+ - b_q)$$

which gives

$$H' = \bar{\varepsilon}_c d^+ d + \sum_q \omega_q b_q^+ b_q$$

$$\bar{\varepsilon}_c = \varepsilon_c - \Sigma$$

$$\Sigma = \sum_q M_q^2 / \omega_q$$

The quantity Σ is the self energy of the hole, which it gets from the interaction with phonons. The Greens function for the hole is defined conventionally as

$$G(t) = -i < Td(t) \; d^+(0) >$$

However, in the transformed variables it is

$$G(t) = -i\theta(t) \; e^{-it\bar{\varepsilon}_c} < e^{\sum_q \frac{M_q}{\omega_q} (b_q^+ e^{i\omega_q t} - b_q e^{-i\omega_q t})} \; e^{-\sum_q \frac{M_q}{\omega_q}(b_q^+ - b_q)} >$$

This can be solved exactly to give[48-51]

$$G(t) = -i\theta(t) \; \exp[-it \; \bar{\varepsilon}_c - \phi_{ph}(t)]$$

$$N_q = (e^{\beta\omega_q} - 1)^{-1} \; , \phi_{ph}(t) = \sum_q \frac{M_q^2}{\omega_q^2} \; [(N_q+1)(1-e^{-i\omega_q t})$$

$$+ N_q(1-e^{i\omega_q t})]$$

(15)

The spectral distribution function for the core hole is

$$A(\omega) = -2\,\mathrm{Im}G(\omega) = \int_{-\infty}^{\infty} dt \; e^{it(\omega-\bar{\varepsilon}_c) - \phi_{ph}(t)}$$

It would be a delta function $A(\omega) = 2\pi\delta(\omega-\bar{\varepsilon}_c)$ except for the factor $\phi_{ph}(t)$ which provides the spectra shape due to phonons. This may be evaluated by obtaining all of the matrix elements M_q, and then sum over the phonons in the band to obtain $\phi_{ph}(t)$. There is a short-cut which is sometimes useful.[52] In the limit that M_q/ω_q is large, then many phonons are involved. This is the strong coupling limit. Then only the short time response is needed, which is

$$\lim_{\omega_q t << 1} \phi_{ph}(t) = it\Sigma + \frac{t^2}{2} \sum_q M_q^2 \; (2N_q+1)$$

In this limit the spectral function is a gaussian, with a width

$$g = [\tfrac{1}{2} \sum_q M_q^2 \; (2N_q+1)]^{\tfrac{1}{2}}$$

Fortunately the strong coupling limit usually always applies in XPS spectra--in the following sense. If the coupling is weak, then phonon effects are small and can be neglected entirely. They only show up if they are strong. This is because the other

broadening mechanisms--Auger and electron-electron--are not small.
Experimentalists identify phonon effects by looking for a
gaussian broadening which is temperature dependent. [53,54] Note
that for $\beta\omega_q = \omega_q/k_T \ll 1$

$$g = [(k_B T)\Sigma]^{\frac{1}{2}}$$

This provides a simple relationship between two quantities which
can be measured--the gaussian width, and the energy shift Σ.

B.   Electron-Electron Interactions

There are two important terms in the coulomb interaction.
One is between the core hole and the conduction electrons, and
the other is between the conduction electrons

$$H = d^+ d^+ \sum_{kq} V_q \, C_k^+ C_{k-q} + \frac{1}{2} \sum_{qkp} V_q \, C_{k+q}^+ C_{p-q}^+ C_p C_k$$

$$V_q = 4\pi e^2/q^2$$

The general solution to this problem has not been accomplished.
There is an approximate solution due to Langreth which has been
adopted by most theorists. The two important types of excitations
of the electron gas are plasmons and low energy electron-hole
pairs. They both behave as bosons. It is customary to treat
them as bosons, and follow the discussion of the prior section
for phonons. The hole Greens function is

$$G(t) = -i\theta(t) \, \exp[-it\varepsilon_c - \phi_{el}(t)]$$

$$\phi_{el}(t) = \int_{-\infty}^{\infty} \frac{d\omega}{\omega^2} \alpha(\omega) [1 - i\omega t - e^{-i\omega t}]$$

$$\alpha(\omega) = -\frac{1}{\pi} \sum_q \frac{V_q^2}{V_q} \, \text{Im}(1/\varepsilon(q,\omega))$$

This has a form similar to $\phi_{ph}(t)$. They appear to differ by the
temperature factors $N(\omega)$. These are actually present in the
electron-electron case also, but are hidden in the factor
$\text{Im}(1/\varepsilon(q,\omega))$.

Plasmons are the high energy excitation of the electron gas.
In Sec. IV-B we showed that, at long wave length

$$-\text{Im } 1/\varepsilon = \frac{\pi}{2}\omega_p \delta(\omega-\omega_p)$$

so that

$$\alpha = \alpha_o \omega_p^2 \delta(\omega-\omega_p)$$

$$\alpha_o = \frac{1}{2\omega_p} \int \frac{d^3q}{(2\pi)^3} V_q^2/v_q$$

$$\phi_{e\ell}(t) = \alpha_o(1 - i\omega_p t - e^{-i\omega_p t})$$

The factor $\exp(-i\omega_p t)$ in the exponent can be expanded in a series

$$G(t) = -i\theta(t) e^{-it(\varepsilon_c - \alpha_o \omega_p)} e^{-\alpha_o} \sum_{\ell=0}^{\infty} \frac{\alpha_o^{\ell}}{\ell!} e^{-i\omega_p t\ell}$$

and the spectral function is just a poisson distribution

$$A(\omega) = 2\pi e^{-\alpha_o} \sum_{\ell=0}^{\infty} \frac{\alpha_o^{\ell}}{\ell!} \delta(\omega-\varepsilon_c - \alpha_o \omega_p - \ell\alpha_o)$$

This poisson distribution is just the spectral function of the hole. Of course, as was remarked in Sec. IV, one can never make just a hole, but always an electron-hole pair. Thus the poisson distribution which we just computed is not the same as the intrinsic part of Sec. IV. This hole term α_o is related to the intrinsic factor β we had before, but the total intrinsic factor had interference terms between the electron and hole as well.

The other important excitation of the electron gas are the low energy electron-hole pairs which occur at large wave vector. By using the randon phase approximation to the dielectric function, one can show that [51,52,55]

$$-\text{Im } \frac{1}{\varepsilon(q,\omega)} = \omega b [N(\omega)+1] \qquad\qquad -E_F \leq \omega \leq E_F$$

$$b = \frac{2m^2}{\hbar^3} \int \frac{d^3q}{(2\pi)^3} \frac{1}{q} \left| \frac{V_q}{\varepsilon(q,\omega)} \right|^2$$

We shall neglect temperature effects, and set $N(\omega) = 0$. This leaves us with the evaluation of

$$\phi_{e\ell} = ibE_Ft + b \int_0^{E_F} \frac{d\omega}{\omega} (1 - e^{-i\omega t})$$

An approximate evaluation of this integral is

$$\phi_{e\ell} = b[iE_Ft + \ln(itE_F)]$$

The fourier transform for $A(\omega)$ has a cut, and the integral must be evaluated by contour integration. This gives the result [52]

$$A(\omega) = \frac{1}{\pi E_F} \left(\frac{E_F}{\omega - \bar{\epsilon}_c} \right)^{1-b} \sin(\pi b) \ \Gamma(1-b), \ \omega > \bar{\epsilon}_c$$

Thus the spectral function is no longer a delta function, a gaussian, or even a lorentzian. Instead, it is a power law distribution. The hole spectral function is skew on the high energy side of the line, but the XPS spectra is skew on the low energy side. This is energy conservation, since whatever extra energy the system --hole plus electronic excitations--the out-going electron in photoemission has less.

This feature was first predicted by Doniach and Sunjic,[56] and is observed in all XPS spectra from metals. It is related to the edge singularity theory of Mahan-Anderson-Nozieres and deDominicis. [52,57-59] In fact, the latter two authors showed that a better estimate of the factor b is called α

$$\alpha = \frac{2}{\pi^2} \sum_\ell (2\ell+1) \ \delta_\ell^2$$

where δ_ℓ is the phase shift for electron-hole scattering. Citrin, Baer, and Wertheim have measured these exponents for XPS spectra of a number of simple metals.[53] They find them insensitive to core state, so that the same factor α is for the 1s, 2s, or 2p state of the same metal ion. Thus the electron-hole interaction is not sensitive to which hole is in the core. These results are shown in Table I, along with FWHM Auger and phonon widths.

Thus electron-electron interactions have two major effects. They add plasmon satellites, and skew the central line and each plasmon satellite with a power law distribution. A simple approximation is to take the total $\phi_{e\ell}(t)$ for electron-electron interactions as the sum of these two simple terms. Minnhagen [55] has done a much better calculation, in which he evaluates explicitly the RPA result for $\phi_{e\ell}(t)$, and obtains the correct spectra by numerical means. His results show that it is a reasonable approximation to represent the results by these two simple effects.

C. Auger Decay

Two processes contribute to the lifetime of the core hole. One is a radiative decay, wherein a more energetic electron drops to occupy the hole, and emits a photon. The second is an Auger decay, where the dropping electron gives up its energy to another electron, thereby exciting it into the continuum. The Auger process is usually more important, and dominates the lifetime. Some of these processes are shown in Fig. 18. There are three core states, which could be the 1s, 2s, and 2p states of sodium, magnesium or aluminum. The shaded region is the occupied conduction band. The first three processes contribute to the lifetime of the core hole in the 1s state. In (a), a conduction electron falls into the hole, and excites another conduction electron; in (b), a 2s electron falls into the hole and excites another 2s electron; in (c), a 2p electron falls into the hole and excites a conduction electron. These are not the only possible combinations. There is good agreement between theory and experiment for <u>atomic</u> Auger transitions, [18] and metal results differ little from atomic values. [53,]   In general, the inter shell processes, such as (b) are most important. An outer shell electron, as in (d), only has conduction band electrons to provide its Auger transitions, and these processes are slow. For example, in magnesium, the reported values for the lifetime widths (FWHM) in both metal and atom are: 2p is $0.03 \pm .02$eV, and 1s is $0.35 \pm .03$eV. If we estimate a lifetime τ by equating these values to $\hbar/\tau$, the 2p hole lives ten times longer.

These lifetimes are calculated from the Golden Rule. The important question is why the resulting lineshape is lorentzian. After all, none of the other lineshapes we have computed are lorentzian. Let us define the general spectral function for Auger decay

$$\Gamma(\omega) = \pi \sum_{\substack{\text{final} \\ \text{states}}} |M|^2 \, \delta(\omega + E_i - E_f)$$

The measured lifetimes give the energy width

$$2\Gamma(\omega=0) \equiv 2\Delta$$

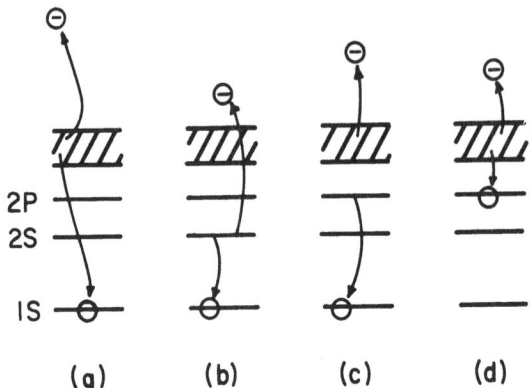

Fig. 18   Different Auger processes in a metal.

The matrix element M is the coulomb interaction between the
orbitals which cause the transitions in Fig. 18, and the
summation is over all allowed transitions. One can show, using
a linked cluster expansion, that the spectral function of the
hole from Auger decays can be represented as

$$G(t) = -i\theta(t) \, e^{-it\varepsilon_c} \, e^{-\phi_A(t)}$$

$$\phi_A(t) = \int_{-\infty}^{\infty} \frac{du}{u^2} \, \Gamma(u) \, (1-iut-e^{-iut})$$

This has the same form as was previously used for phonons and
electron-electron interactions. Although this has the same
mathematical form, the spectral shape of $\Gamma(u)$ is quite different
than the prior cases. Now $\Gamma(u)$ is a smooth, broad band distribution, whose width is of the order of the core hole binding
energy E_B. The phonon distribution was quite narrow, while
the electron-electron distribution had peaks at ω_p and other
structure near zero. When the above integral is evaluated for
a broad band distribution, we obtain the approximation

$$\phi_A(t) \simeq \Delta|t| \qquad\qquad\qquad t > \hbar/E_B$$

For example, one functional form which can be integrated exactly is

$$\Gamma(u) = \Delta\, e^{-|u|/E_B}$$

which gives

$$\phi_A(t) = \frac{2\Delta}{\pi}\, t\, \tan^{-1}(tE_B) - \frac{\Delta}{\pi E_B}\, \ln(1+t^2 E_B^2)$$

The factor $\Delta|t|$ implies a lorentzian line shape. The function $\phi_A(t)$ has to be an analytic function of time, which $\Delta|t|$ is not. However, we now see that $\Delta|t|$ is really an approximation to some function such as

$$\Delta t\,[\frac{2}{\pi}\, \tan^{-1}(tE_B)]$$

which is analytic.

D. Auger and Phonon

The conventional way of interpreting XPS spectra is to assume that each of the three above processes acts independently. Thus the three spectral shapes are convoluted with each other. [53,59] This is equivalent in time space to assuming that the three systems—phonon, electron-electron, and Auger—develop independently with time. Thus, in this approximation, the spectral function is

$$A(\omega) = \int_{-\infty}^{\infty} dt\, \exp[i\omega t - \phi_{ph}(t) - \phi_{e\ell}(t) - \phi_A(t)]$$

When interpreting the experimental data, each of these three terms has been represented by the approximation

$$\phi_{ph}(t) = t^2 g^2$$

$$\phi_{e\ell}(t) = b\, \ln(itE_F)$$

$$\phi_A(t) = \Delta|t|$$

This procedure was questioned by Šunjić and Lucas, [47] and

their remarks have started a lively debate. They focussed on the
interference between the Auger lifetime and the phonon spectra.
However, the remarks also apply to other types of interference,
such as between Auger and electron-electron. However, the
Auger-phonon combination has been chosen as the battleground,
and present debate is on this combination.

Šunjić and Lucas observed that if the core hole is very
shortlived, then the phonons do not get time to act. The
phonons respond on a time scale $\sim\hbar/\omega_q$ and if the core hole is
not around that long, then the phonons should not influence
the broadening. They suggested that equation (15) should be
modified to

$$\phi_{ph}(t) = \sum_q \frac{M_q^2}{\omega_q^2+\Delta^2} \ [\ (N_q+1)(1-e^{-i\omega_q t}) + N_q(1-e^{i\omega_q t})\ ]$$

The factor ω_q^2 has been changed to $\omega_q^2+\Delta^2$. This is an important
alteration since typically $\omega_q\sim.02eV$ while in some cases $\Delta\sim0.4eV$.
For large values of Δ the phonon term is reduced to a negligible
contribution.

This idea has been attacked by experimentalists, who point
out that they see phonon broadening--i.e., a temperature dependent gaussian line width--even in cases where Δ is very large.[60]

I believe that the suggestion of Šunjić and Lucas is essentially correct, although the theoretical formula is different
from the one they suggest. The two processes do interfere. But
we derive an effect which is much smaller, although experimentally significant.

A key fact in understanding the physics is to realize that
there are two or more core hole lifetimes which are involved.
First there is the lifetime of a particular core hole, such as
the magnesium 1s state. This is quite short, since the linewidth is 0.4eV. This is not what is relevent to the phonons.
The lifetime in the Šunjić-Lucas interference is the length
of time for the hole to leave the core of an atom. That is, if
the core hole merely changes its internal state, from the 1s to
the 2s, it is still in the atom insofar as phonons are concerned.
Thus the transition such as Fig. 18c, would contribute to the
1s lifetime, but not to the lifetime of the interference phenomena. We need to define another lifetime, $\tau' = \hbar/\gamma$, which
is based upon how long the core hole remains in the atom. This
is probably governed by the slow process illustrated in Fig. 18d.
It is no longer obvious that $\omega_q \ll \gamma$. In fact, if γ is taken

to be the 2p core hole linewidth mentioned earlier, then $\omega_q \sim \gamma$.

There is another feature to the interference which has been overlooked. The Auger process in Fig. 18b contributes to the 1s lifetime, but it makes two core holes from one. Thus this decay step will increase the coupling to phonons, rather than end it. Thus one should really write down a multichannel process, where the core hole Auger transition either ends the hole-phonon inter-action process(Fig. 18a), increases it (Fig. 18b), or leaves it unaffected (Fig. 18c).

We will ignore the events which make two holes, and concentrate on the process where the hole vanishes from the atom in a time $\hbar/\gamma 2$. Minnhagen [61] has derived one formula for this case, and we shall derive another. The basic procedure is to write down a set of diagrams, or equations of motion, stick in a hole lifetime, and solve the equations. These proce-dures are approximate and _ad hoc_, but at the moment represent the only theories available. We shall try to explain how our result differs from Minnhagen, [61] without urging that either is better.

Puff and Whitfield [62] derived an equation of motion for polarons

$$\left(\frac{id}{dt} - \varepsilon_c\right) G(t) = \delta(t) - i \sum_q M_q^2 \int_o^{-i\beta} dt_2 \, D(q,t-t_2)$$

$$< Td(t) \, d^+(t_2) \, d(t_2) \, d^+(0)>$$

$$D(q,t) = -i[(N_q+1) \, e^{-i\omega_q t} + N_q \, e^{i\omega_q t}]$$

The hole is our polaron. In the absence of a hole lifetime, the time-ordered correlation function

$$< Td(t)d^+(t_2)d(t_2)d^+(0)> = \theta(t-t_2) \; < Td(t)d^+(0)>$$

since the number operator $d^+(t_2)d(t_2)$ is time independent except for its ordering. Thus this factor gives the linear equation for the Greens function

$$[\frac{i\partial}{\partial t} - \varepsilon_c - L(t) \,] \, G(t) = \delta(t)$$

$$L(t) = \sum_q M_q \int_o^{t_2} dt_2 \, D(q,t-t_2)$$

to be the 2p core hole linewidth mentioned earlier, then $\omega_q \sim \gamma$.

There is another feature to the interference which has been overlooked. The Auger process in Fig. 18b contributes to the 1s lifetime, but it makes two core holes from one. Thus this decay step will increase the coupling to phonons, rather than end it. Thus one should really write down a multichannel process, where the core hole Auger transition either ends the hole-phonon inter-action process (Fig. 18a), increases it (Fig. 18b), or leaves it unaffected (Fig. 18c).

Table I: XPS Line Shape Data [a]

Metal		Γ_A(ev) [b]	Γ_{ph}(ev) [c]	α [d]
Li	1s	0.03 ± 0.03	0.37 ± 0.03	0.23 ± .02
Na	2p	0.02 ± .02	0.18 ± .03	0.198 ± .015
	2s	0.28 ± .03	0.20 ± .03	0.205 ± .015
	1s	0.28 ± .03	0.18 ± .04	0.210 ± .015
Mg	2p	0.03 ± .02	0.18 ± .04	0.126 ± .015
	2s	0.46 ± .03	0.19 ± .04	0.130 ± .015
	1s	0.35 ± .03	0.20 ± .05	0.15 ± .02
Al	2p	0.04 ± .02	0.11 ± .02	0.118 ± .015
	2s	0.78 ± .05	-	0.12 ± .015

a. Data from P.H. Citrin, G.K. Wertheim, Y. Baer at room temperature, Phys. Rev. B (to be published).
b. Lifetime of hole, FWHM value
c. Phonon broadening, FWHM
d. Singularity index

Research Supported by a Grant from the National Science Foundation

Reviews

M.L. Glasser and A. Bagchi, to be published

Photoemission from Surfaces, ed. B. Feuerbacher, B. Fitton and
　R.F. Willis (J. Wiley & Sons) to be published

Electronic Structure and Reactivity of Metal Surfaces, ed.
　E.G. Derouane and A.A. Lucas (Plenum, New York, 1976)

X-ray Spectroscopy, ed. L.V. Azaroff (McGraw-Hill, New York, 1974)

N.V. Smith, Crit. Rev. Solid State Sci. $\underline{2}$, 45 (1971)

W.E. Spicer, Comments Solid State Phys. $\underline{5}$, 105 (1973)

D.E. Eastman, in Techniques of Metals Research, VI, ed. E.
　Passaglia (Interscience, New York, 1972)

L. Hedin and S. Lunqvist, Solid State Phys. $\underline{23}$, 1 (1969)

Electron Spectroscopy, ed. D.A. Shirley (North H lland,
　Amsterdam, 1972)

Photoemission, Proceedings of an International Symposium,
　ed. R.F. Willis, B. Feuerbacher, B. Fitton and C. Backx
　(European Space Agency, 1976)

D.C. Langreth, NORDITA Lectures, 1976

Photoelectron Spectroscopy of Solids, ed. M. Cardona and
　L. Ley (Springer-Verlag, Berlin) to be published

References

1. W.L. Schaich and N.W. Ashcroft, Sol. State Comm. $\underline{8}$, 1959
    (1970); Phys. Rev. $\underline{B3}$, 2452 (1971).
2. P.J. Feibelman, Phys. Rev. Letters $\underline{34}$, 1092 (1975).
3. P.J. Feibelman, Phys. Rev. $\underline{B12}$, 1319 (1975).
4. K.L. Kliewer, Phys. Rev. $\underline{B14}$, 1412 (1976); $\underline{B15}$, 3759 (1977).
5. G. Mukhopadhyay and S. Lundqvist, in Photoemission, ed.
    R.F. Willis, B. Feuerbacher, B. Fitton and C. Backx
    (European Space Agency, Noordwijk, 1976) pg. 9
6. C.N. Berglund and W.E. Spicer, Phys. Rev. $\underline{136}$, A1030 (1964).
7. H.Y. Fan, Phys. Rev. $\underline{68}$, 43 (1945).
8. L. Apker, E. Taft, and J. Dickey, Phys. Rev. $\underline{74}$, 1462 (1948).
9. E.O. Kane, Phys. Rev. $\underline{127}$, 131 (1962).
10. G.W. Gobeli and F.G. Allen, Phys. Rev. $\underline{127}$, 141 (1962).
11. R.Y. Koyama and N.V. Smith, Phys. Rev. $\underline{B2}$, 3049 (1970).
12. G.D. Mahan, Phys. Rev. Letters $\underline{24}$, 1068 (1970).
13. G.D. Mahan, Phys. Rev. $\underline{B2}$, 4334 (1970).
14. P.J. Feibelman and D.E. Eastman, Phys. Rev. $\underline{B10}$, 4932 (1974).
15. C. Caroli, D. Lederer-Rozenblatt, B. Roulet, and D. Saint-
    James, Phys. Rev. $\underline{B8}$, 4552 (1973).
16. I. Adawi, Phys. Rev. $\underline{134A}$, 788 (1964).
17. J. Cooper and R.N. Zare, J. Chem. Phys. $\underline{48}$, 942 (1968).
18. W. Bambynek, et al, Rev. Mod. Phys. $\underline{44}$, 716 (1972).
19. H.P. Kelly, in Photoionization and Other Probes of Many
    Electron Interactions, ed. F.J. Wuilleumier (Plenum,
    New York, 1976) pg 83-109.
20. see ref.16
21. D.R. Fredkin and G.H. Wannier, Phys. Rev. $\underline{128}$, 2054 (1962).
22. T. Gustafsson, P.O. Nilsson, and L. Wallden, Phys. Lett.
    $\underline{37A}$, 121 (1971).
23. H. Becker, E. Dietz, U. Gerhardt, and H. Angermuller, Phys.
    Rev. $\underline{B12}$, 2084 (1975).
24. P.O. Nilsson and L. Ilver, Sol. State Comm. $\underline{17}$, 667 (1975);
    ibid, $\underline{18}$, 677 (1976).
25. L. Wallden and T. Gustafsson, Physica Scripta $\underline{6}$, 73 (1972).
26. F. Wooten, T. Huen, and H.V. Winsor, Phys. Lett. $\underline{36A}$,
    351 (1971).
27. N.V. Smith and M.M. Traum, Phys. Rev. Letters $\underline{31}$, 1247
    (1973); Phys. Rev. $\underline{B11}$, 2087 (1975).
28. G.D. Mahan, J. Phys. Chem. Solids $\underline{31}$, 1477 (1970).
29. W.L. Schaich, Phys. Stat. Sol. (b) $\underline{66}$, 527 (1974).
30. D.J. Spanjaard, D.W. Jepsen, P.M. Marcus, Phys. Rev. $\underline{B15}$,
    1728 (1977).
31. P.O. Nilsson and D.E. Eastman, Physica Scripta $\underline{8}$, 113 (1973).
32. K. Siegbahm, et al, ESCA (Amavist and Wiksells, Uppsala, 1967).
33. W.J. Pardee, G.D. Mahan, D.E. Eastman, R.A. Pollock,
    L. Ley, F.R. McFeely, S.P. Kowalczyk, and D.A. Shirley, Phys.
    Rev. $\underline{B11}$, 3614 (1975).

34. R.A. Ferrell, Phys. Rev. 101, 554 (1956); 107, 450 (1957).
35. R.H. Ritchie, Phys. Rev. 106, 874 (1957).
36. O. Sueoka, J. Phys. Soc. Japan 20, 2203 (1965).
37. H. Bethe, Ann. Physik 5, 325 (1930).
38. C.J. Powell, Surface Sci. 44, 29 (1974).
39. G.D. Mahan, phys. stat. sol. (b) 55, 703 (1973).
40. J.J. Chang and D.C. Langreth, Phys. Rev. B5, 3512 (1972);
    B8, 4638 (1973).
41. M. Sunjic and D. Sokcevic, J. Elec. Spec. 5, 963 (1974);
    Sol. State Comm. 15, 165 (1974); 18, 373 (1976).
42. T. McMullen and B. Bergersen, Can. J. Phys. 52, 624 (1974).
43. A.A. Lucas and E. Kartheuser, Phys, Rev. B1, 388 (1970)
    A.A. Lucas, E. Kartheuser, and R.G. Badro, Phys. Rev. B2,
    2488 (1970).
44. S.A. Flodstrom, L.G. Peterson, and S.B.M. Hagstrom, Sol.
    State Comm. 18, 257 (1976).
45. S.A. Flodstrom, R.Z. Bachrach, R.S. Bauer, J.C. McMenamin,
    and S.B.M. Hagstrom, J. Vac. Sci. & Tech. 14, 303 (1977).
46. R.S. Williams, S.P. Kowalczyk, P.S. Wehner, G. Apai,
    J. Stohr, and D.A. Shirley (to be published)
47. M. Sunjic and A. Lucas, Chem. Phys. Letters 42, 462 (1976).
48. M. Lax, J. Chem. Phys. 20, 1752 (1952).
49. R.P. Feynman, Phys. Rev. 80, 440 (1950).
50. C.B. Duke and G.D. Mahan, Phys. Rev. 139, A1965 (1965).
51. D.C. Langreth, Phys. Rev. B1, 471 (1970).
52. G.D. Mahan, Solid State Physics 29, 75 (1974).
53. P.H. Citrin, G.K. Wertheim, and Y. Baer, Phys. Rev. Lett.
    35, 885 (1975); 37, 49 (1976).
54. P.H. Citrin, P. Eisenberger, and D.R. Hamann, Phys. Rev.
    Lett. 33, 965 (1974).
55. P. Minnhagen, Phys. Letters 56A, 327 (1976).
56. S. Doniach and M. Sunjic J. Phys. C3, 285 (1970).
57. G.D. Mahan, Phys. Rev. 163, 612 (1967).
58. P.W. Anderson, Phys. Rev. Letters 18, 1049 (1967).
59. P. Nozieres and C.T. deDominicis, Phys. Rev. 178,1097 (1969).
60. P.H. Citrin and D.R. Hamann, Phys. Rev. B15, 2923 (1977).
61. P. Minnhagen, J. Phys. F6 1789 (1976).
62. R. Puff and G.D. Whitfield, Polarons and Excitons,
    ed. C.G. Kuper and G.D. Whitfield (Plenum, New York, 1963)
    pg. 177.

THE USE OF SYNCHROTRON RADIATION IN UPS: THEORY AND RESULTS

W. E. Spicer

Department of Electrical Engineering

Stanford University, Stanford, California

I. INTRODUCTION

The use of photoemission as a tool for the investigation of the electronic structure of matter has grown at an explosive rate in the last 15 years. For those of you who have become interested in the field in the last few years, it is probably difficult to realize that fifteen or twenty years ago only a small number of papers (approximately 10-20) were being published throughout the world on photoemission per year. Today there are many individual laboratories or groups which publish more than this number per year and the total number of papers are in the hundreds per year. To properly appreciate the growing effect of the use of synchrotron radiation on photoemission, one must view the opportunities opened by synchrotron radiation in terms of the continuous development of photoemission as a tool for exploring the electronic structure of matter. One aspect of this development would be the study of atoms or molecules in the gaseous state. We will not attempt to cover this aspect here but will concentrate on the use of photoemission to study the electronic structure of solids or gases sorbed on or in solids.

It is useful to divide the use of photoemission into two parts – that involved with exploring the bulk electronic structure of solids and that involved with the surface electronic structure. The surface is defined as the last few layers of the solid in which the electronic structure may be different from that of the bulk because of the termination of the lattice. Conversely, the bulk is defined as that part of the solid which is not affected by the surface.

54

By electronic structure, we mean not only the valence band structure but also the core levels which were studied prior to synchrotron radiation primarily by x-ray Photoemission Spectroscopy (XPS).[1] Dr. Wertheim will cover XPS in these proceedings. When we speak of surface electronic structure, we also include the core states of the surface atoms as well as the valence and surface states.

If one looks at the development of photoemission for probing the the electronic structure of solids, one finds that two types of problems had to be overcome. One was experimental. It included developing electronics which would allow energy distributions of the emitted electrons to be measured rapidly and accurately with a minimum of difficulty and expense. This condition was met in the early 1960's with the application of phase sensitive detection to the problem.[2] Another experimental problem was that of obtaining energy distribution curves (EDC's) of the photoemitted electrons over a wide hν range. The use of vacuum monochromators as well as surfaces with thresholds of photoresponse lying in the visible in the 1950's and early 1960's began to solve this problem and to demonstrate the potential of photoemission for probing the electronic structure of solids.[3]

The second key problem was that of interpreting photoemission studies. A timely development here was the development of the three-step model of photoemission. This allowed for relatively easy and definitive interpretation of many photoemission experiments. As this model was developed[4] and systematically applied[3,5,6] with success in metals as well as semiconductors, the potential of photoemission began to be more generally appreciated andthe number of papers began increasing markedly.

A major breakthrough in the use of x-ray sources[1] to probe the core, as well as the valence levels, was the establishment of the chemical shift concept according to which the binding energy of a core level of an atom can be used to determine the chemical state of that atom.[1]

The importance of ultraviolet photoemission spectroscopy (UPS) was established by the use of vacuum ultraviolet spectrometers and LiF windows with special seals to insure an ultrahigh vacuum environment for the samples under study.[7] However, the LiF window limited such studies to hν < 12 eV. This field received an important advancement with the application of resonance arcs[8] which provided a few discrete spectral lines at higher hν (eg the He 21.2 and 41 eV lines).

By the early 1970's photoemission spectroscopy was a well

established discipline divided into two parts, UPS (hν < 42 eV) and
XPS (hν ⪆1 keV). The appearance of synchrotron radiation has
changed this whole picture by the introduction of a continuous
spectral source spanning the regions from the visible into the hard
x-ray region. Thus photoemission spectroscopy is now being practiced
over an increasingly wide range of photon energy allowing for core
and valence; surface and bulk studies can be made on the same sample
by varying the wavelength of radiation[9] over a wide range.

II. CHARACTERISTICS AND BENEFITS
OF SYNCHROTRON RADIATION

Figure 1 indicates the intensity and spectral distribution of
synchrotron radiation from SSRP.[10] As can be seen, only synchrotron
sources offer continuous radiation from the visible to the hard
x-ray region. Previously the only continuum source consistant with
ultrahigh vacuum (UHV) was the vacuum monochromator – LiF window
combination which cuts off for hν > 12 eV. We will give a number
of examples of the application of synchrotron radiation to photoemission here and in each case we will indicate how the ability
to select the optimum wavelength or wavelengths for a given study
was essential to that study; however, we should first examine some
of the attributes of synchrotron radiation which are essential to
its usefulness.

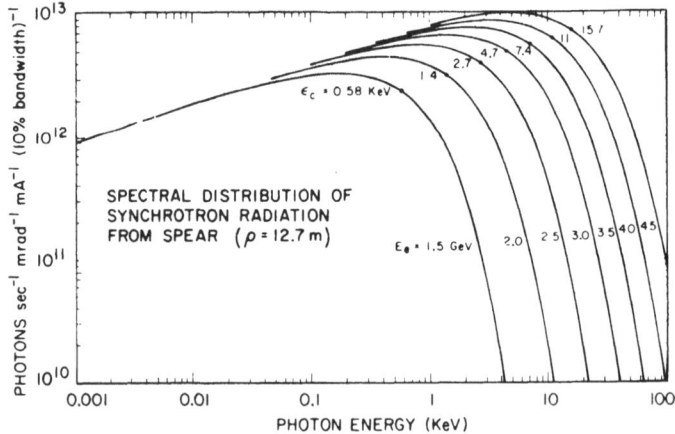

Fig. 1   The spectral distribution of the synchrotron radiation from
         the storage ring SPEAR at Stanford University for various
         energies of stored electron beam. The energy is indicated
         on each curve.

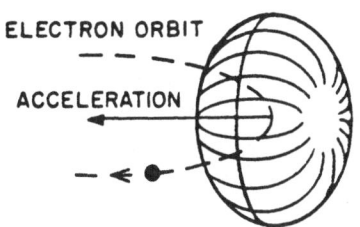

CASE I : $\frac{v}{c} \ll 1$

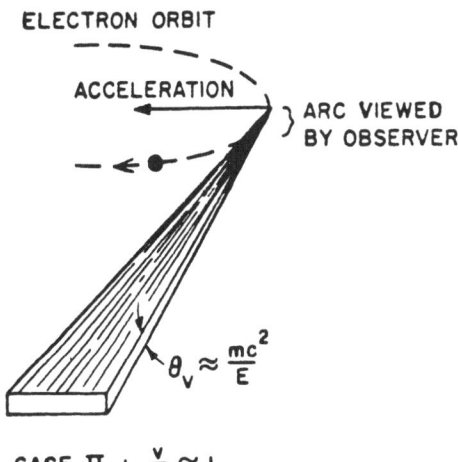

$\theta_v \approx \frac{mc^2}{E}$

CASE II : $\frac{v}{c} \approx 1$

Fig. 2   Radiation emission patterns by electrons in circular
motion at classical and relativistic velocities.

The first of these attributes is the ability to achieve high
intensity after the radiation is sent through a monochromator.[9]
One reason for this is the very large intensity of the emitted
radiation; however, another very important attribute of synchrotron radiation is its natural collimation. This is indicated by
Figure 2.

Figure 2 gives the radiation patterns of nonrelativistic and
relativistic electrons due to their being bent as in a storage ring.
The relativistic effects transform the classical, "donut" shape
pattern into a pattern very strongly collimated in the forward
direction.[12] Thus, as the relativistic electrons are accelerated
by the bending magnets of the storage ring or accelerator, a flat
"pancake" of radiation is produced with very little dispersion away
from the plane of the electron orbit. This lack of dispersion is
very important because it allows the radiation to be collimated
with much less loss than that which would be given by point sources
of radiation. This is particularly important for higher energy
storage rings such as SPEAR[9,11] and Doris[9] since, because of the
radiation hazard, the experimental area must be placed many meters
from the electron beam.

Another important characteristic of the radiation is its
natural polarization (linear) in the plane of the electron orbit.
This can be particularly important for studies involving optical
selection rules and/or angularly resolved photoemission. Also very
important for optics is the small source size and high emittance.
An additional helpful feature is the high-vacuum environment (which
is essential for long electron storage times) of the storage ring.
More details on the characteristics of storage rings can be found
elsewhere.[11,12]

III. FUNDAMENTALS OF THE PHOTOEMISSION PROCESS

A. Introduction

The three-step model[4,6,9,13] is an approximate description
of the photoemission process which has proven extremely useful and
has played a key role in the development of photoemission as a
scientific tool. The reason for the success of this model has
been its simplicity. It divides the photoemission process up into
three successive events each of which can be studied and treated
independently. Most importantly the essential parameters for each
step can be determined and then applied to photoemission.

The weakness of the three-step process is in its treatment of
each step as being independent. Ideally, they should be coupled
and, in the extreme limit of photoelectrons originating only very

close to the surface, the three processes must be treated as simultaneous, completely coupled events.[14] In fact it is exactly in this
limit, that of very short escape depth, that the three-step model
should be used with extreme caution, if at all. However, as the
escape depth of electrons gets longer, e.g. several unit cells or
more, the three-step model appears to be a very useful approximation.

B.   The Three-Step Model

The photoemission is divided into three successive events which
are treated independently: 1) the optical excitation event, 2) the
transport or movement of the excited electron to the surface and,
3) escape over the potential barrier at the surface.

There are two quantities which can be measured in photoemission:[7] 1) the distribution in energy of the photoemitted
electrons and 2) the quantum yield, i.e. the electrons emitted per
absorbed photon. If the energy distribution is given on an absolute
basis, i.e. electrons emitted per eV energy range per photon absorbed,
the yield is the integral of the energy distribution over all energy.
We will concentrate on the energy distribution; however, it should be
recognized that the yield can be obtained from the energy distribution.

In the three-step model, the optical excitation event is assumed
to be governed by the bulk optical constants (in cases where it is
not, a different formulization of the optical excitation process is
often possible). Thus, the probability of an electron being excited
at depth x from the surface (see Figure 3) by a photon of energy $h\nu$
to a final state energy E is:

$$\xi(E, h\nu)dEd\ dx = \sigma(E,h\nu)(1-R(h\nu))e^{-\alpha(h\nu)x}dEdx. \qquad (1)$$

Where $\sigma(E,h\nu)$ is the differential optical absorption coefficient,
(i.e. cross section, for excitation to a final state between E and
E + dE). $R(h\nu)$ is the optical reflectivity at the surface, $\alpha(h\nu)$ is
the optical absorption coefficient (for excitation to all final state
energies), and x is the distance from the surface.

Both theoretically and experimentally,[4,6] it has been shown that
the probability of the excited electron reaching the surface without
appreciable energy loss can be approximated by:

$$Q(E) = e^{-\frac{x}{L(E)}}. \qquad (2)$$

Where L(E) is the characteristic scattering probability of an electron

excited to a final energy E. Table I gives some details of the
principal scattering mechanisms. In Fig. 4 we present escape lengths
against combined electron-electron and electron-plasmon scattering
for a number of materials.[15] Additional data and theoretical treatment can be found in the literature.[16]

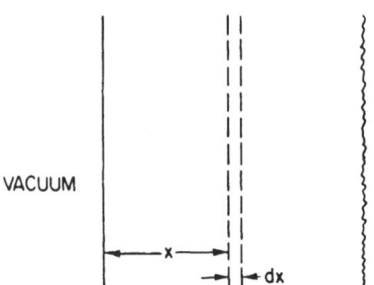

VACUUM

Fig. 3   The photoemission due to excitation from a slab
         of material of thickness dx located a distance x from the
         surface.

Table I

Type of Scat-tering	Threshold	Typical Energy Loss (per event)	Typical Scatter-ing Lengths
Electron-electron	Metal – none non–metal– bandgap	Large fraction of excitation energy (electron-volts)	See Fig. 4 can be few Å
Electron-plasmon	$>$ plasmon energy, $h\nu_p$	$h\nu_p$ or $nh\nu_p$ (n = integer)	Order of 10 Å
Electron-phonon	none	Small (tens of meV)	10 – 100 Å

     The probability of escape over the surface barrier is B(E).
As Fig. 5 shows B(E) is zero if the electron energy is less than
the height of the potential barrier at the surface. B(E) usually
rises monotonically as energy increases above the potential
barrier.

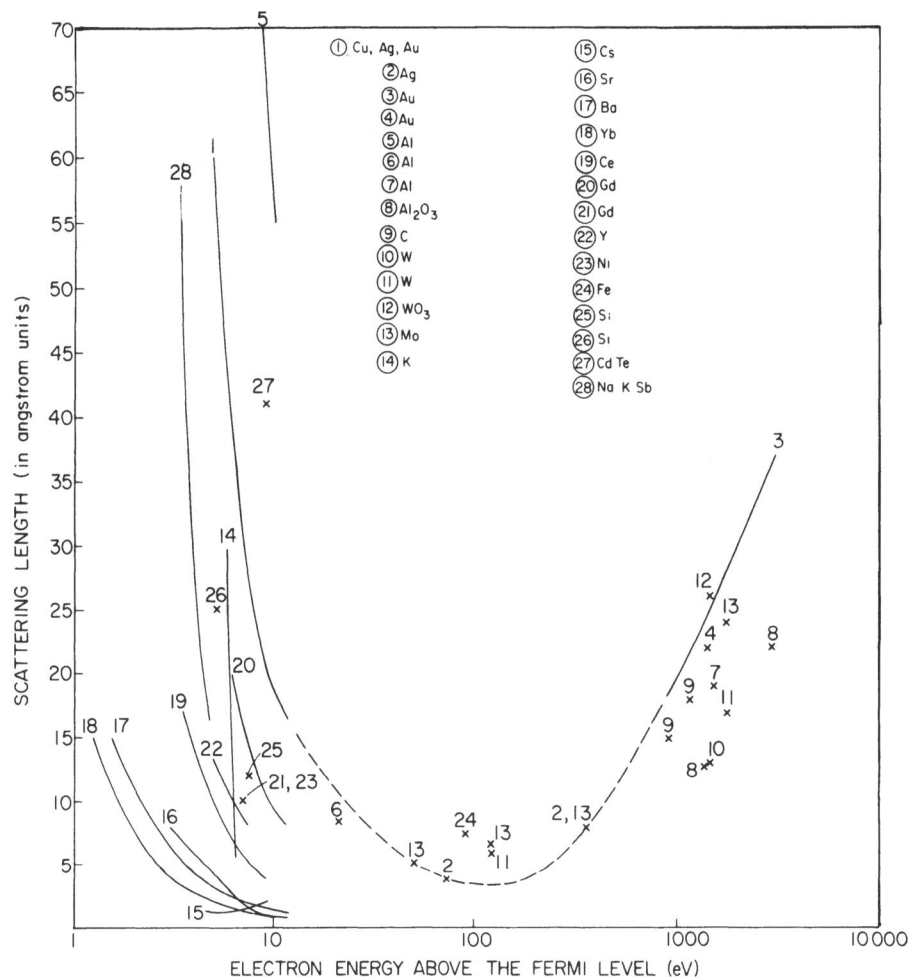

Fig. 4 A summary of data on electrons escape depth versus hν.
The solid curve on the right hand side of the figure is
curve number 3. This figure is taken from Ref. 15 where
references for all data is given.

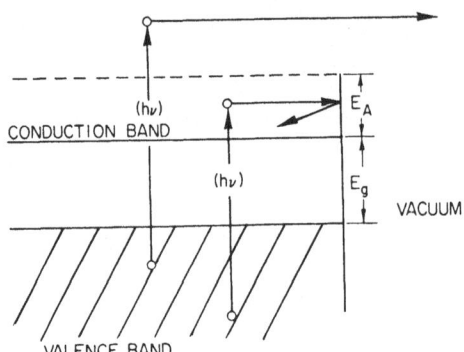

Fig. 5   Photoemission from a semiconductor. Note that only those
          electrons excited into states above the potential barrier
          at the surface can possibly escape from the solid. E_A is
          the electron affinity and E_g the bandgap.

The probability of a photoelectron being excited to energy E
and escaping without appreciable energy loss is given by the product
of probabilities of each of the three steps:

$$P'(E,h\nu,x) = \xi(E,h\nu)dEdxQ(E)B(E). \qquad (3)$$

Substituting in Eqs. (1) and (2) one obtains:

$$P'(E,h\nu,x)dEdx = \sigma(E,h\nu)(1-R(h\nu))e^{-\left[\alpha(h\nu)+\frac{1}{L(E)}\right]x}B(E)dEdx \qquad (4)$$

Assuming a semi-infinite photoemitter and integrating x from 0 to ∞
one obtains:

$$P'(E,h\nu)dE = \frac{\sigma(E,h\nu)(1-R(h\nu))B(E)dE}{\alpha(h\nu) + \dfrac{1}{L(E)}} \qquad (5)$$

Defining a probability, $P(E,h\nu)$ per absorbed photon one obtains:

$$P(E,h\nu) = \frac{\sigma(E,h\nu)B(E)dE}{\alpha(h\nu) + \dfrac{1}{L(E)}} \qquad (6)$$

C.   Effect of Various Scattering Events
     and Contribution of Scattered Electrons

Equation 6 gives the probability of escape without appreciable

energy loss. Here appreciable should be equated with measurable
since the experimental energy resolution is rarely better than
0.1 eV, an electron can suffer several electron-phonon collisions
without appreciable energy loss. However, one electron-electron
collision will normally result in energy loss of many eV.

The experimentally measured energy distribution, $N(E,h\nu)dE$,
contains not only the electrons which escape without appreciable
energy loss but also the distribution produced by the scattering
process, $S(E,h\nu)dE$. Thus,

$$N(E,h\nu)dE = P(E,h\nu)dE + S(E,h\nu)dE. \qquad (7)$$

To first order, it is easy to separate the $P(E,h\nu)$ and
$S(E,h\nu)$ in experimental data since $P(E,h\nu)$ usually gives strong
structure which moves to higher energy as $h\nu$ is increased. In
contrast $S(E,h\nu)$ usually varies monotonically with energy without
sharp structure (except for plasmon production).[4,13] Normally, the
distribution of scattered electrons tends to build up and peak near
low energies. This is shown schematically in Fig. 6. Much more
detail of the scattering processes can be found in the literature.[17]
Since synchrotron radiation is usually used for $h\nu > 12$ eV, the
electron-electron (including plasmon creation) scattering event
usually is dominant and will be of most importance for the work
discussed here.

IV.   BENEFITS OF SYNCHROTRON
      RADIATION IN PHOTOEMISSION STUDIES

A. Introduction

As emphasized earlier, the key benefit of synchrotron radiation
is that it produces the first continuous source of radiation extend-
ing from the infrared to the hard x-ray region. As mentioned in
Section II, the self-collimation and small source size of the
radiation helps in the development and application of monochromators
to produce tunable monochromatic radiation sources. However, full
utilization of the potential of synchrotron radiation is hindered
by the availability of suitable monochromators for photon energies
above a few hundred eV. For example, an excellent monochromator
is available at SSRP which is designed to provide a constant resolution of 0.1Å for $h\nu > 32$ eV.[18] This corresponds to a very respect-
able resolution of 0.1 eV at 100 eV but only 0.5 eV at 500 eV.
Monochromators are now being designed and built which will provide
better resolution and intensity in the range up to 800 eV. In addition, research and development is starting to provide monochromators with high resolution for $500 \gtrsim h\nu \gtrsim 2000$ eV.

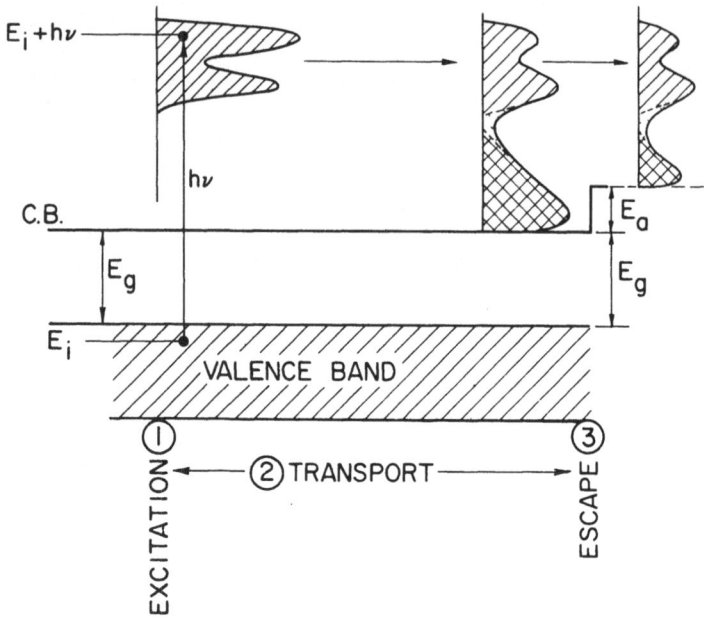

Fig. 6   A schematic diagram illustrating the three-step photoemission process. An initial optically excited distribution is shown on the left. Changes in this distribution
as it approaches the surface (due to inelastic scattering)
and after it has escaped into the vacuum (due to the
potential barrier at the surface) are indicated.

There are at least four primary advantages to be gained by
having photon energies available over a wide energy range.

1) Providing the ability to examine core as well as valence
states.

2) Being able to take maximum advantage of the photon dependence
of the matrix elements. Here we include the dependence of excitation probability in angularly resolved as well as angularly integrated photoemission. An additional benefit of synchrotron radiation is that it is naturally linearly polarized radiation, i.e.
the electric field vector associated with the radiation is uniquely
defined. This, of course, can be of particular importance for making use of or studying matrix elements - particularly in the case
of angularly resolved photoemission.

3) Tuning the photon energy of a given transition in order to
obtain the minimum escape depth (see Fig. 4) and thus maximize

surface sensitivity, or conversely tuning to an energy where the
escape depth is long to emphasize the bulk rather than the surface.

4) Under the proper conditions photoemission can be used to
measure either the bulk optical absorption coefficient[19] or optical
absorption characteristic of the surface.[20,21]

In the following section, we will illustrate these aspects of
photoemission using synchrotron radiation. However, it should be
kept in mind that these advantages are often interrelated. For
example, one may choose a particular $h\nu$ to study adsorption of a
foreign gas on a surface in order to get the proper compromise
between minimum escape depth and maximum matrix element.

The matrix elements for transitions from different core states
may have quite different $h\nu$ dependencies. This is illustrated[21,23]
by Figs. 7 and 8. In these figures the absorption cross section
giving the probability of producing a photoelectron with the energy
characteristic of excitation from 3d (Fig. 7) or 4d and 5d (Fig. 8)
core levels is presented. As can be seen from Figs. 7 and 8, the $h\nu$
dependence of the 3d and that of the 4 or 5d cross section are quite
different. Spectra depending on these characteristic cross sections
are given in Section V-B and VIB.

V. TUNING SYNCHROTRON RADIATION TO MINIMIZE ELECTRON
    ESCAPE DEPTH AND THUS MAXIMIZE SURFACE SENSITIVITY

.A. Introduction

In Section III, we introduced the concept of escape depth and
in Fig. 4 indicated how escape depth varies as a function of electron
energy for a number of solids. In this section, we will illustrate
how the ability to tune the photon energy in order to minimize the
escape depth can give very high surface sensitivity for core as well
as valence states. The examples will be taken from studies of
GaAs[24,26] which was one of the first materials for which these
techniques were applied.

In Fig. 9, we present the escape depth versus electron energy
as determined for GaAs. As can be seen, the escape depth goes
through a shallow minimum at about 60 eV. The minimum falls at
about one and a half molecular layers; thus, roughly half the
emission will come from the first molecular layer and half from
deeper in the sample.

We will examine two cases of the use of the surface sensitivity
given by a short escape depth. The first will be for the Ga and
As 3d core states in which the shifts observed in these due to

adsorption of oxygen is studied. The second will be the GaAs
surface valence states which show very interesting effects which
we associate with the rearrangement of surface atoms on the surface.

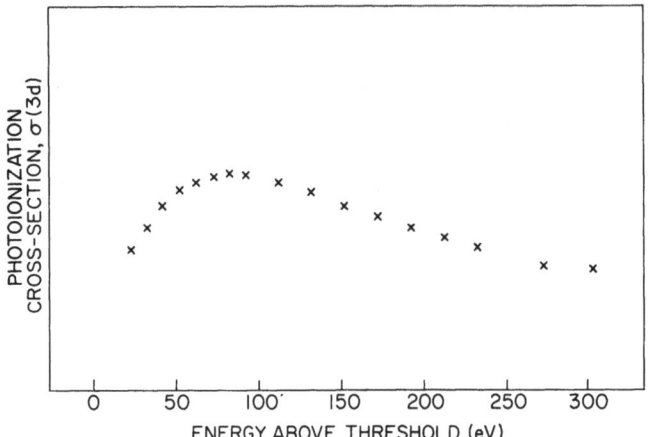

Fig. 7 Photoionization cross section versus hν for exciting an
electron from a 3d (Ga) core level without intrinsic
energy loss. An intrinsic energy loss is one which is
inherent to the excitation process.

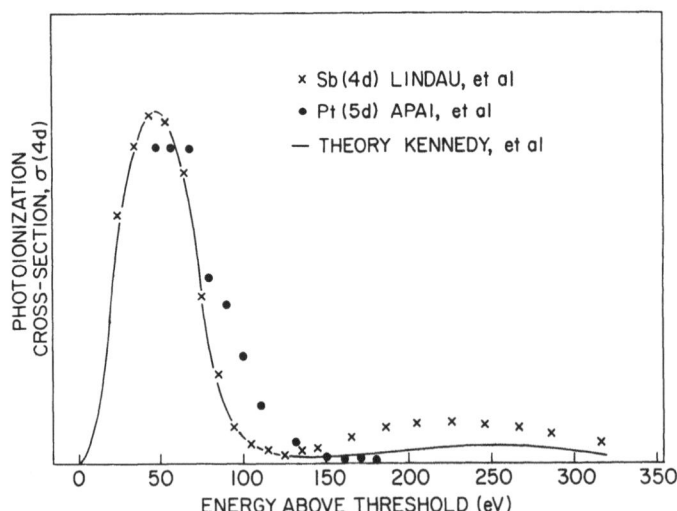

Fig. 8 Photoionization cross section versus hν for exciting an
electron from 4d and 5d levels without intrisic energy loss.
The strong minimum at ~150 eV is due to the Cooper minimum.

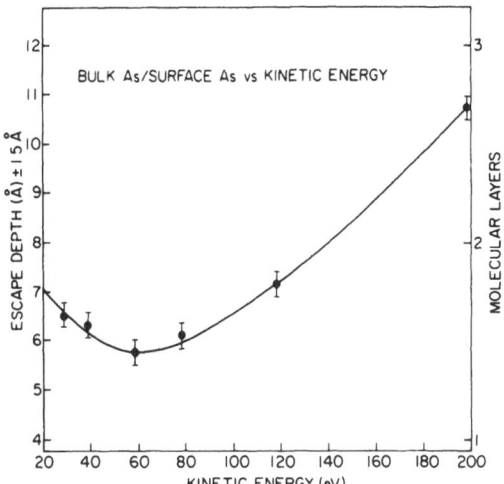

Fig. 9   The escape depth for photoelectrons from GaAs. In order to
         optimize matrix elements, as well as escape depth, it is
         optimum to work near 21 eV for valence band studies and
         near 100 eV for studies of the Ga and As 3d levels as
         well as the oxygen 2p levels.

        B.   GaAs Core Shifts Due to Oxygen Adsorption

     In Fig. 10a, we present spectra taken from the (110) GaAs face
at $h\nu$ = 100 eV as a function of oxygen exposure. (The exposure is
indicated in terms of Langmuirs, L, on each curve. 1 L = 10^{-6}
torr-sec.) An atomically clean (110) surface was obtained by cleaving in a vacuum of 10^{-10} torr. In order to investigate the bonding
of oxygen on an atomic scale, one wishes to see simultaneously the
surface core states of both the Ga and As atoms. $h\nu$ = 100 eV was
chosen because the kinetic energy of the photoexcited As 3d electrons
would be about 60 eV (which places it at the minimum of the escape
depth curve); whereas, the kinetic energy of the Ga 3d's is about
80 eV - close to the escape depth minimum. In addition, as can
be seen from Fig. 3, the photoionization cross section, i.e. excitation probability for the 3d peak is reasonably close to its maximum value for each of these transitions. Whereas, if we had used
a small $h\nu$ (to get shorter escape depth for Ga 3d's) the As 3d
cross section would have been noticeably decreased.

     The benefit of examining the core levels at the surface is well
illustrated by Fig. 10a. As can be seen, there is a single new As
peak shifted to higher binding energy which grows with oxygen exposure; whereas, no shift is observed in the Ga peak. In Fig. 10a,
at approximately monolayer coverage (10^{12}L), the shifted and unshift-
ed As peaks are about equal. This shows that about half of the

photoelectrons are coming from the surface As atoms. If the escape
depth had been longer (as for example with $h\nu \simeq 1.5$ KeV for XPS)
it would have been much harder to see the chemical shift at low
coverages, e.g. 10^6L since a much smaller fraction of the electrons
would have come from the surface.

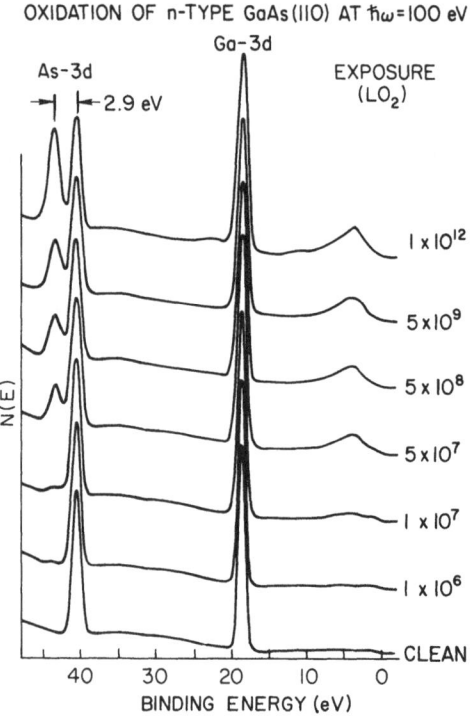

OXIDATION OF n-TYPE GaAs(110) AT $\hbar\omega$=100 eV

Fig. 10a The 3d core levels of Ga and As in GaAs as a function of
         exposure to oxygen in its ground state. Note the strong
         (2.9 eV) As 3d shift and the lack of measurable shift of
         Ga 3d levels. This data is interpreted in terms of chemi-
         sorption of oxygen. The structure at about 6 eV which
         grows with increasing exposure is due to oxygen 2p levels.

The data of Fig. 10a gives definitive evidence that the oxygen
removes electrons from the As in chemisorbing on the surface of the
GaAs. Based on the chemistry of elemental As and Ga, some people had
expected the oxygen to preferentially remove electrons from the Ga
rather than the As. In fact, as Fig. 10b shows, this is what
happens when one passes the chemisorption stage and breaks the GaAs
covalent bonds in order to form bulk Ga and As oxides.

In the studies of the oxygen on GaAs, it has also become apparent that one must control the vacuum and GaAs surface with great
care.[24] One necessary condition is that the radiation source not
compromise the vacuum. This is almost automatically satisfied if a
storage ring provides synchrotron radiation since the storage ring
requires good vacuum in order to achieve long storage times for its
electron beam. As will be pointed out later, it is also important
not to have any source of excitation of the oxygen (for example,
hot filaments or high voltages) and these are not inherent in a synchrotron source as they are in some other sources.

The explanation of the detailed nature of the chemisorption on
(110) GaAs is discussed in detail elsewhere.[21,24,25] Suffice it to
say here that each Ga and As atom on the (110) face is covalently
bonded to three neighbors which would leave one "dangling" bond
on each atom. However, the As core has two more positive charges
than the Ga. As a result, both "dangling" electrons tend to
reside on the As surface atoms - leaving two As valence electrons
not involved in the covalent bonds which are available for the
oxygen chemisorption.

Finally, a little should be said about the details of oxygen
exposure - again details are available from the literature. In
order to limit the oxygen to chemisorption (Fig. 10a) great care
had to be taken[24] that there is no excitation of the oxygen. This
was done by turning or valving off all hot filaments, ion guage or
or other possible sources of excitation. This is necessary even
when the source of excitation is well removed from (and well out
of line of sight of) the sample under study. As can be seen from
Fig. 10a only the As shift of 2.9 eV (which we associate with chemisorption) takes place even when the GaAs is exposed to an atmosphere
of pure, unexcited oxygen for about 15 minutes (10^{12} L).

In contrast (see Fig. 10b) the bulk oxides are formed by orders
of magnitude lower exposure ($\sim 10^6$ L) by simply turning on an ion
guage which is well out of line of sight of the GaAs, i.e. any
oxygen excited by the ion guage must be scattered from several
metal surfaces before it can reach the GaAs. In order to understand
this behavior, it is important to realize that oxygen has a very
long lived (~ 15 min) excited state. Thus, even if oxygen ions
or atoms cannot reach the sample, the long lived oxygen excited
states almost certainly can.

By tuning the photon energy to 21 eV one can look at the valence
band within the last few layers of GaAs. As can be seen from Fig. 9,
21 eV does not give the minimum escape depth. However, the optical
matrix elements for excitation from the valence band drop very fast

as hν is increased. As a result, hν = 21 eV represents an optimum compromise between minimizing the escape depth and obtaining a large enough matrix element to make the measurements feasible.

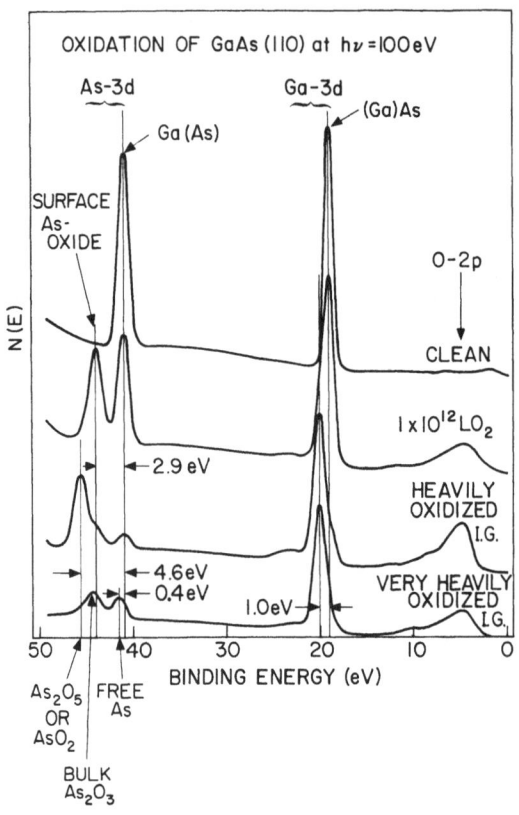

Fig. 10b   The chemical shifts after heavy oxidation of GaAs. Note
           the various As shifts due to different chemical shifts
           as well as Ga shift corresponding to Ga_2O_3.

Figure 10c shows the change in the valence band (VB) structure associated with the surface as a function of time after cleaving to form the clean surface and as a function of exposure to un-excited molecular oxygen as in Fig. 10a. Note first the difference in the VB between the dashed and solid clean curves. This took place due to allowing the crystal to stand at room temperature for about 12 hours after the cleave. The dashed curve was taken just after cleaving and the solid curve after the room temperature "anneal." Note that the Fermi level, E_f, of this n-type sample,

was pinned just after cleaving. The pinning was removed by the
annealing and then brought back by oxygen exposure. It is
established that the pinning is due to extrinsic surface states;
thus, these must be removed by the annealing.

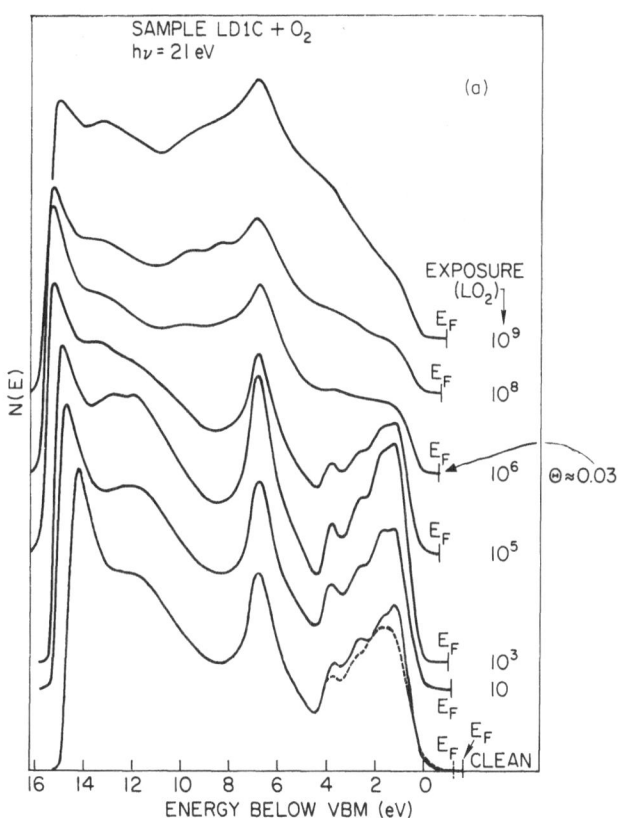

Fig 10c.   The valence band at the surface of GaAs as a function of
           exposure to unexcited molecular oxygen as in Fig. 10b.
           Note the strong, phase-like change in electronic structure
           for a coverage θ = 0.03, i.e. 3% of a monolayer.

     Up to 10^5L oxygen exposure, there is little change in the
valence band; however, between 10^5 and 10^6L there is a very strong
and abrupt change. It has been suggested that this is due to dis-
order induced in the valence band by a relatively small oxygen over-
age (~ 3% of a monolayer). A detailed discussion of possible causes
for this is given elsewhere.[21,26]

VI.   USE OF PHOTON ENERGY DEPENDENCE OF ABSORPTION
CROSS SECTION TO SEPARATE PHOTOEMISSION FROM
TWO SOURCES WHICH ARE DEGENERATE IN ENERGY

A.   Introduction

If one examines Fig. 9, one notices that there is very little
photoemission from the valence band. This is because for 100 eV,
the matrix elements for excitation from the valence band are very
small in agreement with the discussion in the last section. However, as the oxygen exposure is increased, a strong peak appears
at about 5 eV binding energy. This is due to emission from the
2p levels of the adsorbed oxygen. The area under this 2p peak is
important because it gives a measure of the oxygen coverage.[26]
However, it should be noted that it overlays the valence bands and
is only clearly visible because the matrix elements for the GaAs
valence bands are so weak in this photon energy range. This illustrates an important point. When two sets of levels are
degenerate in energy, they can be separated if a photon energy can
be found at which the matrix elements for one transition are
dominant.

In order to emphasize the potential of photoemission based on
synchrotron radiation to determine the amount of adsorbed gas, it
is useful to make a comparison with Auger techniques. In the last
few years, Auger electron spectroscopy has been used increasingly
to monitor the amount of a gas adsorbed on a surface. However,
recently, there is increasing evidence that the adsorption itself
can be strongly perturbed by the electron beam needed for the Auger
measurement. In contrast, photoemission techniques such as that
described above appear to perturb the adsorbed species much less
than the electron beam. Thus, photoemission techniques may be used
increasingly to determine quantitatively the amount of adsorbed
gas. As synchrotron sources are able to produce monochromatic
radiation at higher energies so that the deep core states of the
of the gas, e.g. the oxygen $1s^2$ states can be probed, this use of
photoemission will become increasingly important.

In this section we will illustrate a use of the matrix element
or cross section difference between the adsorbing solid and the
adsorbed gas which is somewhat different from that discussed for
GaAs. This is the use of the photon energy dependence of the
matrix element to suppress the substrate emission so that weaker
features of the spectra due to the adsorbed gas can be seen.[26]

B.   "Missing" Bonding Orbitals of CO Adsorbed on Metal

The UPS spectra of CO adsorption on various metals has perhaps

been studied more than that of any other adsorbed gas. Fig. 11
indicates rather symbolically, the spatial distribution of the CO
valence orbitals. For CO in the gas phase, the 4σ, 1π, and 5σ
levels have binding energies of 20 eV or less; whereas, the 3σ has
a binding energy of about 38 eV. When CO is adsorbed on a metal,
the 5σ levels bond to the surface and are moved downward in energy
so that they merge in energy with the 1π orbitals, whereas, the

4σ levels retain their independent identity.[27] However, prior to
the work to be described below, the 3σ levels had never been clearly
observed for the adsorbed molecule. There were at least two reasons
for this. First, the matrix elements for the 3σ transition are
weak in the ultra-violet range where most of the previous work was
done. Second, the secondary electrons produced by scattering of
valence band electrons tended to obscure the weak 3σ excitation.
Here we will show how, by using the continuously tunable radiation
from SSRP, these difficulties were overcome.

In Figs. 7 and 8, we have presented data on the photon dependence of the cross section for 3d, 4d, and 5d states plotted in terms
of the adsorption cross section, σ, versus photon energy above
threshold. To the first approximation the results appear to be
independent of whether the electrons come from valence or core
states. The most striking characteristic of this data is the difference between the 3d states on the one hand and the 4 and 5d states
on the other. The cross section for the 3d states changes rather
monotonically with photon energy. In contrast the 4 and 5d states
have a strong maximum about 50 eV above the excitation threshold
and then σ goes through a very low and shallow minimum about 150 eV

above the threshold. This is known as the Cooper minimum[28] after
the physicist who first predicted it. The minimum is due to destructive interference between the initial and final states involved
in the transition. Such interferences are only possible when there
are nodes in the initial states. The 4 and 5d wave functions have
such nodes but the 3d does not.

The Pt valence bands obtained for a photon energy lying in the
Cooper minimum ($h\nu$ = 150 eV) are shown in Fig. 12 for the clean
surface and after adsorption of CO. Note that the CO spectra dominates that of Pt. This is strikingly different from the usual
spectra of chemisorbed species where the emission from the substrate dominates that from the adsorbate. Due to the reduction
in substrate emission, it is possible to see not only the 3σ orbital
at 28.5 eV but also the "shake up" structure which lies between
the 3σ and 4σ peaks.

In Fig. 13 we present the difference curve obtained by subtracting the emission of the clean Pt surface from that of the CO
covered surface. Also included in Fig. 13 is the CO spectrum

obtained by x-ray photoemission.[29]   The two spectra have been
aligned so that the 4σ orbitals coincide.   The intensity of the
3σ orbital at 150 eV is depressed with respect to that in the x-ray
spectrum because of reduced matrix elements.[30]

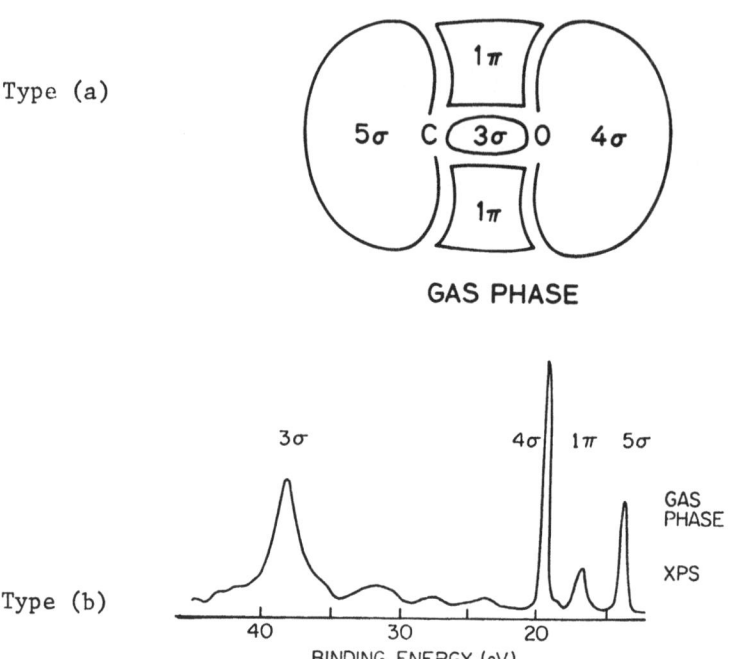

Type (a)

GAS PHASE

Type (b)

Fig. 11   (a) A schematic sketch of the filled valence orbitals of
          the free (gaseous) CO molecule.   (b) An x-ray photoemission
          spectrum (Ref. 29) of gas phase CO in which the various
          orbitals are identified.

          The spectra of Fig. 13 differ in three important ways.   The
first is the downward shift of the 5σ orbitals so that they converge
with the 1π orbitals.   This has been observed previously and dis-

cussed in detail.[31] It is due to the bonding between the 5σ orbitals
and the metal. Two new features are revealed in Fig. 13. These are
the change in the shakeup structure and the upward shift of the 3σ
orbitals of the adsorbate with respect to the 4σ level. The former
effect is undoubtedly due to the involvement of the metal electrons
in the shakeup process and the latter different "relaxation shift
for the 4σ and 3σ levels. Both effects are discussed in more

detail elsewhere.[36] Ultimately these data should help provide important information on the metal CO interaction which bind the gas
molecule to the metal surface.

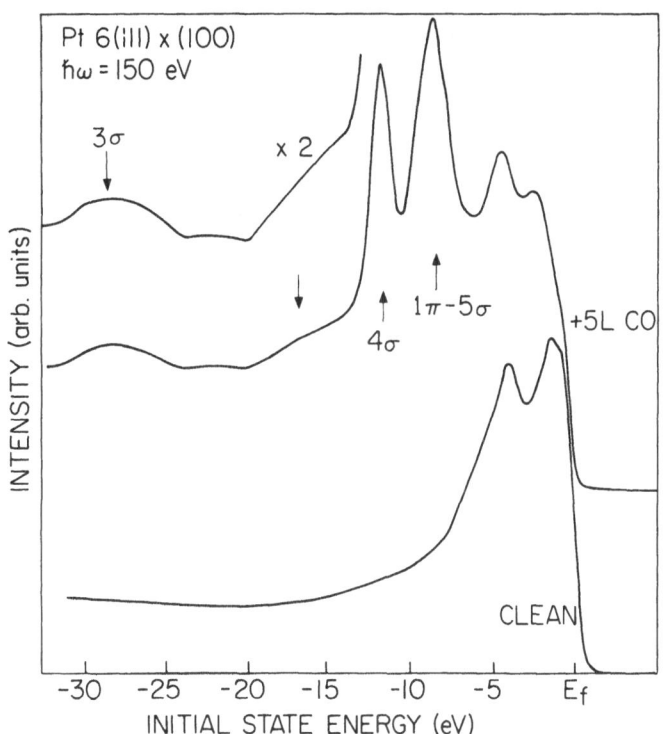

Fig. 12   The valence band spectra from clean Pt for $h\nu$ = 150 eV and
          for Pt with adsorbed CO. The CO spectra dominates that of
          Pt because the photon energy of 150 eV falls in the Cooper
          minimum of the Pt adsorption (the 5d curve of Fig. 8)
          curve. Because the Pt valence emission is suppressed it
          is possible to clearly see, for the first time, the CO
          3σ orbitals and the "shake up" electrons.

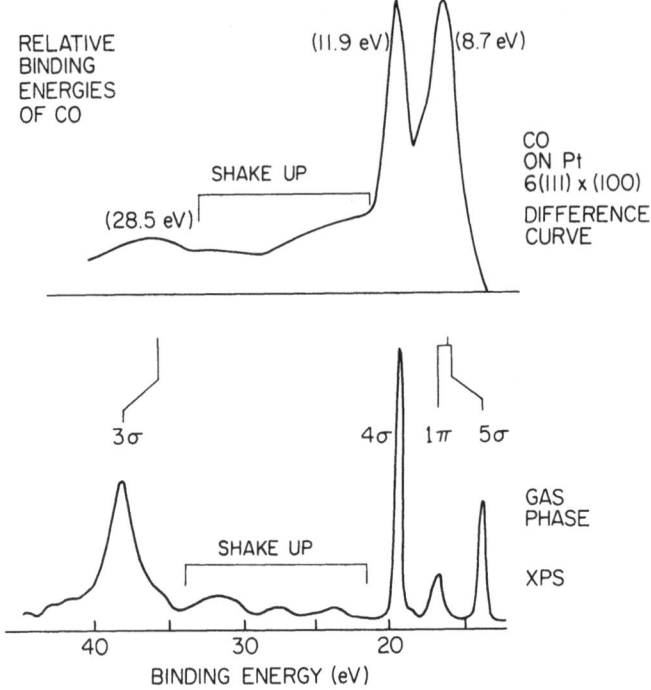

Fig. 13   The difference curves (CO covered Pt minus clean Pt) of CO
          on Pt compared to CO in the gas phase. Note the strong
          changes in the shake up structure and the shift of the 3σ
          relative to the 4σ orbitals.

VII. ANGULARLY RESOLVED PHOTOEMISSION ELECTRON SPECTROSCOPY (ARPES)

A. Introduction

    Additional information is available from the photoemission ex-
periments if emission within restricted solid angles is measured.
Such measurements are usually spoken of as angularly resolved photoemission electron spectroscopy (ARPES).[32]   In such experiments the
direction of polarization of the incident light is usually an important parameter. Thus, one should remember that controlling the
alignment of the $\bar{A}$ of the incident light with respect to the substrate under study may also be an important requirement of the
angular resolved experiment.   Once again, it is the continuous
radiation available from synchrotron sources which can be so important in these experiments. In addition, the synchrotron
radiation has the additional advantage of being naturally polarized.

ARPES is a fast growing and important spectroscopy. This is reflected by the set of lectures being devoted to its theoretical aspects by Dr. Liebsch in these proceedings. Because of his coverage, I will only give two examples of ARPES in conjunction with synchrotron radiation. In the first example, it is used to obtain quite detailed information on the geometry and electronic structure of a gas adsorbed on a crystal.[33] The second involves its use to determine features of the band structure of a metal.[34] Many other applications can be envisioned, for example, to use ARPES to determine the electronic states of a substrate which are involved in bonding a sorbed gas molecule to the solid.

B.   Studies of CO Adsorbed on Ni(100)

For such a young field, the use of ARPES to study adsorbed gases has produced a relatively large body of literature. We will not attempt to review this here. Rather we will discuss a recent study by Allyn, Gustafsson and Plummer of CO on Ni(100)[33] which gives a particularly good example of the way theory and experiment have combined in ARPES to provide strikingly detailed information about an adsorbed gas.

Figure 14 shows EDC's taken by ARPES for four different geometries of the polarization vector of the incident light and emission angle of the photoelectrons. The structure at energies above -5 eV is due to the valence states of the Ni, that below -5 eV to the adsorbed CO. The reader is referred to Fig. 11 and reminded of the discussion of the orbitals of gaseous and adsorbed CO in Section VI since these are also relevant to this discussion.

The photoemission peak due to the 4σ CO orbital is indicated by 4σ in Fig. 14. The peak labeled "P" is due to the 1π and the 5σ CO orbitals which overlap in energy. The latter is strongly shifted due to its involvement in the bonding of the CO to the metal.

The significance of Fig. 14 lies in the strong modulation of the two CO peaks as a function of the direction of the polarization of the radiation and the direction of emission of the emitted electrons. By keeping these angles fixed and varying the photon energy, Allyn et al studied the variation of peak intensities with photon energy in ARPES.

The theoretical work of Davenport[35] and Dehmer and Dill[36] had suggested that a resonance, i.e. a peak in absorption cross section would occur for emission from the 4σ and 5σ (but not $1\bar\pi$) states when the direction of polarization of the radiation, $\bar{A}$, and the direction of electron emissions were along the axis of the CO

molecule. The resonance is associated with scattering of the
excited electrons between the carbon and oxygen atoms and thus its
directional dependence is not surprising. Just such a resonance
is seen in Figure 15b. The geometry of the experiment is shown
by the insert in Fig. 14. The CO molecule is perpendicular to the
surface with the carbon atom next to the surface. The alignment
of the CO molecule had been determined in previous work and was
confirmed as will be related here by the studies of Allyn et al.
Thus the axis of the molecule is along the z axis of the insert
in Fig. 14.

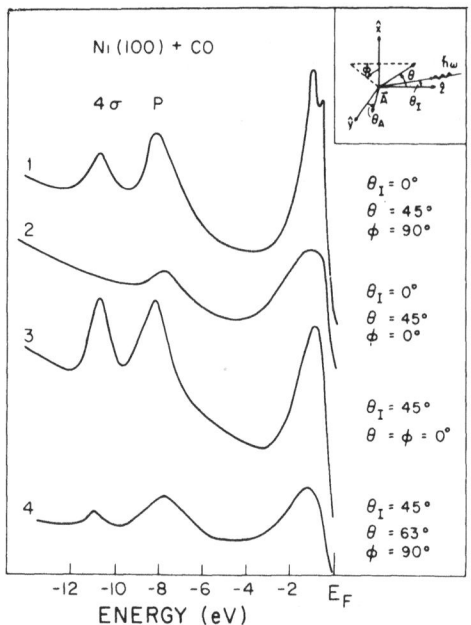

Fig. 14   Angularly resolved energy distributions for hν = 32 eV
          for CO on Ni(100) from Allyn et al (Ref. 33). The
          ordinate is proportional to the number of electrons emitted/
          eV. The structure between −5 eV and E_f is due to the Ni
          valence states. "4σ" indicates the CO 4σ molecular levels.
          "P" indicates the 1π CO molecular levels and the levels
          due to the CO 5σ levels and the bonding to the metal. The
          experimental geometry is shown in the inset. The yz plane
          is in the incident and polarization plane and the z axis
          is the crystal normal. The experimental configuration
          is indicated on the right.

        Note that the resonance is seen in Fig. 15 only when both the
direction of emission is close the CO axis (θ ≈ 0°) and the $\bar{A}$ has a

component along the CO axis ($\theta_I = 45°$ but not $0°$). This provides
independent evidence that the axis of the CO molecule is perpendicular to the surface.

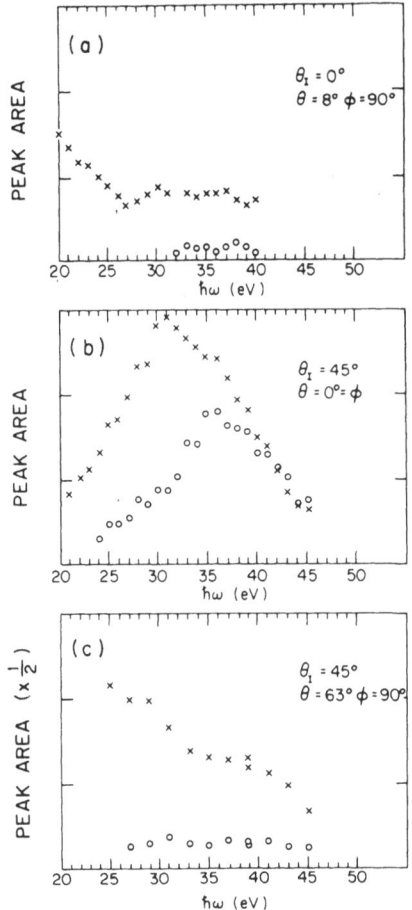

Fig. 15   CO on Ni(100) P(open circles) and 4σ (x's) peak strength
           (obtained by taking the areas under the peaks) as a
           function of hν for various sample geometries (see Fig. 14
           and the insert therein).

    A more quanitative study of the orientation of the CO molecule
with respect to the surface is illustrated by Fig. 16 where the
strength of the 4σ peak (i.e. the area under the 4σ peak) is plotted
versus the polar angle of emission θ for hν = 35 eV (hν of the
resonance peak in Fig. 15b). The solid curve was calculated assuming the molecule to be perpendicular to the surface and the points
are from the experiment. As can be seen, the agreement is excellent.

Based on these studies Allyn et al estimated that the CO molecule could not be tilted by more than 5° from the normal.

One additional experiment by Allyn et al was necessary to confirm that the CO was bonded to the Ni by the carbon atom. The results of this are shown in Fig. 17. As can be seen the experiment agrees much better for the case with carbon "down", i.e. next to the surface than for the other calculated cases. Allyn et al point out that the nice agreement between experimental and theoretical positions for the resonance peak in Fig. 17 is fortuitous and was not obtained for the gaseous CO molecule.[37]

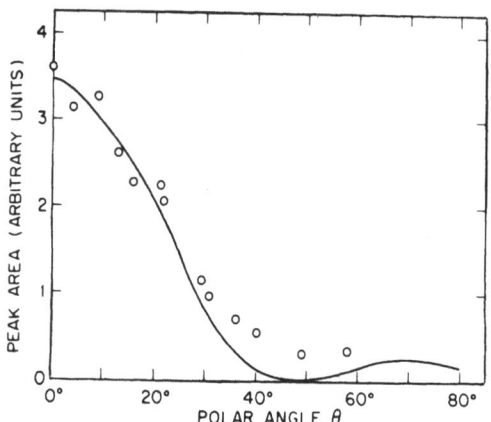

Fig. 16   CO on Ni(100). 4σ peak area (see Fig. 14) versus polar
         angle   measured perpendicular to the incident plane.
         $h\nu = 35$ eV, $\theta_I = 45°$. The theory curve has the shape

         characteristic of the 4σ emission on resonance (i.e. where
         the direction of emission is along the CO molecular axis)
         and is normalized to best fit data at small θ.

Having established the geometry of the CO molecule with regard to the surface, Allyn et al were able to go back to the data of Fig. 14 and disentangle the positions of the overlapping 5σ and 1π peaks. Since $\theta_I = 0°$ in curve 2 of Fig. 14, the single CO peak should be dominated by transitions from the 1π orbitals. Whereas, for $\theta_I = 45°$ and $\theta = \varphi = 0°$ (curve 3) the 5σ should dominate.

thus the 5σ is found to be 0.5 ± 0.2 lower in energy (i.e. 0.5 ± 0.2 higher binding energy) than the 1π orbital. The $4\sigma - 1\pi$ separation is found to be 3 eV – identical (within experimental accuracy) to the separation in the gas. In contrast the 5σ is found to have moved 3.4 eV closer to the 4σ level due to the bonding on the Ni.

This example illustrates how ARPES experiments can be used
with theory to obtain detailed knowledge of the geometry of a gas
chemisorbed on a metal and, in addition, to obtain the energy of
two peaks which overlap so strongly in energy as to make it difficult
to determine their individual energies by conventional means. The
experimental studies involved continuously tunable radiation for
$20 < h\nu < 45$ eV and were only made possible by the use of synchrotron
radiation.

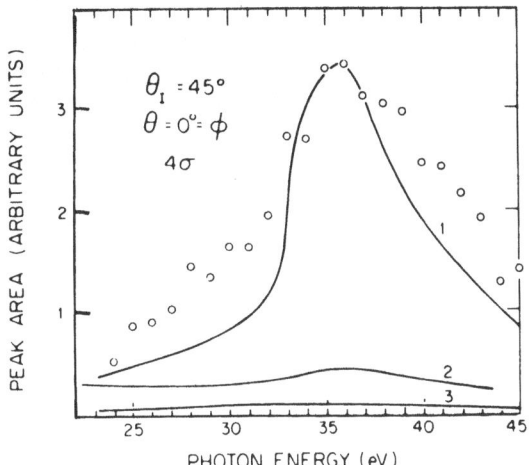

PHOTON ENERGY (eV)

Fig. 17. CO on Ni(100). 4σ peak area (see Fig. 14) for electron
emission along axis of the molecule, θ_I = 45°, versus hν.

Solid curves are calculated for (1) CO axis normal to the
surface, carbon end down, (2) CO axis normal to surface,
oxygen end down, and (3) CO axis in surface plane. Experimental data normalized to theory at hν = 36 eV.

C.   Use of ARPES to Obtain Information
of the Bulk Band Structure

If electrons are excited in a solid and escape without scattering, the angle of emission will be, in principle, directly related
to the crystal momentum, nk, of the state into which the electron
is optically excited. If the optical excitation process is direct,
i.e. k is conserved between the initial and final states, the
angular resolved photoemission spectra can be used to obtain information on the band structure in a very restricted volume of k -
space within the Brillioun zone.

One complexity of this approach lies in relating the momentum
of the electron outside the crystal to the crystal momentum, hk,

inside the crystal. (Because of periodic variations of the crystal
potential in the solid, momentum is not a constant of motion inside
the crystal; however, $\hbar k$ is a constant of motion.) Traditionally,
$\hbar k_{\parallel}$ (i.e. parallel to the surface) is set equal to $p_{\parallel}$ (i.e.

momentum parallel to the surface outside the crystal) and $p_{\perp}$

(momentum perpendicular to the surface) is set equal to the difference between $\hbar k_{\perp}$ and the momentum loss due to overcoming the

potential barrier at the surface. This momentum loss, in general,
produces a refraction so that the direction of motion in vacuum
is different from that in the crystal. One can avoid the complications of this refraction by studying the electrons emitted perpendicular to the surface. Since $p_{\parallel}$ is zero for these electrons,

it is normally assumed that one only is concerned with $p_{\perp}$ and one
can then simply relate the energy of the emitted electrons to their
energy in the solid. In this way the EDC's for normal emission
are directly related to features of the band structure in the direction of $\bar{k}$ space perpendicular to the surface.

        In Fig. 18, data from a study by Smith, Anderson and Lapeyre[34]
of emission in the (001) direction from tungsten for $10 \gtrsim h\nu \gtrsim 30$ eV
using synchrotron radiation with light polarized parallel to the W
surface is shown.

        This work was closely related to that of Feuerbacher and
Christensen[38] (FC) who had made a very extensive theoretical[39] and
experimental[38] study of the photoemission and surface and bulk
band structure of W. The work of FC was very extensive except for
certain restrictions imposed by their radiation sources for
$h\nu > 13$ eV, where they were only able to make measurements at 16.8
and 21.2 eV. Their experimental geometry gave light incident on
the sample at almost grazing incidence; as a result, there was a
large component of light polarized perpendicular to the sample.
Because of the lack of continuously tunable radiation, it was
impossible to make measurements at certain wavelengths where
critical transitions were predicted by the theoretical work of
Christensen and Feuerbacher (CF).

        Smith, et al[34] used synchrotron radiation in order to study
the transitions of interest. Their light was polarized in the
plane of the surface and only electrons emitted within a 4° cone
normal to the (001) crystal face were collected. One might argue
that surface states will be excited only by radiation polarized
normal to the surface and thus that the experiment of FC[38] would
be much more sensitive to surface states than would that of Smith
et al. Smith et al argue that this has been confirmed experimentally,[3,4] however, additional confirmation would be satisfying.

Figure 19 compares some of the experimental results from Smith et al to the band calculations of CF.[39] On either side of the band diagram are plots used by Smith, et al to locate important structure in the band structure. It is important that some care be taken in explaining those diagrams.

At the top of each diagram are plots labeled CIS. These are data taken by changing the final energy, E_f, as hν is changed so that one is always measuring the electrons emitted from a fixed initial state energy, E_i. Thus, $E_i = hν - E_f$. For that reason, the plots are called CIS (constant initial state) plots. In order to obtain CIS plots the EDC's are normalized to each other so that the height of the EDC at any final energy is proportional to the emission at that electron energy. As a result, a CIS gives the emission probability from a fixed initial state energy, E_i, as a function of final state energy, E_f. Thus, in the CIS curves, maxima indicate final state energies for which the transition probability has peaked for a given initial state energy. A similar method was earlier developed by Shay[40] using energy distribution curves alone.

Three CIS curves are presented in Fig. 19. Each corresponds to an initial state (-0.4, -1.3, and -5.7 eV respectively) for which a peak appears in the EDC's of Fig. 18. In the bottom halves of the left and right hand panels of Fig. 19 are EDC's in which the initial state structure of interest is strong (hν = 16 eV for E_i = -0.6 and -1.3 eV; hν = 15 eV for E_i = -5.7 eV). From the right hand panel it is clear that the -5.7 eV transition peaks for E_f = 9.3 eV; and from the left hand panel, that $E_f \cong 14.6$ is the peak for transitions from the initial states of -0.6 and -1.3 eV. Detailed analysis of Smith, et al, using more closely spaced hν than those in Fig. 18, located the initial states at ~ -0.4, ~1.3, and -5.7 eV and the final state for the transition maximum at 14.6 and 9.3 eV. These values were found to be in agreement with the band theory of CF within 0.2 eV. The transitions from E_i = -0.4 and -1.3 eV to E_f = 14.6 eV were assigned to optical transitions near Γ and that from -5.7 to 9.3 eV to states near the H point in the Brillouin zone (see the band structure diagram of Fig. 19).

Other band structure information was obtained from the EDC's of Fig. 18. The peak near -1.5 eV for hν = 18 eV was studied for values of hν spaced 0.1 eV apart and found to correspond to a transition from -1.5 to 16.8 eV. This was associated with a transition in the band structure 1/4 of the distance from Γ to H, i.e. to the structure resulting from the crossing of the bands near k = 1/4 $|\bar{H}|$ at E = 16.8 eV in the band structure.

   The small peak near −3.3 eV for 18 < hν ≲ 15 eV was identified
by Smith et al with transitions near the middle of the zone to
final energies associated with the band minimum at 13.8 eV; however,
the maximum was found experimentally at a final state 1.1 eV lower,
i.e. at 12.7 eV.

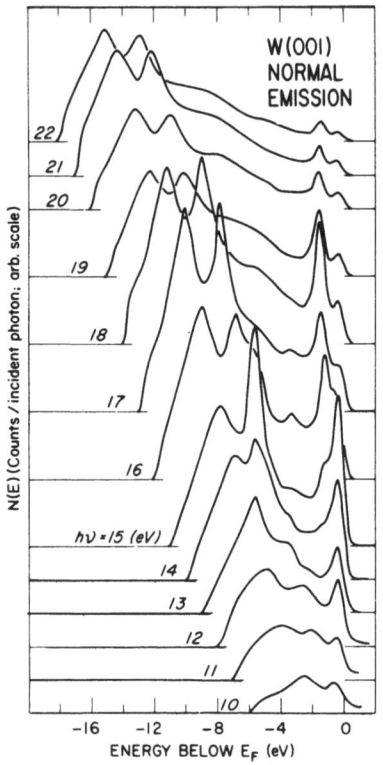

Fig. 18   Angular resolved energy distribution of Smith, et al.
           (Ref. 34) from atomically clean W(001) for various photon
           energies. As the ordinate indicates, the curves are
           normalized on an absolute basis. The Fermi level, E_f, is

           taken as the zero of energy and the curves are plotted (on
           the abscissa) versus the initial state from which the
           electron is excited.

   Long experience in interpreting photoemission data suggests that
certain caution should be taken in treating these specific assign-
ments with surety until detailed calculations have been made based
on the band structure; however, the overall agreement between theory
and experiment is very encouraging.

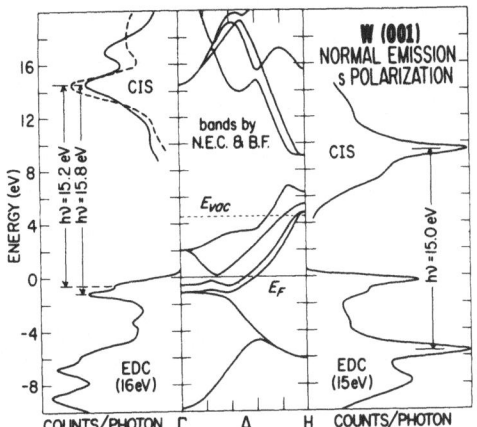

Fig. 19   The center panel indicates the W energy bands in the
          direction of emission (001) from Ref. 39. Normal
          emission angular resolved energy distribution curves (EDC)
          and constant initial state spectra (CIS) are given in the
          left and right hand panels. The hν for each EDC is indi-
          cated on the EDC. The dashed and solid curves respectively
          in the left hand panel were generated by the shoulder at
          −0.6 eV and the peak at −1.2 eV in the hν = 16 eV EDC.
          Similarly, the CIS of the right hand panel was produced by
          the peak at −5.7 eV in the hν = 15 eV EDC.

     Several other features of the EDC's of Fig. 18 are worthy of
notice. For high values of hν, there is a doublet at a constant
final state energy near the low energy cutoff of the EDC's. Smith
et al assign these peaks quite reasonably to electrons scattering
inelastically (due to electron – electron scattering) into the high
density of final states associated with the band maximum and minimum at H (i.e. at about 5.5 and 9 eV in the band diagram). However,
note that, although there is an absolute bandgap of about 2 eV, in
the band structure there is relatively strong emission between
the peaks where band theory predicts no emission. Likewise, there
is reasonably strong emission for hν < 15 eV; however, examination
of the band diagram of Fig. 19 indicates that there is no way to
account for these in terms of direct transitions. In
fact, there should be no transitions possible to final states in
the range $7 < E_f < 9$ eV (i.e. roughly 3 to 5 eV above the low energy
cutoff of the EDC's of Fig. 18) since there are no states in that
energy range. In contrast, as can be seen from Fig. 18, for all
values of hν the emission in this range of final energy is always
quite strong. Clearly the simple direct transition – bulk band
structure model is insufficient to explain all features of the EDC's.
Such disagreements between the simple theory and experimental results
are not unknown[41]; however, they are often ignored. In the view

of the present author, this is a pity since understanding these
differences can lead to much deeper knowledge of photoemission,
optical excitation, and the electronic structure of solids.

D. Discussion of ARPES

The preceding discussion of ARPES using synchrotron radiation
should in no way be considered as comprehensive. Rather we have
used two examples of experiments which lend themselves to fairly
straightforward interpretation of the experimental data. Many
other examples can be found in the literature, for example the work
of Smith, Tramm, Knapp, Anderson, and Lapeyre[42] on polarization
effects in angle-resolved photoemission from the layered material
TaS_2 in which they show that the polarization dependence of ARPES
may possibly be used to determine both the plane-wave decomposition
of the final states and the shapes of the atomic orbitals of the
states involved in the photoemission.

Another potentially important aspect of ARPES is possible
multiple scattering effects, such as those in LEED. An example
might be scattering of electrons excited from an adsorbed atom by
the atoms of the substrate. Such scattering was successfully ignored
in the treatment of CO on Ni report in Section VII B; however, there
may be other cases where this is important. Above all, the reader
is referred to Dr. Liebsch's contribution to these proceedings for
a more complete treatment of ARPES. Since Professor Mahan was one
of the first theorists to recognize the importance of angular
resolved photoemission, it is likely also that valuable material
on this will be found in his section of these proceedings.

VIII. CONCLUSIONS

We have attempted to show here some of the advantages of using
synchrotron radiation for photoemission spectroscopic investigation;
however, the examples used to illustrate these examples are in no way
all inclusive. Undoubtedly, advantages which have not yet been
demonstrated will become important in the future.

Assuming that the reader is impressed with the advantages of
synchrotron radiation, what are some of the disadvantages? The
most obvious one is that there are only a few synchrotron sources
suitable for photoemission studies throughout the world. Even
though the number of suitable sources will probably double by 1980,
most workers will have to travel some distance to find the closest
source. If one is contemplating use of synchrotron sources for
photoemission studies, what should be the governing considerations?
The first is the radiation intensity available at the sample. Even
though storage rings are basically high intensity sources, the
manner in which the radiation is extracted from the ring and

delivered to the sample determines both the intensity and wavelength range available. It should be recognized that photoemission
experiments may require several orders of magnitude – more intensity
than optical absorption or reflection measurements. Of all photoemission experiments, measurements of the total yield of photoemitted electrons require the smallest intensity and angular resolved energy distributions the highest intensity.

It should also be recognized that the optical system used to
extract the radiation from the storage ring and provide monochromatic radiation determines the $h\nu$ range and polarization which is
available to the experimentalist. If photoemission experiments
were not a prime consideration in the design of a given optical
system (including monochromator), it may not be adequate for photoemission measurements. The experimenter should satisfy himself
as to these parameters before committing effort to use a given
facility.

A second consideration is the availability of suitable experimental chamber(s). If an experimenter has to travel many miles
to make use of a synchrotron facility, it is probably impractical
for him or her to carry a large and complex vacuum chamber back
and forth. Rather, it is necessary that a chamber be kept at the
facility. Since this is expensive, it can prove very advantageous
for several groups to share common equipment. This has been done
in an exemplary way by users of the Wisconsin storage ring which
has perhaps the oldest tradition of strong and continuous photoemission work of any synchrotron facility. It is typical to find
names of workers from two or more different laboratories on papers
from Wisconsin. This usually is a result of collaborative effort
to solve the equipment problem.

Despite the difficulties of doing work at most synchrotron
facilities, most workers find the rewards of such work very much
worth the effort.

ACKNOWLEDGEMENT

The support by the National Science Foundation Grant No. DMR 74-22230 and by NSF (Contract No. DMR 73-07692 A02) in cooperation with the Stanford Linear Accelerator Center and the U.S. Energy Research and Development Administration; by the Advanced Research Projects Agency of the Department of Defense and monitored by Night Vision Laboratory, U.S. Army Electronics Command, under Contract No. DAAK 02-74-C-0069; and by the U.S. Office of Naval Research (Contract No. N00014-75-C0289), is gratefully acknowledged.

REFERENCES

1. D.A. Shirley, Ed. Electron Spectroscopy, North Holland –
    Amsterdam, Amer. Elsevier, N.Y. 1972.
    Kai Siegbahn, et al, ESCA Atomic Molecular and Solid State
    Structure Studied by Means of Electron Spectroscopy, Almquist
    and Wiksells Boktryckeri AB, Uppsala, 1967.

2. W.E. Spicer and C.N. Berglund, Rev. Sci. Inst., $\underline{35}$, 1665 (1964).
    R.E. Eden, Rev. Sci. Inst. $\underline{41}$, 252 (1970).

3. See for example, F.Abeles, Ed., Proc. of Intem. Colloq. on
    Optical Properties and Electronic Structure of Metals and
    Alloys, Paris, 1965 (North-Holland, Amsterdam; John Wiley,
    New York).

4. W.E.Spicer, Phys. Rev., $\underline{112}$, 114 (1958).
    C.N. Berglund and W.E. Spicer, Phys. Rev., $\underline{136}$, 1030, 1044 (1964).

5. L.H. Bennett, Ed., Electronic Density of States, U.S. Govern-
    ment Printing Office, Washington, DC. (SD Catalog No. C13:323)
    1971.

6. W.E. Spicer, in Proc. Intem. Colloq. on Optical Properties
    and Electronic Structure of Metals and Alloys, 1965, Paris,
    Ed. F. Ables (North-Holland, Amsterdam; John Wiley, New York)
    pp. 296-315.

7. G.F. Derbenwick, D.T. Pierce, and W.E. Spicer, in Methods of
    Experimental Physics Vol 11, Solid State Physics, Ed. R.V.
    Coleman, Acad. Press., N.Y. and London, 1974, p. 67.

8. See for example, J.W. Rabalais, Principles of Ultraviolet
    Photoemission Spectroscopy, J. Wiley and Sons, 1977, p. 19-28
    and references therein.

9. E.E. Koch, R. Haensel, and C. Kunz, Ed., Proc. of IV Inter-
    national Conference on Vacuum Ultraviolet Radiation Physics;
    Hamburg, 1974; Pergamon, Vieweg. 1974.

10. Robert W. Morse, "An Assessment of the National Need for
    Facilities Dedicated to the Production of Synchrotron Radiation,"
    National Acad. of Sciences, Washington, DC, 1976. (Available
    from Solid State Sciences Committee, NRC, 2101 Constitution
    Ave., Washington, D.C. 20418.)

11. S. Doniach, I. Lindau, W.E. Spicer, and H. Winick, J. Vac.
    Sci. Technol. $\underline{12}$, 1123 (1975).
    I. Lindau, "Photoemission Studies of Semiconductor and Metal

Surfaces Using Synchrotron Radiation," in course of Synchrotron Radiation, Int. College of Applied Phys, Ed. A.N. Mancini and I.F.Quercis, 1976, p. 321.

12. K.O. Hodgson, H. Winick and G. Chu, Synchrotron Radiation Research, SSRP Report # 76/100 Stanford University, Stanford, CA 94305.

13. W.E. Spicer, "Bulk and Surface Ultraviolet Photoemission Spectroscopy," in Optical Properties of Solids - New Developments, Ed. B.O. Seraphin, North Holland, 1975.

14. P.J. Feibelman and D.E. Eastman, Phys. Rev. B10, 4932 (1974).

15. I. Lindau and W.E. Spicer, J. Elect. Spectroscopy 3, 409 (1974).

16. D.R. Penn, Phys. Rev. 13, 5248 (1976).
    C.J. Powell, Surf. Sci., 44, 29 (1974); Rev. Mod. Phys. to be published.

17. T.H. DiStefano and W.E. Spicer, Phys. Rev. B7, 1554 (1973); R. Powell and W.E. Spicer, G.B. Fisher, and P. Gregory, 8, 3987 (1973).

18. F.C. Brown, R.Z. Bachrach, S.B.M. Hagstrom, Proc. IV International Conference on Vac. Ultraviolet Rad. Phys., Hamburg, 1974, p. 785.

19. N. Schwentner, M. Skibowski, and W. Steinmann, Phys. Rev. B8, 2965 (1973).
    W. Gudat and C. Kunz, Phys. Rev. Lett. 29, 169 (1972).

20. P.W. Chye, I.A. Babalola, T. Sukegawa, and W.E. Spicer, Phys. Lett., 34, 1624 (1975).
    G.J. Lapeyre and J. Anderson, Phys. Rev. Lett., 35, 117 (1975) D.E. Eastman and J.L. Freeouf, Phys. Rev. Lett., 33, 1601 (1974).

21. W.E. Spicer, I. Lindau, J.N. Miller, D.T. Ling, P. Pianetta, P.W. Chye, and C.M. Garner, Physica Scripta, in press.

22. I. Lindau, P. Pianetta, and W.E. Spicer, Phys. Lett., 54A 225 (1976).

23. G. Apai, P.S. Wehner, J. Stohr, R.S. Williams, and D.A. Shirley, Solid State Comm. 20, 1141 (1976).

24. P. Pianetta, I. Lindau, C.M. Garner, and W.E. Spicer, Phys.
     Rev. Lett., $\underline{37}$, 1166 (1976), $\underline{35}$, 1356 (1975).

25. W.E. Spicer, I. Lindau, P.E. Gregory, C.M. Garner, P. Pianetta,
     P.W. Chye, J. Vac. Sci. Technol. $\underline{13}$, 780 (1976).

26. W.E. Spicer, I. Lindau, P.E. Gregory, C.M. Garner, P. Pianetta,
     and P.W. Chye, J. Vac. Sci. Technol., in press.

27. C.R. Brundle, in Electronic Structure and Reactivity of Metal
     Surfaces, Ed., E.G. Derouane and A.A. Lucas, Plenum Press,
     New York, 1975), p. 389.

28. J.W. Cooper, Phys. Rev. $\underline{128}$, 681 (1962).

29. U. Gelius, E. Basilier, S. Svensson, T. Bergmark, and K.
     Siegbahn, J. Elec. Spectroscopy, $\underline{2}$, 405 (1973).

30. M.S. Banna and D.A. Shirley, J. Elec. Spectroscopy, $\underline{8}$, 255
     (1976).

31. H. Hochst, S. Hufner, A. Goldman, Phys. Lett. $\underline{57A}$, 265 (1976).

32. A.L. Robinson, Science, $\underline{194}$, 1306 (1977).

33. C.L. Allyn, T. Gustafesson, and E.W. Plummer, Chem. Phys.
     Lett., $\underline{47}$, 127 (1977) and references therein.

34. R.J. Smith, J. Anderson, J. Hermanson, and G.J. Lapeyre,
     Solid State Comm., $\underline{19}$, 976 (1976); G.J. Lapeyre, R.J. Smith,
     J. Anderson, J. Vac. Sc. Technol. $\underline{14}$, 384 (1977).

35. J.W. Davenport, Phys. Rev. Lett. $\underline{36}$, 945 (1976).

36. J.L. Dehmer and D. Dill, Phys. Rev. Lett., $\underline{35}$, 213 (1975).

37. E.W. Plummer, T. Gustafesson, W. Gudat, and D.E. Eastman,
     Phys. Rev. A, $\underline{15}$, 2339 (1977).

38. B. Feuerbacher and N.E. Christensen, Phys. Rev. $\underline{B10}$, 2373
     (1974).

39. N.E. Christensen and B. Feuerbacher, Phys. Rev. $\underline{B10}$, 2349
     (1974).

40. J.L. Shay, Ph.D. Dissertation, Stanford University, 1966,
     copies can be obtained from University Microfilms, Ann Arbor,
     Mich. USA, or High Wycomb, England.

J. Shay and W.E. Spicer, Phys. Rev. <u>161</u>, 799 (1967).

41.   W.E. Spicer and R.E. Eden, 1968, Proc. 9th Intern. Conf. on
      the Phys. of Semicond. Moscow, USSR, p. 65.

42.   N.V. Smith, M.M. Traum, J.A. Knapp, J. Anderson, and
      G.J. Lapeyre, Phys. Rev. B, <u>13</u>, 4462 (1976).

ANGLE RESOLVED PHOTOEMISSION:

THEORETICAL INTERPRETATION OF RESULTS

Ansgar Liebsch

Institut für Festkörperforschung der Kernforschungsanlage Jülich, D-5170 Jülich, Germany

I. INTRODUCTION

During the past few years, photoelectron spectroscopy has
emerged as a widely used technique in the study of the electronic
properties of a variety of physical systems.[1] Experimentally, a
wealth of new information has become available through the use of
electron analyzers with high angular resolution as well as continuously tunable light sources producing well polarized radiation. On
the theoretical side, a much better understanding has been reached
concerning the single-particle aspects of the excitation process. In
fact, consensus appears to exist today with regard to their quantitative formulation. Schemes have been developed to evaluate the
differential photoionization cross section for such different systems as clean surfaces, [2-5] molecules[6,7] and chemisorption com-
plexes.[8,9] Although the entire field is still at an early stage,
applications to specific cases in each of these areas have been
carried out which show encouraging agreement between measured and
calculated spectra. These examples represent evidence that the one-
electron properties play a significant role in determining the characteristics of the photocurrent and that they can be used to obtain
information about the electronic structure of the system.

In principle, excitation of electrons from the bound states
of a system by means of photons represents a highly complex physical process whose complete theoretical description is not yet possible today.[10] Somewhat arbitrarily, three main aspects may be
distinguished: single-electron features, the vector field induced
by the incident radiation,[11] and many-particle processes.[12] These
aspects are not unrelated and their relative significance may depend greatly on experimental parameters (e.g. photon frequency),

on the sample, and on the type of bound state under consideration.
The theoretical models to be discussed below are all based on the
independent-particle approximation. Interactions of the excited
electron and of the created hole[3] with the remaining electrons are
taken into account only to the extent that they can be described in
terms of a finite lifetime. Furthermore, the photon field is taken
to be spatially constant. As the severity of these assumptions remains largely unknown, the results obtained with existing schemes
should be treated with caution. Quasi-elastic scattering of the
emitted electron by phonons or diffuse scattering at imperfect surfaces may   also lead to sizeable momentum redistributions which
are generally not included so far. Finally, consideration of the
finite acceptance angle of the electron spectrometer might represent a significant aspect in a correct interpretation of experimental data.

The scope of these notes is limited to a selection of those
theoretical studies whose aim is the analysis of absolute intensity
variations of the photocurrent with emission angles, photon frequency and polarization. This excludes a wide class of so-called
band structure investigations[13] which try to correlate observed
peak positions with the momentum dispersion of the energy bands of
two- and three-dimensional systems. These latter schemes can be
viewed as special limiting cases of the more general and complete
theoretical models which we discuss here. Reference will be made,
however, to related spectroscopies in order to illustrate specific
points and to draw attention to common phenomena.

The structure of these notes is as follows. In Sec. II, the
one-electron theory of the photoionization process is reviewed.
Applications to specific systems and, wherever possible, comparisons
to experimental observations are given in Secs. III to VI. Discussed
are, respectively, results obtained for atoms, molecules, adsorbates
and clean surfaces. Emphasis is placed upon the particular aspects
of the electronic structure to which the photocurrent is sensitive.
A more detailed outline of the formal procedure of evaluating the
differential cross section is given only in the case of emission
from clean surfaces (Sec. VI. 1).

II. THE EXCITATION PROCESS

Within the one-electron theory,[14-17] the wave function of the
outgoing photoelectron at the position R of the detector may be
expressed as

$$\Psi(R) = \int d^3r \; G^r(R,r) \; p \cdot A \; \Psi_i(r)$$

$$= G^r \; p \cdot A \; |\Psi_i> \tag{2.1}$$

where Ψ_i represents the initial state wave function of the electron and $\underset{\sim}{p}$ the momentum operator. (We refer in this section mainly to the emission from solids. The following derivation, however, is entirely general and applies to atoms or molecules as well.) The photon field $\underset{\sim}{A} = A\; \hat{\epsilon}$ is assumed to be constant in space, with its phase properties given by the (complex) polarization vector $\hat{\epsilon}$. The quantity G^r denotes the retarded one-particle Green's function at the final energy E_f. It describes the propagation of the electron from the point of excitation towards the detector. This includes the transport of the electron to the boundary of the system and the transmission into the vacuum. These two "steps", together with the excitation,are treated in (2.1) as a coherent unit indicating the "one-step" nature of the emission process. In Sec. VI, we discuss the conditions under which this general approach reduces, in the case of emission from clean surfaces, to the more approximate "three-step" model.[18-20] In this limit, the three steps - excitation, transport and transmission - enter as separate factors into the expression for the cross-section. For molecule or adsorbate photoionization, only the one-step model is meaningful since excitation, transport and transmission all take place in the same region in space.

The propagator G^r may formally be expanded[21] in terms of the free electron Greeen's function G_o and the lattice potential V by using the Dyson equation

$$G^r = G_o + G_o\; V\; G_o + \ldots$$

$$= G_o(1 + T\; G_o) \tag{2.2}$$

where the T matrix of the entire system is defined as

$$T = V + V\; G_o\; T\; . \tag{2.3}$$

Since the detector can be assumed to be at infinity relative to the sample, we may replace the first G_o in Eq. (2.2) by its asymptotic form for $R \rightarrow \infty$:

$$G_o(\underset{\sim}{R}-\underset{\sim}{r}) \sim \frac{e^{i\kappa|\underset{\sim}{R}-\underset{\sim}{r}|}}{|\underset{\sim}{R}-\underset{\sim}{r}|}$$

$$\rightarrow \frac{e^{i\kappa R}}{R}\; e^{-i\underset{\sim}{k}_f \cdot \underset{\sim}{r}} \tag{2.4}$$

where $\kappa = \sqrt{E_f}$ and the vector $\underset{\sim}{k}_f \equiv \kappa\; \hat{R}$ represents the momentum of the outgoing electron. ($\hat{R}$ is a unit vector pointing toward the detector.) Inserting this expression into (2.2) and (2.1) we obtain:

$$\Psi(\underset{\sim}{R}) \rightarrow \frac{e^{i\kappa R}}{R} <\underset{\sim}{k}_f | (1 + T\ G_o)\ \underset{\sim}{p} \cdot \underset{\sim}{A} | \Psi_i >$$

$$= \frac{e^{i\kappa R}}{R} <\Psi_f | \underset{\sim}{p} \cdot \underset{\sim}{A} | \Psi_i > \qquad (2.5)$$

with

$$<\Psi_f | = <\underset{\sim}{k}_f | + <\underset{\sim}{k}_f | \ T\ G_o\ . \qquad (2.6)$$

Thus, at the detector, the photoelectron is described by an outgoing spherical wave[14] whose amplitude and phase are determined by the transition matrix element

$$M_{fi}(E_f, E_i; \underset{\sim}{k}_f, \underset{\sim}{A}) \equiv <\Psi_f | \underset{\sim}{p} \cdot \underset{\sim}{A} | \Psi_i >\ . \qquad (2.7)$$

The boundary conditions that are satisfied by the final state wave function Ψ_f which appears in the matrix element may easily be found by inserting the asymptotic expression for G_o, Eq. (2.4), into the second term of (2.6):

$$\Psi_f^*(\underset{\sim}{r}) = e^{-i\underset{\sim}{k}_f \cdot \underset{\sim}{r}} + \iint d^3r' d\ r''\ e^{-i\underset{\sim}{k}_f \cdot \underset{\sim}{r}'}\ T(\underset{\sim}{r}', \underset{\sim}{r}'')\ G_o(\underset{\sim}{r}'' - \underset{\sim}{r})$$

$$\rightarrow e^{-i\underset{\sim}{k}_f \cdot \underset{\sim}{r}} + f^*(E_f, \underset{\sim}{k}_f)\ \frac{e^{i\kappa r}}{r} \qquad (2.8)$$

for $r \rightarrow \infty$. Thus, the complex conjugate form of Ψ_f consists of an incoming plane wave plus outgoing spherical waves[22], i.e. Ψ_f satisfies the same boundary conditions as the time-reversed LEED wave function![14,15,17] The important point here is that this kind of final state wave function is required for the evaluation of the transition strength. The actual state of the emitted electron is clearly an outgoing wave as shown in Eq. (2.5). The quantity $f(E_f, \underset{\sim}{k}_f)$ in (2.8) represents a complex scattering amplitude, which essentially determines the LEED reflectivities. Using (2.3), the second term in Eq. (2.6) can be viewed as a perturbation series in the one-electron potential V. It ensures that Ψ_f is an eigenstate of the system, which the first term, the plane wave, is not.

If Ψ_f is approximated by the outgoing plane wave, the matrix element assumes a particularly simple form[23]

$$M_{fi} \rightarrow \int d^3r\ e^{-i\underset{\sim}{k}_f \cdot \underset{\sim}{r}}\ \underset{\sim}{p} \cdot \underset{\sim}{A}\ \Psi_i(\underset{\sim}{r})$$

$$\rightarrow \cos(\underset{\sim}{k}_f, \underset{\sim}{A})\ \Psi_i(\underset{\sim}{k}_f) \qquad (2.9)$$

i.e. it is proportional to the Fourier transform of the initial state Ψ_i. This expression demonstrates the point that the angle de-

pendence of the photocurrent is intimately related to the spatial
distribution of the electronic states of the system. For an adequate interpretation of spectra, however, this plane wave limit
turns out not to be generally sufficient. Instead, as will become
evident from the results discussed in the following sections, Ψ_i
and Ψ_f need to be treated in a consistent manner. This has been
found to be true for emission from molecules as well as for clean
surfaces. In addition, going to higher photon frequencies in the
far ultra-violet range (kinetic energies of the order of 200 eV)
does not essentially change this conclusion.

If both initial and final state wave functions are obtained
from the same potential, they are orthogonal and the following
alternative expression for the matrix element (2.7) may be derived:

$$M_{fi} = (E_f - E_i)^{-1} <\Psi_f | \underset{\sim}{A} \cdot [H, \underset{\sim}{p}] | \Psi_i>$$

$$= (E_f - E_i)^{-1} i\hbar <\Psi_f | \underset{\sim}{A} \cdot \nabla V | \Psi_i> . \qquad (2.10)$$

This form is particularly convenient if the potential is of muffin-
tin form since in this case the constant-potential region between
atomic spheres does not contribute to the matrix element.

The outgoing electron may undergo strong inelastic scattering
by creating particle-hole pairs or collective excitations. These
processes are the origin of the surface sensitivity of the photocurrent since they imply that an electron can propagate only relatively short distances without loosing some of its energy. The
escape depth varies with energy and depends on the material. Generally, it is of the order of a few lattice spacings. In the evaluation of the matrix element (2.7), the presence of electronic
damping is taken into consideration by means of an optical potential, i.e. by inserting a complex self-energy $\Sigma = V_o - i\Gamma$ into the
single-particle propagator G_o in Eq. (2.6). Thus, in momentum
space, G_o is of the form[24]

$$G_o(E, \underset{\sim}{k}) = (E - \underset{\sim}{k}^2 - V_o + i\Gamma)^{-1} \qquad (2.11)$$

where V_o is the inner potential and Γ is inversely proportional to
the mean free path:

$$\lambda = \frac{2}{\Gamma} \sqrt{E - V_o} . \qquad (2.12)$$

Due to the optical potential acting on the excited electron, the
identity (2.10) is no longer strictly valid since the Hamiltonian
of the system is energy dependent. Corrections to (2.10) may
arise[25] which are related to the spatial variation of the optical
potential near the solid-vacuum interface.

Within a solid angle $d\Omega$ at the detector, the current due to the emitted wave $\Psi(\underset{\sim}{R})$ is given by the standard expression

$$j \sim d\Omega \lim_{R \to \infty} R^2 \text{ Im } \Psi^*(\underset{\sim}{R}) \nabla_R \Psi(\underset{\sim}{R}) \tag{2.13}$$

The differential cross section is obtained by summing over all possible initial states which satisfy energy conservation

$$\frac{d\sigma}{d\Omega}(E_f,\hbar\omega,\underset{\sim}{k}_f,A) \sim \kappa \sum_i |<\Psi_f|\underset{\sim}{p}\cdot\underset{\sim}{A}|\Psi_i>|^2 \delta(E_f-E_i-\hbar\omega). \tag{2.14}$$

The prefactor κ defines the density of final states. The theoretical results which are discussed in the subsequent sections are all based on this golden rule expression.[14-17] The above derivation shows that this formula is compatible with the one-step nature of the emission process. It enters the matrix element in two ways, (a) through the boundary conditions imposed on Ψ_f (time reversed LEED state), and (b) through the proper incorporation of the surface and the electronic damping in the evaluation of Ψ_f.

Using Eqs. (2.1) and (2.13), the differential cross section can alternatively be written as[10]

$$\frac{d\sigma}{d\Omega} \sim \lim_{\substack{R,R' \to \infty \\ \hat{R}=\hat{R}'}} R^2 \nabla_R \text{ Im } \iint d^3r \, d^3r' \, G^r(E_f,\underset{\sim}{R},\underset{\sim}{r}) \, \underset{\sim}{p}\cdot\underset{\sim}{A} \frac{1}{2\pi i}$$

$$\times G^+(E_f-\hbar\omega,\underset{\sim}{r},\underset{\sim}{r}') \, \underset{\sim}{p}\cdot\underset{\sim}{A} \, G^a(E_f,\underset{\sim}{R}',\underset{\sim}{r}') \tag{2.15}$$

where

$$G^+(E,\underset{\sim}{r},\underset{\sim}{r}') = 2\pi i \sum_i \Psi_i(\underset{\sim}{r}) \, \delta(E-E_i) \, \Psi_i^*(\underset{\sim}{r}')$$

$$= -2i \text{ Im } G^r(E,\underset{\sim}{r},\underset{\sim}{r}') \tag{2.16}$$

and

$$G^a(E,\underset{\sim}{r},\underset{\sim}{r}') = G^r(E,\underset{\sim}{r},\underset{\sim}{r}')^*. \tag{2.17}$$

The quantity G^a denotes the advanced Green's function of the system and $G^+(E,\underset{\sim}{r}=\underset{\sim}{r}')/2\pi i$ is the local density of states. Eq. (2.15) has a simple diagrammtic representation as shown in Fig. 1. The solid lines include the interaction with the crystal lattice and, via a uniform optical potential, the effective interaction with the remaining electrons of the system. The wavy lines include the refraction of the incident photon field at the surface. This diagram describes the excitation process in the single-particle approximation. Caroli et al.[10] have discussed higher-order diagrams which illustrate various corrections to this term due to many body interactions.

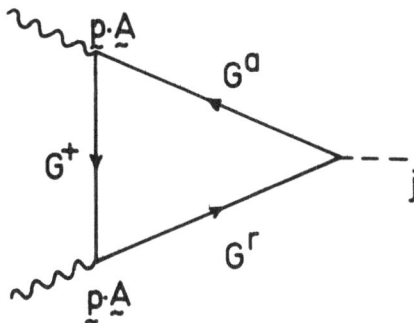

Fig. 1. Golden rule expression (2.15) in diagrammatic form.

III. ATOMS

We consider first the photoionization of isolated atoms in
order to illustrate the basic quantities that determine the variation of the differential cross section with emission angles, photon
frequency and polarization. This simple case also serves as a convenient starting point for the subsequent sections since the matrix
elements for molecules or extended systems can essentially be viewed
as a coherent superposition of atomic transitions. Let us assume
that the initial state is characterized by a particular angular
momentum $L_i \equiv (\ell_i, m_i)$:

$$\Psi_i(\underset{\sim}{r}) = R^i_{\ell_i}(r) \, Y_{L_i}(\hat{r}) \tag{3.1}$$

where $R^i_{\ell_i}$ is a radial wave function and Y_{L_i} a spherical harmonic.
As a result of the interaction of the outgoing wave with
the atomic potential, the final state is a superposition of solutions at the final energy[26] :

$$\Psi^*_f(\underset{\sim}{r}) = 4\pi \sum_L Y^*_L(\hat{k}_f) \, i^{-\ell} \, e^{i\delta_\ell} \, R^f_\ell(\kappa r) \, Y_L(\hat{r}) \tag{3.2}$$

The exponential factor containing the phase shift $\delta_\ell(E_f)$ ensures
that the radial function has the asymptotic form

$$R_{\ell}^f(\kappa r) = \cos\delta_{\ell}\, j_{\ell}(\kappa r) - \sin\delta_{\ell}\, n_{\ell}(\kappa r)$$

$$= e^{i\delta_{\ell}}\, h_{\ell}^{(1)}(\kappa r) + e^{-i\delta_{\ell}}\, h_{\ell}^{(2)}(\kappa r) \tag{3.3}$$

for $r \to \infty$, i.e. there is a phase shift of $2\delta_{\ell}$ between the incoming and outgoing waves. Eq. (3.2) can be derived by expanding the expression (2.6) for Ψ_f^* into partial waves. If the potential is switched off by setting all phase shifts equal to zero and by replacing R_{ℓ} by j_{ℓ}, Eq. (3.2) reduces to the standard angular momentum representation of the incoming plane wave, $\exp(-i\underset{\sim}{k}_f \cdot \underset{\sim}{r})$, i.e. the first term of Eq. (2.6).

Inserting Eqs. (3.1) and (3.2) into the $A \cdot \nabla$ form of the matrix element, the differential cross section may be written as[26]

$$\frac{d\sigma}{d\Omega} \sim \left| \sum_{m_{\varepsilon}=-1}^{1} Y_{1m_{\varepsilon}}(\hat{\varepsilon}) \sum_{L_f} Y_{L_f}^*(\hat{k}_f)\, i^{-\ell_f}\, e^{i\delta_{\ell_f}}\, R_{\ell_f \ell_i}\, I(1m_{\varepsilon};L_f,L_i) \right|^2 \tag{3.4}$$

where we have used the identity

$$\hat{\varepsilon} \cdot \hat{r} = \frac{4\pi}{3} \sum_{m_{\varepsilon}=-1}^{1} Y_{1m_{\varepsilon}}(\hat{\varepsilon})\, Y_{1m_{\varepsilon}}^*(\hat{r}). \tag{3.5}$$

The quantities $R_{\ell_f \ell_i}$ denote the radial matrix elements

$$R_{\ell_f \ell_i}(E_f, E_i) = \int_0^{\infty} dr\, r^2\, R_{\ell_f}^f(r)\, R_{\ell_i}^i(r)\, \frac{\partial}{\partial r}\, V(r) \tag{3.6}$$

and the $I(1m_{\varepsilon};L_f,L_i)$ are the Gaunt coefficients

$$I(1m_{\varepsilon};L_f,L_i) = \int d\Omega_r\, Y_{1m_{\varepsilon}}^*(\hat{r})\, Y_{L_f}(\hat{r})\, Y_{L_i}(\hat{r}). \tag{3.7}$$

Eq. (3.4) shows that the radial matrix elements determine the variation of the atomic cross section with the energies of the initial and final state and, therefore, with photon frequency. Depending on the character of the radial functions, the transition strength can vary by many orders of magnitude. In general, more localized core states have better overlap with final sate wave functions at higher kinetic energies.[27] This is the origin of the so-called delayed onset of the 4f levels in the rare earths.[28] At low photon frequencies, the spectra show only the p-type valence levels, while at higher energies ($\hbar\omega \gtrsim 40$ eV) they are dominated by the 4f emission. Also, as Cooper[29] has shown, the radial matrix elements exhibit zeros as function of frequency if the initial state radial function has nodes. This is the case for all states other than 1s, 2p, 3d, etc. The above arguments indicate that the relative intensity with which a particular adatom level, for example, appears in a spectrum is more strongly influenced by the transition strength than by the escape depth of the excited electron. Except at very low kinetic energies, below approximately 10 eV, the mean free path varies far more slowly as function of frequency than the matrix elements.[27]

The variation of the cross section with emission angles and polarization of the incident radiation is determined by the Gaunt integrals which specify the selection rules of the transition process:

$$\ell_f = \ell_i \pm 1 \qquad\qquad\qquad (3.8a)$$

$$m_f = m_\varepsilon - m_i \qquad\qquad\qquad (3.8b)$$

where m_ε defines the orientation of the polarization vector (see (3.5), e.g. $m_\varepsilon = 0$ for $\hat{\varepsilon}_\parallel = 0$). The amplitudes and phases with which the various allowed transitions from a given initial state are superposed, depend on the radial matrix elements and the phase factors in Eq. (3.4).

The selection rules (3.8) obeyed by the magnetic quantum numbers are of considerable practical importance for angle resolved photoemission measurements since they can give specific information about the symmetry of the initial state without the aid of a calculation. For example, emission along the z-axis ($m_f = 0$) and s-polarized light ($m_\varepsilon = \pm 1$) imply $m_i = \pm 1$. For this configuration, therefore, initial states that are odd with respect to reflections in the x-y plane (p_x,p_y or π-type states) are observable while all others (e.g. s, p_z or δ-type states) give zero intensity. Further below, we discuss cases where these kinds of symmetry arguments have been used to identify adsorbate levels.

If the final state (3.2) is approximated by a plane wave, the various contributions to a transition are related to one another in such a manner that the matrix element is of the form (2.9). The polarization dependence appears as a separate factor, and in contrast to the conditions (3.8), it is not related in any way to the orbital symmetry of Ψ_i. Instead, the polarization dependence is simply given by the cosine of the angle between the polarization vector and the detector direction as specified by $\underline{k}_f$. Thus, the matrix element in this limit is zero whenever $\hat{\varepsilon}$ and $\underline{k}_f$ are perpendicular. This is clearly in contradiction to the selection rules described above.

For a p_z orbital and $\hat{\varepsilon}_{\parallel} = 0$, we have $L_i = (1,0)$ and $m_\varepsilon = 0$, so that transitions may take place to the $L_f \overset{\_}{=} (0,0)$ and $L_f = (2,0)$ partial wave components of the final state[30]:

$$\frac{d\sigma}{d\Omega} \sim |\, e^{i\delta_s} R_{sp} - e^{i\delta_d} R_{dp}(3\cos^2\theta - 1)\,|^2 \qquad (3.9)$$

where θ denotes the polar angle. If, on the other hand, the interaction of the outgoing wave with the potential of the emitting atom is ignored, it is easily shown that

$$\frac{d\sigma}{d\Omega} \sim \cos^4\theta \qquad (3.10)$$

regardless of E_f. This result follows from Eq. (2.9) since both the polarization term and the Fourier transform of the p_z orbital contribute a factor $\cos^2\theta$ to the cross section. Eq. (3.9) shows that the actual θ dependence may deviate strongly from this limit depending on the interference of the two transition channels. If $\exp(i(\delta_s -\delta_d)) R_{sp}/2R_{dp}$ were to approach unity, for example, we would obtain instead of (3.10)

$$\frac{d\sigma}{d\Omega} \sim \sin^4\theta . \qquad (3.11)$$

This simple case illustrates that the final state interaction redistributes the plane wave intensity into different directions. In particular, we may observe appreciable emission at angles for which the polarization vector and the detector direction are perpendicular.

Similar intra-atomic final state partial wave interferences have recently been discussed by Gadzuk[31] for the case of atoms with filled d shells adsorbed on a wide gap insulator such as LiF. Here, the substrate provides a large crystal field which splits the various d quantum numbers. If these levels fall within the gap, they should remain sharp and thus might be resolvable in a photoemission experiment. For normally incident, unpolarized light ($m_\varepsilon = \pm 1$), transitions from the $d_{x^2-y^2}$ orbital ($L_i = (2, \pm 2)$), for example, can occur to different final state f channels, namely $L_f = (3,\pm 3)$ and $L_f = (3, \pm 1)$. Ignoring for the moment the d-p transition which is also allowed, the contribution to the cross section arising from the interference of these two transitions leads to an angular distribution of the form

$$\frac{d\sigma}{d\Omega} \sim \sin^2\theta \,|\sin^2\theta\, e^{4i\phi} - (\cos^2\theta - \frac{1}{5})|^2 . \qquad (3.12)$$

Three azimuthal patterns at different polar angles θ are shown in Fig. 2 together with the Fourier transform of the $d_{x^2-y^2}$ orbital. The latter is proportional to $(1 + \cos 4\phi)$ regardless of the polar

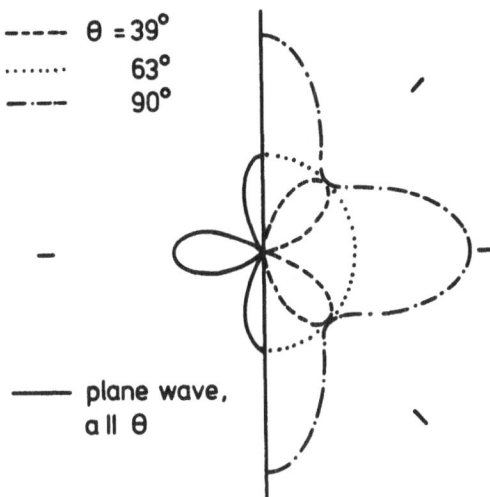

Fig. 2. Azimuthal emission distribution from $d_{x^2-y^2}$ orbital using
normally incident, unpolarized light. Left half plane:
plane wave final state approximation; right half plane:
actual cross section, Eq. (3.12) for $\sin^2\theta = 2/5$, $4/5$ and
1, respectively. All curves are divided by common factor
$\sin^2\theta$. (Ref. 30)

angle. In contrast, the actual cross section above undergoes a reversal of extrema at $\cos^2\theta = 1/5$ or $\theta = 63^0$. We note that this effect is not related to transitions into different ℓ channels whose
strength would depend on the respective radial matrix elements.
Instead, it is caused entirely by the selection rules obeyed by the
magnetic quantum numbers within the same channel. The identification
of these angular distributions (including possible modifications
due to backscattering from the substrate) might give information regarding the crystal field and the adatom geometry.[31,32]

So far, we have assumed a fixed orientation of the initial
state orbital. In the gas phase, averaging over all possible configurations leads to the following general expression for the differential cross section:[33]

$$\frac{d\sigma}{d\Omega} \sim \frac{\sigma}{4\pi} \left(1 + \beta/2(3\cos^2\gamma - 1)\right) \qquad (3.13)$$

where σ is the integrated cross section and γ the angle between

detector direction and polarization vector. The so-called asymmetry
parameter β gives a measure for the deviation of the final state
from a plane wave. Typical values lie between −1 and 2, with β = 2
in the plane wave limit or, in the absence of spin-orbit coupling[34],
for an initial s orbital. Eq. (3.13) is also valid in the case of
emission from gas phase molecules[25,33] which are discussed in the
following section.

IV. MOLECULES

1. Gas Phase Molecules

Dill and Dehmer[6,35] have proposed a scheme for determining the
differential cross section for photoionization from isolated, random-
ly oriented molecules. K-shell absorption spectra of several first
row diatomic molecules exhibit resonance-type behavior within a few
Rydberg above the edge. Such enhancements are not observed for the
constituent atoms, and they have been identified as one-electron
phenomena originating in the molecular nature of the final state.
Although excitation from a K-shell produces a p-type atomic state,
repeated scattering of this wave within the molecular potential
region causes coupling to all other angular momentum states. In the
case of N_2, CO and NO, this effect leads to a predominantly f-like
resonance which is responsible for the observed enhancement of the
photoelectric current.[35]

The scheme for evaluating the continuum states is based on the
so-called scattered-wave method[36] which is frequently used to determine the bound state energies and charge density of molecules. The
molecular potential is represented as a cluster of non-overlapping
spherical potentials centered on the atomic sites. At energies above
the vacuum, the correct asymptotic behavior of the states is that
of an outgoing plane wave plus incoming spherical waves as shown
in Sec. II. Dill and Dehmer have emphasized the close relationship
between these boundary conditions and those appropriate for an elec-
tron-molecule scattering problem. The excited states are orthogonal
to the ground state wave function since both are derived from the
same one-electron potential.

Fig. 3 shows the theoretical angle-integrated cross section
for K-shell ionization in the case of N_2 with

$$\sigma_{tot} = \sigma_g + \sigma_u \qquad (4.1)$$

and σ_g, σ_u represent the partial cross sections for transitions into even (g) and odd (u) final state channels, respectively. These
results are compared with the atomic K-shell cross section (dashed

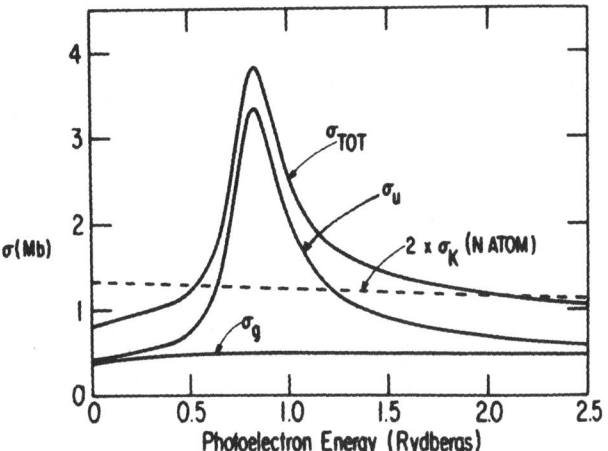

Fig. 3. K-shell photoionization cross section for N_2. Twice the K-shell photoionization cross section for atomic nitrogen is given by the dashed line for comparison. (Ref. 35)

line) which does not exhibit any structure. The overall resonance behavior of σ_{tot} agrees closely with observed spectra[37] for N_2, CO and NO.

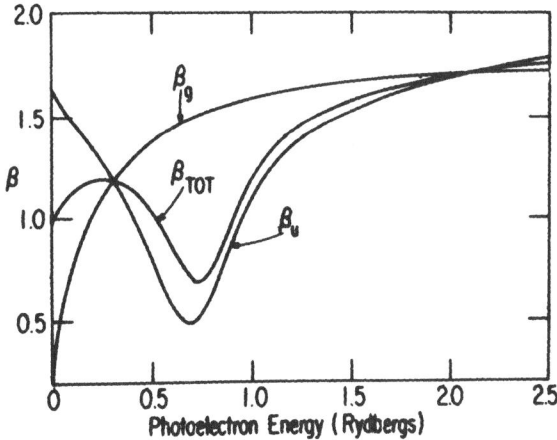

Fig. 4. Asymmetry parameter for K-shell photoionization in N_2. In the case of atomic nitrogen, the corresponding value is equal to 2 at all energies. (Ref. 35)

As indicated above, the differential cross section for emission from gas phase molecules is generally of the form (3.13). In the energy region where the integrated cross section exhibits resonance behavior, the asymmetry parameter β also varies rapidly, decreasing to 0.5 in the case of N_2 as shown in Fig. 4, where

$$\beta_{tot} = (\sigma_g \beta_g + \sigma_u \beta_u)/\sigma_{tot} \qquad (4.2)$$

Experimentally, this structure in β has not yet been verified. The strong deviation from the atomic result ($\beta = 2$) indicates that the angular distributions can be greatly influenced by the multiple-scattering of the outgoing electron off the surrounding atomic potentials.

The interferences of these various scattered waves depends, of course, also on the interatomic distances. In principle, therefore, the variation of the cross section with final state energy and with emission angles should reflect the local atomic geometry. For completeness, we note here that this kind of structural information is the primary aim of the extended X-ray absorption fine structure spectroscopy[38] (EXAFS). This technique is closely related to photo-emission since the absorption coefficient is equivalent to the 4π angle integrated ionization cross section. Plotted as function of photon frequency or kinetic energy of the excited electron, it exhibits characteristic oscillations whose separations depend on the internuclear spacings as well as the phase shifts of the various atomic potentials. As an example that illustrates this technique, a recent measurement[39] on gas phase Br_2 molecules is reproduced in Fig. 5 together with the corresponding theoretical spectrum.

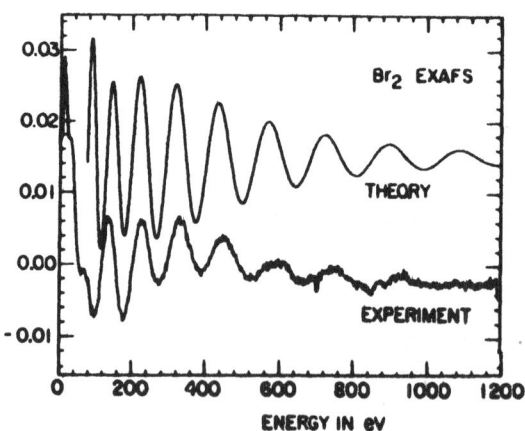

Fig. 5. K-edge X-ray absorption coefficient for Br_2 gas as function of kinetic energy of the excited electron. A smooth background is subtracted from the experimental data and the theoretical curve is displaced upwards for clarity. (Ref. 39)

2. Oriented Molecules

Davenport[7] has applied the scattered-wave method in order to determine the differential cross section for the valence states of oriented molecules. The underlying motiviation for this approach is the fact that a substrate surface tends to preferentially orient chemisorbed molecules along a particular axis.[40] Those molecular orbitals that do not participate in the bonding to the substrate should closely resemble their gas phase analogs. The ground state of these levels is, therefore, assumed to be that of the isolated molecules. The excited state in this scheme is also that of the free molecule. It is derived from the same one-electron potential as the initial state. In principle, corrections may arise[21] as a result of the interaction of the outgoing wave with the substrate atoms (see following section). It is well possible, however, that this is of minor importance if the final state scattering is dominated by intra-molecular resonances such as those discussed in the preceding subsection.

Theoretical[7] and experimental[41] results obtained for the gas phase ionization cross-section for CO indeed show the same kind of resonance behavior for the valence levels as for the K-shell.[35] In the case of the 4σ and 5σ levels, maxima are obtained at photon frequencies of about 36 eV and 28 eV, respectively (see Fig. 6).

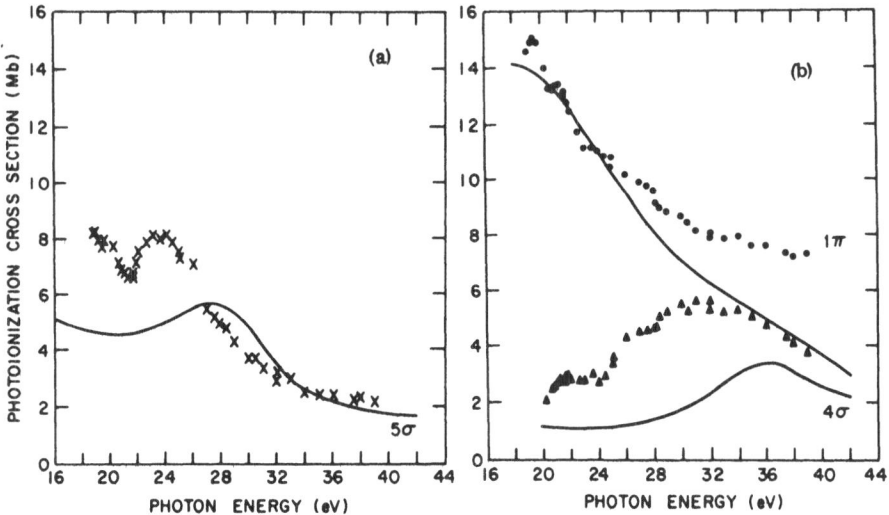

Fig. 6. Photoionization cross sections for gas phase CO. (a) 5σ level and (b) 1π and 4σ levels. The calculated spectra are indicated by the solid lines. (Ref. 7)

Since the theoretical ionization energies are -19.7 eV and -14.0 eV,
this implies that the peaks occur at different final-state energies,
indicating that the initial states are important as well. The
1π level does not couple to the final state resonance as a consequence of matrix element effects. The calculated spectra in Fig. 6
agree considerably better with the data than those obtained by previous methods which do not treat initial and final state on the
same footing.

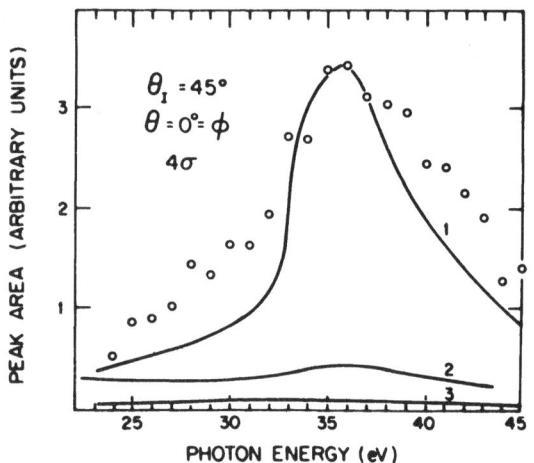

Fig. 7. 4σ peak area for CO on Ni(001) versus photon energy at nor-
        mal exit and 45^0 angle of incidence. Solid curves are cal-
        culations for (1) CO axis normal to the surface, carbon end
        down, (2) CO axis normal to surface, oxygen end down, and
        (3) CO axis in surface plane. Experimental data normalized
        to theory at $\hbar\omega$ = 36 eV. (Ref. 42)

     For CO chemisorbed on Ni(001), two peaks are observed at -10.7
eV and -8.0 eV below E_F. It is now generally accepted that the lower
corresponds to the 4σ state while the upper is a superposition of
the 1π and 5σ levels. The latter is shifted downwards from its gas
phase position as a result of the binding to the substrate. Fig. 7
shows the differential cross section of the 4σ level as function of
photon energy.[42] P polarized light is incident at 45^0 and emission
is along the surface normal. The measured spectrum (circles) exhibits a pronounced maximum near $\hbar\omega$ = 36 eV, similar to the one seen
in the gas phase. These data are compared to calculated spectra[43]
for three distinct orientations of the isolated molecule. Striking
agreement is obtained if the CO axis is parallel to the surface

normal, with the carbon end towards the Ni surface. The remaining
configurations (oxygen end down or molecular axis parallel to surface plane) can clearly be ruled out as possible alternatives.

The above identification receives additional support from comparisons of the variation of the 4σ intensity with polar emission
angle.[42] This is illustrated in Fig. 8 for the photon frequency at
resonance and p polarized light incident at 45^0. The emission direction is perpendicular to the incidence plane in order to minimize the interference of the components of the polarization vector

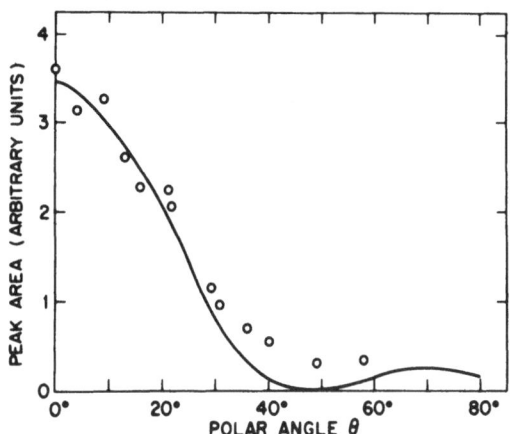

Fig. 8. 4σ peak area for CO on Ni(001) versus polar angle Θ measured
perpendicular to the incidence plane at $\hbar\omega$ = 35 eV and 45^0
angle of incidence. The theory curve has the shape characteristic of the 4σ emission on resonance (where $\Theta = 0^0$ is
the direction of the CO axis) and is normalized to best fit
the data at small angles. (Ref. 42)

parallel and normal to the surface. The angular distribution is seen
to be strongly peaked in a rather narrow cone of about $\pm 25^0$ along
the surface normal. The excellent agreement between the experimental
and theoretical[43] results suggests that the molecular axis is perpendicular to the surface plane with an accuracy of approximately
5^0. The measurements are performed at saturation coverage of CO
where it is known that the overlayer is not in registry with the
substrate.[44] This clearly tends to average out any modifications of
the angular distribution due to interaction of the excited electron
with the crystal lattice.

Calculations[43] show that the angular profiles of the CO levels
vary considerably, both in shape and in absolute intensity, as
function of photon energy as well as polarization. This fact under-

lines the usefulness of variable frequencies and polarizations in
order to optimize the observable effects. It is important to focus
on prominent spectral features (such as the final state resonance
in the case of CO) since their existence and their essential behavior as function of experimental parameters should be predicted
by an adequate theoretical model.

The above identification of the CO bonding configuration evidently requires a fairly detailed theoretical treatment of both
initial and final state. The orbital symmetry of the observed levels, on the other hand, can be determined experimentally simply by
choosing the appropriate polarization. An example of this has
recently been given by R.J. Smith et al.[45] (see also Ref. 42) for
CO adsorbed on Ni(001). At finite angles of incidence, using p polarized light, both CO induced peaks are observed for several emission directions. At normal incidence (s polarization), the lower
peak at -10.7 eV disappears for emission near the surface normal
while the -8.0 eV peak remains finite. This behavior does not depend on the detailed spatial distribution of the states. It is a
direct consequence of their symmetry and the selection rules (3.8)
obeyed for this particular polarization. Both the 4σ and 5σ states
have even parity for reflections in the x-y plane while the 1π
state is odd. $\underset{\sim}{A}$ itself is also odd in this case, i.e. the former two
levels give zero normal emission whereas the 1π remains finite. This
implies that the 1π level must be assigned to the upper peak at
-8.0 eV. The above result unambiguously confirms earlier identifi-
cations of CO levels using their photon energy dependence at low
frequencies[46] ($\hbar\omega \leqslant 100$ eV) and their relative intensity at X-ray
energies.[47] (In addition, these measurements give the correct assignment of the 4σ and 5σ states which is not possible using the
above symmetry arguments alone.) It should be noted that this polarization effect can equally well be observed for unpolarized light
at normal incidence. An example will be discussed in the following
section for O on Ni(001).

V. ADSORBATES

1. Core Levels

In addition to providing information on orbital symmetries and
molecular orientations, photoemission can also be used to investigate adsorption geometries and the electronic properties of chemical bonds at surfaces.[21,23,40] With regard to the atomic structure
determination, photoemission is complementary to LEED; it is, to
some extent, more general in that it is not limited to ordered overlayers and relatively high coverages. Information on bonding states,
on the other hand, is not accessible, as directly, through any other
surface sensitive technique. The theoretical difficulties in the

description of emission from adsorbed species are considerable. The
neglect of many-body effects and the assumption of a spatially constant photon field in the vicinity of the surface are particularly
questionable. Even within the independent-electron picture, self-
consistency of the potential is most likely required to obtain
reasonably accurate one-particle wave functions.

Conceptually, the simplest case is the emission from adatom
core levels.[21,48-52] Here, the initial state is given in terms of
a known atomic wave function and it is the excited state which
contains structural information. The outgoing wave can be thought
of as a superposition of two coherent contributions:[51] (i) a
"direct" wave whose angular symmetry and complex amplitude are determined by the intra-atomic transition at the emitting site, and
(ii) an "indirect" wave which is caused by the repeated scattering
of the direct wave by the surrounding atomic potentials within overlayer and bulk (see Fig. 9). The interference of these contributions
leads to structure in the energy and angle distributions that is
sensitive to the adsorption geometry.

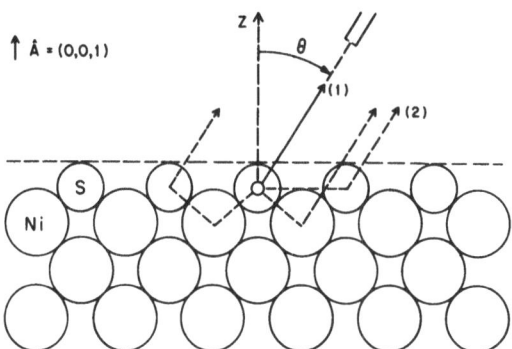

Fig. 9. Illustration of two processes contributing to photoemission
        from a localized adsorbate s orbital: (1) direct emission
        into atomic p-type final state and (2) indirect emission
        via backscattering of the direct wave from the surrounding
        atoms. Only two single and one multiple scattering event
        are indicated. (Ref. 51)

A rough estimate of the magnitude of these interferences can be
obtained by comparing the core level ionization to related spectroscopies in which the same kind of scattering processes play an
important role. As indicated already at the end of Sec. IV.1, the
absorption coefficient in EXAFS corresponds to the ionization cross

section, summed over all possible final states at a particular
energy. Typical fine structure due to scattering events is of the
order of 10% of the mean signal.[38] This implies (i) that the angle
integrated photoemission cross section should exhibit, as function
of final electron energy, structure of similar magnitude, and
(ii) that at a particular emission direction, the interference effects have to be considerably larger than 10% since much of this
structure is averaged out upon angular integration. Also, the
above ionization experiment can be viewed as a LEED-type problem
with a spherical rather than a plane wave source. Since the direct
and indirect wave are, however, detected coherently (see Eq. (2.5)),
the main correction to the direct emission intensity is proportional
to the cross term between these two contributions. For a typical
LEED reflectivity $(\sim|f(E_f,k_f)|^2)$ of 4%, the scattering amplitude
$f(E_f,k_f)$ in Eq. (2.8) is of the order of 0.2. Thus, the cross term
is roughly 2 x 0.2 or 40%, in qualitative agreement with the preceding estimate. The scattering amplitude needs to be small at all
scattering angles for the cross term to be negligible at some energy
E_f.

The matrix element (2.7) can be evaluated by expanding the scattering potentials as well as the one-electron propagators in Eq.
(2.6) in terms of their angular momentum components. For the direct
wave, one obtains

$$M^o_{fi}(E_f,E_i;\underset{\sim}{k}_f,\underset{\sim}{A}) \sim \sum_L Y^*_L(\hat{k}_f)\, M_{LL_i}(E_f,E_i)\qquad\qquad (5.1)$$

with

$$M_{LL_i}(E_f,E_i) \sim <R^f_\ell Y^*_L\,|\underset{\sim}{A}\cdot\nabla V|\,R^i_{\ell_i}\, Y_{L_i}> .\qquad\qquad (5.2)$$

(The same notation is used as in Sec. III.) The final state radial
functions in (5.2) are the result of the interaction of the outgoing wave with the potential at the emitting site. To first order
in the scattering from the surrounding potentials, the indirect
contribution to the matrix element is given by:

$$M^1_{fi} \sim \sum_{LL'} Y^*_{L'}(\hat{k}_f)\, \sum_R e^{-i\underset{\sim}{k}_f\cdot\underset{\sim}{R}}\, t_{L'}(\underset{\sim}{R})\, G^o_{L'L}(\underset{\sim}{R}-\underset{\sim}{R}_o)\, M_{LL_i}\qquad (5.3)$$

where $\underset{\sim}{R}$ denotes the lattice vectors and $\underset{\sim}{R}_o$ the position of the emitting atom. The quantities $t_{L'}$ and $G^o_{LL'}$ are the partial wave com-
ponents of the single-site scattering matrix and of the propagator
G_o, respectively. G_o contains a complex self-energy (see Eq. (2.11))
to account for the inelastic processes that give the excited electron its short mean free path. Because of the translational symmetry
of the system, the summation over sites parallel to the surface can
be converted into a sum over two-dimensional reciprocal lattice

vectors $\underset{\sim}{g}$ of the surface Bravais net. The remaining sum over the
planes of the crystal is a geometric series, i.e. Eq. (5.3) can be
expressed in the following simple form:[51]

$$M_{fi}^l \sim \underset{\underset{\sim}{g}}{\Sigma}\ t(\underset{\sim}{k}_f, \underset{\sim}{K}_g^-)\ \frac{\exp(i(\underset{\sim}{K}_o^+ - \underset{\sim}{K}_g^-)\cdot\underset{\sim}{d}_o)}{1 - \exp(i(\underset{\sim}{K}_o^+ - \underset{\sim}{K}_g^-)\cdot\underset{\sim}{d})}\ \frac{8\pi^2}{a^2 i\kappa(g)}$$

$$\times \underset{L}{\Sigma}\ Y_L^*(\hat{\underset{\sim}{K}}_g^-)\ M_{LL_i}(E_f, E_i) \tag{5.4}$$

where

$$t(\underset{\sim}{k}_f, \underset{\sim}{K}_g^-) = \underset{L}{\Sigma}\ Y_L^*(\hat{\underset{\sim}{k}}_f)\ t_\ell\ Y_L(\hat{\underset{\sim}{K}}_g^-) \tag{5.5}$$

$$\underset{\underset{\sim}{g}}{K}^+ = (\underset{\sim}{k}_{f\parallel} + \underset{\sim}{g}, \pm\ \kappa(g)) \tag{5.6}$$

$$\kappa(g) = (E_f - V_o + i\Gamma - (\underset{\sim}{k}_{f\parallel} + \underset{\sim}{g})^2)^{1/2} \tag{5.7}$$

and $\underset{\sim}{d}$, $\underset{\sim}{d}_o$ specify the relative position of neighboring bulk layers
and of overlayer and first substrate plane, respectively. (Only
backscattering from the substrate is included in (5.4).) Due to
the evanescent nature of the large g terms in (5.4), only a small
number of contributions are of importance. This shows that only a
discrete set of backscattered waves is coherently added to the
direct wave.

Complex phase and amplitude of the indirect wave are determined
by the scattering factors t and by the exponents containing the
structural information. In kinematic LEED theory (first order in t),
the peak positions are given by the well known Bragg conditions,
which are related only to the atomic geometry. In the above case,
however, they depend sensitively on the phase properties of the
scattering factors. A feature that might, therefore, be of considerable practical importance is the fact that at normal emission
($\underset{\sim}{k}_{f\parallel}$ = 0) many terms in (5.4) are degenerate. It is likely that such
spectra are easier to reproduce than those at finite angles where a
far more complicated superposition of many different contributions
takes place.

Fig. 10 shows theoretical results[51] for the emission from the
2s core level of sulfur adsorbed on a Ni(001) surface. Plotted is
the ratio of total intensity (including multiple scattering) and
direct emission intensity as function of final energy at $\underset{\sim}{k}_{f\parallel}$ = 0.
For illustrative purposes, only the normal component of the pola-
rization vector is used. The overall magnitude of the interference
effects is in good agreement with the qualitative estimates discussed
above. The comparison of the centered (a) and top (b) position

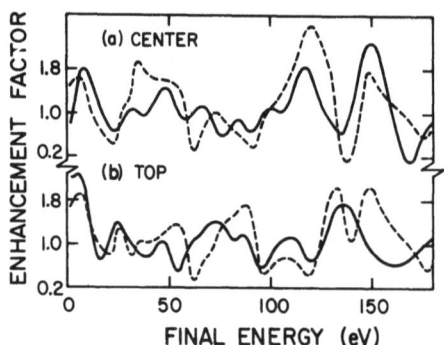

Fig. 10. Enhancement of direct emission intensity from 2s core level
of sulfur on Ni(001) due to backscattering from substrate.
(a) Adsorption in center position, (b) in top position.
Solid lines: multiple scattering, dashed lines: single
scattering limit. (Ref. 51)

demonstrates the influence of the adsorption geometry on the emission spectra. Some similarities exist between the multiple scattering results (solid lines) and those in the single scattering
limit (dashed lines). The agreement, however, does not appear sufficient for the latter to be adequate. Similar conclusions apply
to polar and azimuthal angle distributions.

Tong and Van Hove[52] have recently performed analogous calculations for emission from core levels of oxygen on Ni(001). They
obtain final state effects of the same overall magnitude and similarly strong multiple scattering interferences. Large differences
are found for various adsorption geometries.

Azimuthal emission spectra for CO on Ni(001) by R.J. Smith
et al.[45] show no discernable anisotropy which could be attributed to
the above described phenomena. However, in the case of emission from
molecular orbitals, a more complicated superposition of outgoing
waves results than for localized core levels. This might well lead
to an effective smoothing of interferences. Besides, at high coverages, the CO overlayer is not in registry with the substrate,[44]
while at low coverages, more than one binding site appears to exist
and the molecules might not be aligned preferentially along a particular axis.[53] Both effects cause considerable averaging of possible structure due to backscattering.

The final state scattering effects should, of course, be also
present in the case of core level emission from clean surfaces.
This has, in fact, recently been observed by N.V. Smith et al.[54]

for the spin-orbit split In $4d_{3/2}$ and $4d_{5/2}$ core levels of the layer
compound InSe which are located about -18 eV and -16 eV below
the vacuum, respectively. Fig. 11 shows an azimuthal spectrum of the
$4d_{5/2}$ level taken at 33 eV photon energy. P polarized light is incident at -45° in the x-z plane. The detector is fixed at a polar
angle of 60° in the x-z plane (solid curve) or in the y-z plane
(dashed curve). The spectrum is obtained by rotating the crystal
about the surface normal. The asymmetry of the dashed spectrum
is caused by the relative orientation of polarization vector and
detector direction.[54] It is a direct indication of the non-plane
wave nature of the final state: In this limit, the two curves should
each be proportional to the Fourier transform of the initial state
since the factors $\cos(\underline{k}_f, \underline{A})$ in Eq. (2.9) are constant as the crystal is rotated. The observed azimuthal structure could be caused by
the intra-atomic transition at the In atom and by the interference
of the outgoing wave with surrounding sites. An indication that
the latter scattering events play a role are spectral variations
with photon frequency. Increasing $\hbar\omega$ by only 1 eV changes the fine
structure considerably.[54] This effect is most likely not related to
the intra-atomic matrix element which usually varies more slowly
with $\hbar\omega$.

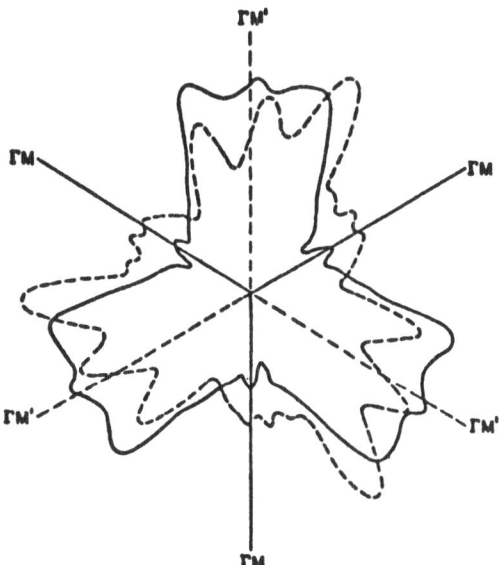

Fig. 11. Azimuthal photoemission spectrum from In $4d_{3/2}$ core level
of InSe at 33 eV photon energy. Detector and photon
beam are in same plane (solid curve) or in perpendicular
planes (dashed curve). (Ref. 54)

2. Surface Molecules

We now turn to the discussion of the photon excitation of
valence electrons of adsorbed species. It is evident that this requires a reasonably accurate description of the electronic structure of the chemisorption state.[55] In addition, the results shown in
the preceding sections suggest that proper treatment of the final
state (including the electronic mean free path) is also important.
Two schemes, which try to incorporate both initial and final state
in a consistent manner, have recently been proposed. The first is
based on the surface-molecue picture, i.e. the initial state is
derived from a cluster involving the adatom and several neighboring
substrate atoms.[9,81] The second approach applies to the emission
from chemisorbed overlayers;[8] it is discussed in the following subsection.

Scheffler et al.[9] have performed calculations of the differential cross section for p(2x2) oxygen on Ni(001), using the surface-
molecule approach. Previous angle integrated spectra for this system
exhibit[56] a single, rather broad oxygen induced peak at about 6 eV
below E_F, i.e. below the Ni d bands. To interpret angle resolved
spectra[9,82] the photocurrent is calculated in the following manner:
(i) The oxygen coverage for the p(2x2) structure is considered sufficiently low so that direct and indirect interactions between adatoms can be neglected. Also, the oxygen induced states are assumed
to be rather concentrated near the adatom and its neighboring atoms
in the substrate. The initial state is, therefore, represented as
a linear combination of atomic orbitals, with radial functions and
expansion coefficients determined by a $O(Ni)_5$ scattered wave Xα
calculation.[57] Four Ni atoms are arranged in a square corresponding
to the unit cell of the (001) crystal face. The fifth Ni atom is
below, the oxygen atom above the center of the square at distances
as for the p(2x2) structure. The Ni-0 bond length is 1.98 A as determined by LEED studies.[83] For such a cluster, two oxygen derived
states are obtained below the Ni 3d levels, one of p_z-like symmetry
and at slightly lower energies, another (doubly degenerate) of p_x,
p_z-like symmetry. Spatially, the weight of these states is concentrated on the oxygen atom and its four Ni neighbors; only the p_z-
like level shows some weight also on the fifth nickel atom below
the oxygen atom. (ii) As final state, a time reversed LEED wave
function appropriate for the p(2x2) overlayer on a semi-infinite
Ni substrate is used. The muffin-tin constant is the same throughout bulk and overlayer. The outgoing electron is refracted but not
scattered by the potential step at the adsorbate-vacuum interface.
Inside the muffin-tin spheres, the final state is approximated by
a continuation of the plane wave expansion that is valid in the
interstitial region between spheres. (iii) The vector potential
describing the photon field is assumed to be spatially constant. The
complex polarization vector $\hat{\epsilon}$ at the solid-vacuum interface is derived from the clean Ni optical constants $\hat{n} = n - ik$ with n = 0.9

and k = 0.15 for the photon frequency under consideration (40.8 eV).

For unpolarized light, incident in the y-z plane at an angle α with respect to the surface normal, the cross section is of the general form:

$$\frac{d\sigma}{d\Omega} \sim |\varepsilon_x M_x|^2 + |\varepsilon_y M_y + \varepsilon_z M_z|^2 \tag{5.8}$$

where $\underset{\sim}{M}$ represents the vector $<\Psi_f|p|\Psi_i>$. Thus, at normal emission, symmetry arguments require $\underset{\sim}{M} = (0,0,M_z^1)$ and $\underset{\sim}{M} = (M_x,M_y,0)$ at the energy of the p_z and the p_x,p_y level, respectively, with $M_x = M_y$. This implies:

$$\frac{d\sigma}{d\Omega}\bigg|_{p_z} \sim |\varepsilon_z M_z|^2$$

$$\sim |1+r_{\parallel}|^2 \sin^2\alpha |M_z|^2 \tag{5.9}$$

$$\frac{d\sigma}{d\Omega}\bigg|_{p_x p_y} \sim |\varepsilon_x M_x|^2 + |\varepsilon_y M_y|^2$$

$$\sim (|1+r_{\perp}|^2 + |1-r_{\parallel}|^2 \cos^2\alpha) |M_{x,y}|^2 \tag{5.10}$$

where

$$r_{\parallel} = \frac{\hat{n}^2\cos\alpha - (\hat{n}^2 - \sin^2\alpha)^{1/2}}{\hat{n}^2\cos\alpha + (\hat{n}^2 - \sin^2\alpha)^{1/2}} \tag{5.11}$$

$$r_{\perp} = \frac{\cos\alpha - (\hat{n}^2 - \sin^2\alpha)^{1/2}}{\cos\alpha + (\hat{n}^2 - \sin^2\alpha)^{1/2}}$$

are the complex reflectivities (the indices $\parallel$ and $\perp$ refer to the plane of incidence). Fig. 12 shows the dependence of the differential cross section on the incidence angle α at normal emission, as given by the above equations. Both curves include an arbitrary normalization constant. These results clearly demonstrate the usefulness of varying the angle of incidence as a means of determining the orbital symmetry of adsorbate states. The fact that at normal emission the p_z-like level vanishes for $\alpha \to 0$ while the p_x,p_y level remains finite, is a consequence of selection rules (see Eq. (3.8b)) and, thus, model independent. The results also illustrate the importance of including the refraction of the incident radiation for a theoretical interpretation of the polarization dependence. The use of only the incident light as exciting field ($r_{\parallel} = r_{\perp} = 0$) leads to rather different distributions, in particular, at large angles:

$$\frac{d\sigma}{d\Omega}\bigg|_{p_z} \sim \sin^2\alpha |M_z|^2 \tag{5.12}$$

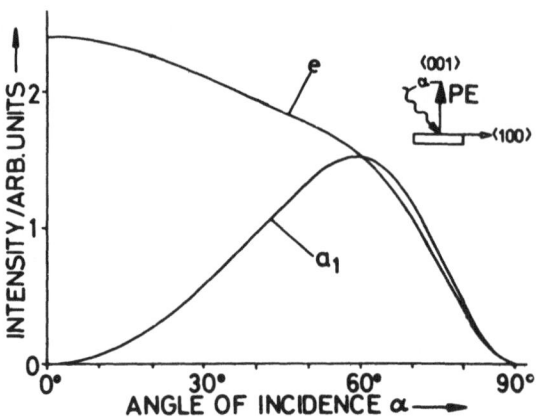

Fig. 12. Calculated photoemission intensities for p(2x2) O on Ni(001) using unpolarized light at $\hbar\omega$ = 40.8 eV. Emission is normal to the surface. The two adsorbate levels have $a_1(p_z$-derived) and $e(p_x,p_y$-derived) symmetry. Ref. (9)

$$\left.\frac{d\sigma}{d\Omega}\right|_{p_x p_y} \sim (1 + \cos^2\alpha) \quad |M_{x,y}| \; . \tag{5.13}$$

Fig. 13(a) shows the calculated differential cross section for the two levels as function of polar emission angle ((100) azimuth) at normal incidence. In panel (b), theoretical results are shown in which the initial state is replaced by atomic oxygen p functions. The pronounced minimum of the p_z-like level at about 45^0 is almost entirely obscured. In panel (c), the LCAO initial state is used and the final state is approximated by an exponentially damped plane wave. The p_x,p_y level now exhibits the wrong behavior at normal emission (as pointed out in Sec. II, the plane wave limit gives vanishing intensity whenever detector direction and polarization vector are perpendicular) and the 20^0 lobe of the p_z-like level is absent. These comparisons illustrate the necessity of deriving both initial and final state from a reasonably detailed model of the electronic structure of the system.

Experimentally, the two oxygen induced levels are not resolvable because of their large intrinsic width. Thus, if unpolarized light is used, only a superposition of the p_z and p_x,p_y distributions in Fig. 12 or 13 can be compared with the measured spectra since both levels are excited simultaneously.[82] With polarized radiation, on the other hand, it is possible to select the individual p components and investigate their variation with emission angles. This will be discussed further in the following subsection.

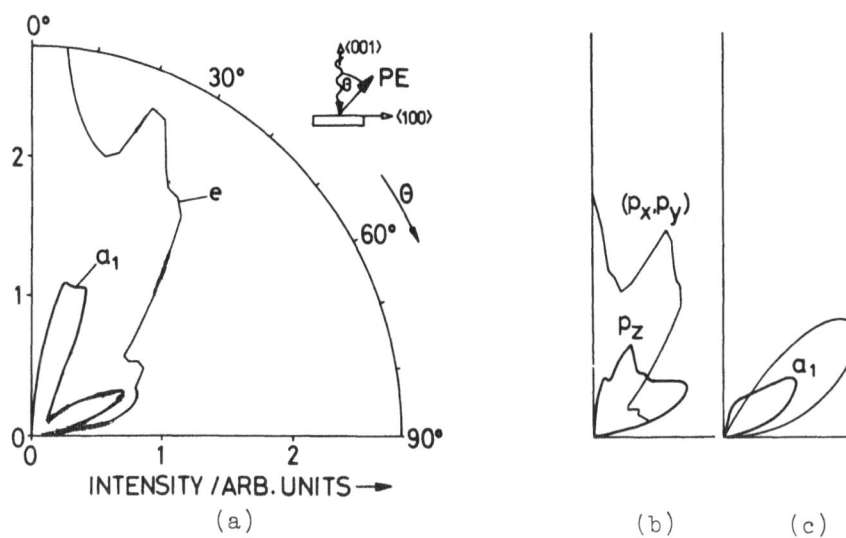

Fig. 13. Polar plot of oxygen induced levels with symmetry a_1 (p_z-
         derived) and e(p_x,p_y-derived). Unpolarized light is
         incident normal to the surface, $\hbar\omega$ = 40.8 eV. The
         curves show calculated spectra using (a) LCAO initial
         state, LEED final state; (b) oxygen atomic 2p initial state,
         LEED final state; (c) LCAO initial state, damped plane wave
         final state. (Ref. 9)

3. Chemisorbed Overlayers

A complementary approach to the emission from valence states
of adsorbed species, applicable in the limit of ordered overlayers
on a semi-infinite substrate, has recently been suggested by
Liebsch.[8] Whereas cluster models focus on the local aspects of the
bonding, the two-dimensional translational symmetry of the overlayer
scheme makes possible a realistic description of both extended and
local features of the surface electronic structure. Furthermore,
experiments are usually performed in the high coverage region where
ordered adsorbate structures are formed. The close spacing between
adatoms can also lead to observable effects[58] on the electronic
properties which are not contained in a cluster approach.

The method of evaluating the differential cross section of the
metal-overlayer system is an extension of the procedure for clean
surfaces which is discussed in some detail in the following section.
The bulk states are obtained by solving for the complex band structure of the substrate. The presence of the overlayer is conveniently

taken into account by means of a transfer matrix which relates the
two-dimensional Fourier coefficients of a wave function and its normal derivative on one side of the plane to those on the other. In the
vacuum, the potential is assumed to depend only on the coordinate
normal to the surface. Formally, three types of bound states of the
metal-adsorbate system may be distinguished. At energies and $k_\parallel$ points
within the range of the substrate conduction bands, the adsorbate
levels are able to hybridize with them unless the interaction is
forbidden on symmetry grounds. In the first case, the bound states
form asymptotically a standing wave whose amplitude is strongly peaked
near the overlayer if the energy approaches the adatom resonance.
The second case is that of a sharp adsorbate resonance whose wave
function falls off exponentially toward the interior. Finally, below
the conduction band or within band gaps, there may exist split-off
states that are also, like the sharp resonances, localized in the
vicinity of the surface. A state that hybridizes with the metal
bands is expressed inside the semi-infinite solid as a superposition
of a propagating Bloch wave incident on the surface and a series of
reflected waves that propagate or decay toward the interior.[59] All
of these Bloch waves are characterized by the same energy and the
same $k_\parallel$ within the overlayer SBZ. The incident wave is normalized
such that it carries unit normal current.[60] In the case of the discrete adsorbate levels, the state in the interior consists only of
evanescent waves. The wave functions of these levels are normalized
to unity over the surface unit cell.[60] The excited state is expressed as a LEED wave function of the metal-overlayer system. It is
derived from the same potential as the ground state.

Calculations based on this scheme have been carried out for the
(1x1)[8] and c(2x2)[61] structures of oxygen on Ni(001). The oxygen
atoms are adsorbed in the four-fold centered positions with a Ni-O
bond length of 1.98 Å as found in LEED studies.[63] The self-consistent
Wakoh potential is used for the Ni substrate up to the surface plane
and a neutral-atom Xα potential (α = 0.744) for the oxygen overlayer.
The muffin-tin model is adopted with a constant of −1.0 Ry in the
interstitial region throughout substrate and adsorbate.

To illustrate the electronic properties of the ground state in
the case of the c(2x2) overlayer, we show in Fig. 14 several partial
wave projections of the local density of states (integrated over
the muffin-tin sphere volume) at $k_\parallel$ = 0 as function of energy,
(a) at an oxygen site, (b)-(d) in the first Ni plane, and (e) in the
second. The following main features can be noticed:

(i) The presence of the oxygen overlayer leads to the formation
of two groups of levels, bonding states below and unoccupied anti-
bonding states above the Ni d bands. This splitting is caused by the
O 2p-Ni 3d interaction. Near the p_z levels, the Ni d_{z^2} and d_{xy} components are strongly enhanced, with the d_{z^2} component having appreciable weight also in the second Ni layer. At the p_x, p_y resonances,

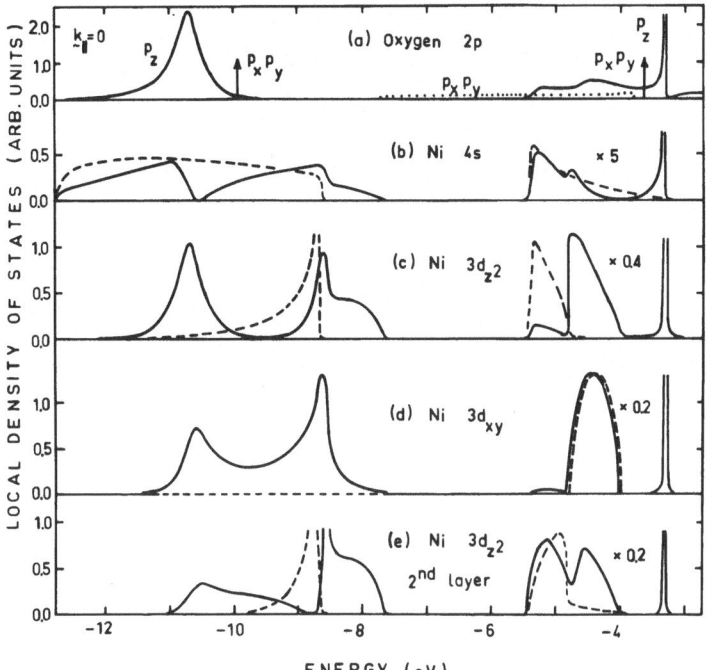

Fig. 14. Various partial wave projections of the local density of
states at $\bar{\Gamma}$ for c(2x2) O on Ni(001), (a) at an overlayer
site, (b)-(d) in the first substrate plane, and (e) in
the second layer. Energies are relative to the vacuum. In
(b)-(e), the dashed lines are for the clean surface, the
solid ones for the adsorbate covered surface. The Ni s-p
band extends approximately from -12.5 eV to -8.5 eV, the 3d
bands from -8 eV to about -4 eV. Notice the different
scales. (Ref. 61)

the Ni density has mainly d_{xz}, d_{yz} character (not shown here). For
the clean surface, the d density at these energies is negligible.
This enhancement indicates that a metal-adsorbate surface molecule
is formed with occupied valence states well below E_F and unfilled
states just above E_F. (The Fermi energy in paramagnetic Ni lies
slightly below the top of the d band.) Both types of states are
predominantly of p character within the overlayer and of d character
in the first two to three substrate planes. This picture is qualitatively consistent with O 1s appearance potential spectra for oxygen chemisorbed on Ni(001) which show[62] relatively narrow unfilled
states involving the oxygen atoms at about 2 eV above E_F. Furthermore, electron energy loss spectra for the same system indicate[62]
that a transition of about 7 eV exists from occupied to unoccupied

nickel-oxygen valence states. In connection to these observations, we also note that recent photoemission data for c(2x2) O on Cu(001) show[63] a small adatom-induced peak between the top of the Cu d bands and E_F. These results indicate that metal-oxygen anti-bonding levels exist similar to those for O on Ni but that they are at least partially filled as a result of the lower-lying Cu d bands.

(ii) The bonding as well as anti-bonding states are split, partly due to the direct coupling between oxygen levels and partly due to their interference with the Ni conduction bands. The p_z bonding level lies at $\underset{\sim}{k}_{\parallel} = 0$ <u>below</u> the doubly degenerate p_x, p_y states.

(iii) The hybridization with the Ni s-p band broadens the p_z bonding level by about 0.5 eV. The p_z anti-bonding level is narrower because of the smaller Ni s-p density at the higer energies. The p_x, p_y states remain sharp at $\underset{\sim}{k}_{\parallel} = 0$ since they cannot couple to the s-p band because of their symmetry. Investigation of the spatial distribution of the local density shows that a charge accumulation in the Ni-O bond region occurs for states in the lower part of the p_z bonding resonance. In the upper part, on the other hand, a depletion takes place indicating the anti-bonding character of the states. This interpretation agrees with what one expects from the interaction of a single level with a continuum of states each of which leads to a small splitting. The result is a broadened peak in which the bonding states accumulate in the lower and the anti-bonding in the upper part. Among the O 2p-Ni 3d bonding states, we can, therefore, identify those that are bonding or anti-bonding with respect to the s-p band.

Fig. 15 illustrates the position and broadening of the oxygen 2p levels at finite $\underset{\sim}{k}_{\parallel}$. Plotted are the three p wave projections of the local density at an adatom site. (The x axis is along the crystallographic (110) direction and k_x is in units of $2\pi \sqrt{2}/a$, where a is the Ni lattice constant.) Below the conduction band minimum, the p levels can exist only as discrete split-off states. Above the minimum, the p_x resonance is broadened due to interference with the s-p band (except at $\underset{\sim}{k}_{\parallel} = 0$) whereas the p_y state remains sharp at all $\underset{\sim}{k}_{\parallel}$ values shown. These results demonstrate that the level width depends sensitively on $\underset{\sim}{k}_{\parallel}$ and on the symmetry of the adsorbate induced state.

The variation of the energy position of the oxygen resonances can be understood partly in terms of the direct interaction between adatoms. The overlap of the p functions leads to the formation of two-dimensional bands within the overlayer. Calculations of the bound states of an isolated oxygen monolayer give a dispersion which is qualitatively similar to that in Fig. 15. The dispersion of the 2p levels in Fig. 15 suggests that the ordering of the oxygen resonances does not remain the same throughout the SBZ. At $\bar{\Gamma}$ ($\underset{\sim}{k}_{\parallel} = 0$), the p_z level lies below the doubly degenerate p_x, p_y states while at $\bar{M}$ ($\underset{\sim}{k}_{\parallel} = (1/2,0)$), it is above the doublet. Recent angle resolved

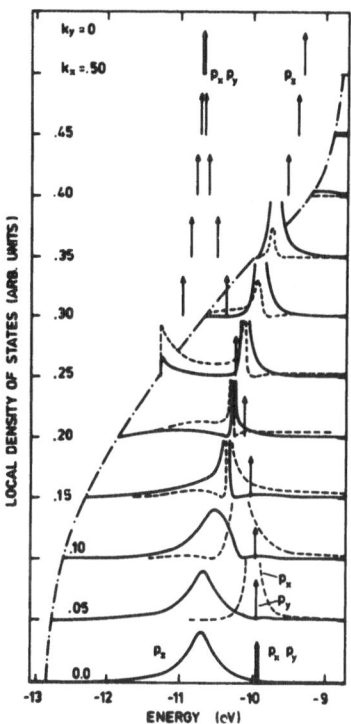

Fig. 15. P wave projections of the local density of states for
c(2x2) O on Ni(001) at an oxygen site for several $\underset{\sim}{k}_{\parallel}$
values along $\bar{\Gamma}\bar{M}$. The arrows indicate sharp adsorbate resonances above or split-off states below the conduction
band minimum (dot-dashed line). (Ref. 61)

spectra for c(2x2) O on Ni(001) indicate[64] that such a reversal of
the level positions does indeed take place. In this experiment, the
individual 2p components are identified by varying the polarization
of the incident radiation. At normal emission ($\underset{\sim}{k}_{\parallel}$ = 0), a mixture of
s- and p-polarized light shows a peak near -6.0 eV below E_F while
pure s-polarization gives a peak at about -5.5 eV. Since normal
emission from the p_z like level with s-polarized light is forbidden
as a consequence of selection rules, the data imply that at $\bar{\Gamma}$ the
energy of the p_z state lies below the p_x,p_y levels. At emission
angles corresponding to $\underset{\sim}{k}_{\parallel}$ close to the $\bar{M}$ point in the SBZ, a
peak near -5.5 eV is observed for s-polarized light if the polari-
zation vector lies within the emission plane. The spectrum shows a
feature at roughly -6.5 eV below E_F if the polarization vector is
perpendicular to the emission plane. Since a transition from the p_z-
like state is not allowed for the latter configuration, the p_z level has to lie above the p_x,p_y doublet at $\bar{M}$ in agreement with the
theoretical results shown in Fig. 15. A quantitative comparison with

the calculated level energies cannot be made, of course, since they
are derived from a non self-consistent one-electron potential and
since they apply only to the ground state of the system. With regard
to angle integrated measurements,[56] the results discussed above suggest that the broadening, splitting and dispersion of the adsorbate
levels are important contributions to the observed width.

Fig. 16 illustrates for the (1x1) overlayer structure the differential cross section as function of initial energy at several
detector angles. The photon energy is 21.2 eV and only the emission
induced by the normal component of the polarization vector is
shown. (In practice, there is always a parallel component present.)
For $\Theta_f = 15^0$, the dashed line illustrates the intensity for normally incident unpolarized light. The low-energy cut-off of these
spectra corresponds to the lower edge of the Ni s-p band. In the
absence of the oxygen layer, this band leads to a low featureless
distribution and is, therefore, not shown here. The dispersion of
the oxygen induced peaks is larger than that in Fig. 15 because of
the smaller spacing between adatoms. The fact that the lower level
gives a finite intensity at normal emission ($\Theta_f = 0$) implies that
it has p_z-like symmetry since both the final state and the polarization vector are even with regard to the reflections in the plane

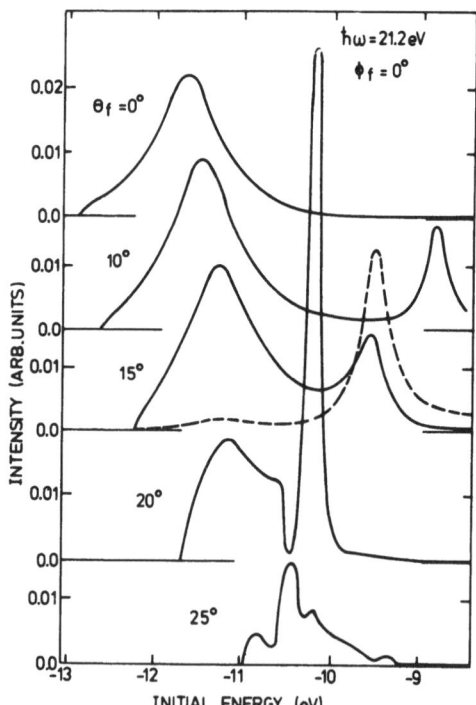

Fig. 16. Intensity as function of E_i for (1x1) O on Ni(001) at
several angles Θ_f. Solid curves: $\hat{\varepsilon} \parallel \hat{n}$; dashed curve at 15^0:
unpolarized, normally incident light. (Ref. 8)

of the surface. The response of the two levels to the change in polarization at $\theta_f = 15°$ agrees with this identification: The p_z-derived state couples most strongly to the normal component of $\vec{\varepsilon}$ while the p_x-like level is excited predominantly by the parallel component. This behavior is qualitatively consistent with the theoretical results in Fig. 12 which show that near normal incidence and normal emission, the p_x, p_y levels give finite intensity whereas the p_z state vanishes. In the plane-wave final state approximation, on the other hand, the polarization dependence of the matrix element is not related to the initial state symmetry (see Eq. (2.9)), i.e. both peaks in Fig. 16 should change by the same proportionality factor.

VI. CLEAN SURFACES

1. Formalism

The purpose of this section is to give an outline of the formal procedure for determining the differential cross section for emission from clean surfaces. The basic features of this scheme are the following: (i) The excitation is considered as a one-step process. (ii) Ground state and excited state of the semi-infinite system are derived from the same one-electron potential. (iii) The finite escape depth of the outgoing electron is included in the evaluation of the final state wave function by means of an optical potential. The method is sufficiently general so that it can be readily extended to the case of emission from chemisorbed overlayers. An application to such a system has already been discussed in the preceding section.

The essential element in the evaluation of the bound states of a semi-infinite system is the so-called complex band structure of the substrate. This technique has originally been developed to describe surface states.[65] Subsequently, it has been applied in theoretical studies of several surface electron spectroscopies, such as low energy electron diffraction[66] and field emission.[59,60] We follow here closely the derivation given by Nicolaou and Modinos.[59] Only the main steps are indicated since the details can be found in the appropriate references.

Because of the translational symmetry along the surface plane, the parallel momentum component $k_{i\parallel}$ is a good quantum number. As solution of the complex band structure problem at a given energy E_i and (reduced) wave vector $\underset{\sim}{k}_{i\parallel}$, one obtains a discrete set of Bloch states that are characterized by perpendicular momentum components $k_{\alpha\perp}^i(E_i, \underset{\sim}{k}_{i\parallel})$. While only the states with real $k_{\alpha\perp}^i$ may exist in the three-dimensional bulk, near a surface it is necessary to consider also those with complex $k_{\alpha\perp}^i$ that decay into the crystal. Both types together form an adequate expansion set for the ground state inside the crystal. At an energy within the range of the bulk bands, this state may be constructed as follows:

$$\Psi_\alpha^i = \psi_\alpha^i - \sum_\beta \psi_\beta^i R_{\beta\alpha}^i \tag{6.1}$$

where $\psi_\alpha^i(E_i, k_{i\parallel}; r)$ denotes a Bloch wave that propagates towards the surface, and $\psi_\beta^i(E_i, k_{i\parallel}; r)$ those waves that either propagate or decay towards the interior. The incident wave ψ_α^i is normalized to unit normal current.[60] Since among the ψ_β^i there is one Bloch wave whose group velocity is opposite ot that of ψ_α^i, the function Ψ_α^i may be interpreted as a standing wave plus a series of reflected waves which ensure continuity along the solid-vacuum interface ($z = 0$).

The number of reflected waves is, in principle, infinite although in practice only a relatively small number of waves (typically 13 or 21) is required to obtain convergence. As a superposition of propagating and evanescent Bloch functions, the ground state (6.1) is obviously not characterized by a single perpendicular wave vector component.

At the surface plane, the Bloch waves can be expressed as a two-dimensional Fourier series in the following manner:

$$\psi_\alpha^i(r_\parallel; z=0) = \sum_g e^{i(k_{i\parallel}+g)\cdot r_\parallel} \phi_{g\alpha}^i \tag{6.2}$$

where g are the reciprocal lattice vectors of the surface Bravais net. The corresponding components of the normal derivative $\psi_\alpha^i{}'$ at $z = 0$ are denoted by $\phi_{g\alpha}^i{}'$.

In the external region ($z \geq 0$), the potential is assumed to be independent of $r_\parallel$. We denote the solutions that decay toward infinity by $u(\sqrt{-E_i}, z)$ and take them to be normalized to unity at $z=0$. An appropriate representation of the initial state in this region is given by

$$\psi_\alpha^i(r_\parallel, z \geq 0) = \sum_g \phi_{g\alpha}^i e^{i(k_{i\parallel}+g)\cdot r_\parallel} u(\kappa_g, z) \tag{6.3}$$

where $\kappa_g^2 = -E_i + (k_{i\parallel}+g)^2$. The condition of continuity of value and slope of ψ_α^i at $z = 0$ leads to the following equations:

$$\phi_{g\alpha}^i = \phi_{g\alpha}^i - \sum_\beta \phi_{g\beta}^i R_{\beta\alpha}^i$$

$$\gamma_g^i \phi_{g\alpha}^i = \phi_{g\alpha}^i{}' - \sum_\beta \phi_{g\beta}^i{}' R_{\beta\alpha}^i \tag{6.4}$$

where γ_g^i are the logarithmic derivatives of $u(\kappa_g, z)$ at $z = 0$. The solution of (6.4) provides both the Fourier components $\phi^i_{g\alpha}$ and the reflection coefficients $R^1_{\beta\alpha}$.

At energies within gaps of the bulk band structure, the ground state consists solely of evanescent functions, i.e. the incident Bloch wave ψ^1_α is absent from (6.1). The discrete energies at which a surface state might exist are specified by the roots of the homogeneous system which one obtains instead of (6.4). In practice, it is convenient to solve for the eigenvalue which vanishes at the surface state energy since the slope with which it goes through zero is equal to the normalization of the wave function.[60]

For the evaluation of the matrix element between initial and final state, it is useful to determine the projections of Ψ^i_α onto the muffin-tin spheres.[2] We denote these by $A^1_{\alpha nL}(E_i, k_{i\parallel})$ where n represents a layer index and $L = (\ell, m)$ the angular momentum numbers:

$$A^i_{\alpha nL} = \int_n d\Omega_r \; \psi^i_\alpha(r) \; Y^*_L(\hat{r}) \tag{6.5}$$

where the Y_L are spherical harmonics and the integral is over the surface of the muffin-tin sphere at the origin of the nth plane. Thus, within a sphere, Ψ^1_α consists of a superposition of solutions $R^1_{n\ell}(r)$ to the spherically symmetric potential with expansion coefficients given by (6.5). The $R^1_{n\ell}(r)$ are normalized to unity at the muffin-tin radius.

As shown in Sec. III, the final state in photoemission is identical to the time-reversed LEED function. Since it is actually Ψ^*_f which is required for the matrix element, we may directly use the LEED state with the convention, however, that the plane wave incident from the vacuum carries a parallel momentum $-k_{f\parallel}$. The detector direction is specified by the unit vector $\hat{R} = (k_{f\parallel}, (E_f - k^2_{f\parallel})^{1/2})/E_f^{1/2}$ and $k_{f\parallel}$ may lie outside of the first Brillouin zone. Several physically equivalent schemes[66,67] exist for the evaluation of LEED intensities. Although it is not the computationally most efficient scheme, we choose here the so-called Bloch wave method[66] since it is closely related to the procedure described above for the initial state. It also has the advantage that it provides more physical insight into the photoemission process, as the transmission factors[15] of the final state Bloch waves across the surface barrier appear explicitly. Furthermore, it is the only scheme which allows the investigation of direct transitions in the limit of weak electronic damping.

In the interior of the crystal, Ψ^*_f consists of a superposition of Bloch waves $\psi^f_\beta(E_f, -k^r_{f\parallel}; r)$ that decay away from the surface (as bulk quantities, the Bloch states depend only on the reduced wave vector, indicated by the superscript r. In contrast, Ψ_f^* depends on the extended vector $\underset{\sim}{k}_{f\parallel}.$):

$$\Psi_f^* = \sum_\beta \psi_\beta^f \; T_\beta^f \tag{6.6}$$

The transmission coefficients $T_\beta^f(E_f,-k_{f\parallel})$ — here of the incident plane wave into the Bloch states ψ_β^f — are determined by continuation of Ψ_f^* across the solid-vacuum interface. Within a muffin-tin sphere, the excited state consists of a superposition of solutions $R_{n\ell}^f(r)$ evaluated at E_f and the expansion coefficients $A_{nL}^f(E_f,-k_{f\parallel})$ are defined in analogy to Eq. (6.5).

It is convenient to evaluate the matrix element in the $\underset{\sim}{A} \cdot \nabla V(\underset{\sim}{r})$ form, since, in the muffin-tin model of the potential, one obtains a sum over atomic matrix elements of the crystal plus an external contribution.[2] Because of the Bloch nature of initial and final state along the surface, the internal contribution may be simplified by summing over sites $\underset{\sim}{P}$ within planes parallel to the surface:

$$\langle \Psi_f | \underset{\sim}{A} \cdot \nabla V | \psi_\alpha^i \rangle_{int} = \sum_{\underset{\sim}{P}} e^{i(\underset{\sim}{k}_i - \underset{\sim}{k}_f^r)_\parallel \cdot \underset{\sim}{P}}$$

$$\times \sum_{n=0}^{\infty} \langle \Psi_f | \underset{\sim}{A} \cdot \nabla V | \psi_\alpha^i \rangle_n \tag{6.7}$$

i.e. $\underset{\sim}{k}_{i\parallel} = k_{f\parallel}^r$. Thus, the parallel component of the Bloch wave vector is conserved during the transition as a result of the translational symmetry of the system along the surface.

The remaining sum over atomic planes can also be simplified by using Eqs. (6.1), (6.6) and the Bloch conditions in the direction normal to the surface, e.g.

$$\psi_\alpha^i(\underset{\sim}{r} + \underset{\sim}{d}) = e^{i(\underset{\sim}{k}_{i\parallel},k_{\alpha\perp}^i)\cdot\underset{\sim}{d}} \; \psi_\alpha^i(\underset{\sim}{r}) \tag{6.8}$$

where $\underset{\sim}{d}$ represents the (usually three dimensional) vector between the origins of two neighboring layers. One obtains:

$$\sum_{n=0}^{\infty} \langle \Psi_f | \underset{\sim}{A} \cdot \nabla V | \psi_\alpha^i \rangle_n = \sum_{\beta'} T_{\beta'}^f \langle \psi_{\beta'}^f | \underset{\sim}{A} \cdot \nabla V | \psi_\alpha^i \rangle_o \; \lambda_{\beta'\alpha}$$

$$- \sum_{\beta}\sum_{\beta'} T_{\beta'}^f \; R_{\beta\alpha}^i \langle \psi_{\beta'}^f | \underset{\sim}{A} \cdot \nabla V | \psi_\beta^i \rangle_o \; \lambda_{\beta'\beta} \tag{6.9}$$

with

$$\lambda_{\beta'\beta}^{-1} = 1 - \varepsilon_o \exp(-i(k_{\beta'\perp}^f + k_{\beta\perp}^i)d_\perp) \qquad (6.10)$$

and similarly for $\lambda_{\beta'\alpha}$. The factor $\varepsilon_o = \exp(-\xi d_o)$ accounts for the attenuation of the incident radiation inside the crystal:[15,16] $A(z) = A e^{-\xi z}$. At photon frequencies in the UV range, the absorption coefficient ξ is of the order of several hundred inverse lattice spacings, i.e. ε_o is close to unity.

The expression (6.9) indicates the various types of transitions between initial and final state Bloch functions which contribute to the matrix element. It also demonstrates the difficulty of separating in a well defined manner terms associated with the surface from those depending on the bulk. The evanescent waves of the ground state may decay rather slowly into the crystal and the inelastic scattering, except at very low final energies, prevents electrons from all but the first few layers to be detected. Apart from (6.9), there is, of course, the external contribution to the matrix element related to the gradient of the surface potential. Since this term depends, in addition, on the amplitudes of the wave functions, it may, in principle, also reflect bulk properties.

The respective weights of the transitions between Bloch states are strongly influenced by the factors $\lambda_{\beta'\beta}$ and $\lambda_{\beta'\alpha}$ whose magnitude may be interpreted as a measure of the effective distance over which an initial and final state Bloch function are coherent. The coherence is most pronounced if both perpendicular wave vector components are real and equal in magnitude but opposite in sign (see Eq. (6.10)). The reduced momentum of the actual final state (time reversed form of (6.6)) in this case is, therefore, equal to that of the initial state. This is the condition of $\underset{\sim}{k}$ conservation for the direct transitions between propagating Bloch states in the traditional three-step model.[18-20] The transition is then dominated by a single term, say,

$$T_{\alpha'}^f \quad <\psi_{\alpha'}^f|\underset{\sim}{A}\cdot\nabla V|\psi_\alpha^i>_o \lambda_{\alpha'\alpha} \qquad (6.11)$$

since all other contributions involving evanescent waves as well as that due to the external region are small by comparison. Because of the finite escape depth of the excited electron, the condition of $\underset{\sim}{k}$ conservation is never truly satisfied as all final state Bloch waves are evanescent. Nevertheless, at low kinetic energies the imaginary wave vector components caused by the damping are generally smaller than those of the Bloch states which are evanescent also in the absence of inelastic scattering. The matrix element (6.9) is in this limit still dominated by a term like (6.11). If we take account of the weak damping in an ad hoc manner by inserting an imaginary component of $k^f_{\alpha'\perp}$ into $\lambda_{\alpha'\alpha}$, we obtain[15]

$$|\lambda_{\alpha'\alpha}|^2 = |(1-e^{-\Lambda d\perp})^2 + 4e^{-\Lambda d\perp} \sin^2(\tfrac{1}{2} Kd_\perp)|^{-1} \qquad (6.12)$$

where

$$A \equiv Im(-k^f_{\alpha'\perp}) + \xi$$
$$K \equiv Re(k^f_{\alpha'\perp} + k^i_{\alpha\perp}) . \qquad (6.13)$$

The differential cross section is accordingly a simple product of three factors:

$$\frac{d\sigma}{d\Omega} \sim |M^{fi}_{\alpha'\alpha}|^2 \ |\lambda_{\alpha'\alpha}|^2 \ |T^f_{\alpha'}|^2 \qquad (6.14)$$

The first involves the matrix element in (6.11) between a Bloch wave ψ^i_α and a (weakly evanescent) Bloch wave $\psi^f_{\alpha'}$, integrated over the volume of the first layer unit cell. The second factor, apart from defining the effective $k_\perp$ conservation ($|\lambda_{\alpha'\alpha}|^2$ reaches its maximum at $K = 0$), gives the total probability for an electron, excited at some point inside the crystal, to reach the surface. The third determines the strength with which the internal final state Bloch wave couples to its outgoing counterpart in the vacuum.

In the above discussion, we have assumed that, in the absence of inelastic scattering, a propagating final state Bloch wave actually exists at the energy E_f and momentum $k_{f\parallel}$ under consideration. This will not be true in the case of a gap in the final state band structure. Ψ^*_f in Eq. (6.6) then contains only evanescent waves which may be viewed as the exponential tails of the vacuum states. This type of "surface emission" appears to have been observed for several materials.[68]

We conclude this section by giving a general expression (i.e. not limited to weak electronic damping) for the differential cross section in terms of the partial wave projections of ground state and excited state onto the individual atomic sites:

$$\frac{d\sigma}{d\Omega} (E_f, k_{f\parallel}, A, \hbar\omega) \sim \sum_\alpha |<\Psi_f|A\cdot\nabla V|\psi^i_\alpha>_{int}$$

$$+ <\Psi_f|A\cdot\nabla V|\psi^i_\alpha>_{ext}|^2 \ \delta(E_f-E_i, \hbar\omega) \qquad (6.15)$$

where

$$<\Psi_f|A\cdot\nabla V|\psi^i_\alpha>_{int} = \sum_{m=-1}^{1} \frac{4\pi}{3} Y_{1m}(\hat{A}) \cdot \sum_{n=0}^{\infty} \sum_{L_f L_i} A^f_{nL_f}(E_f, -k_{f\parallel})$$

$$\times R_{\ell_f\ell_i}(E_f, E_i) \ I(1m; L_f, L_i) \ A^i_{\alpha n L_i}(E_i, k^r_{f\parallel}) \qquad (6.16)$$

The quantities $R_{\ell\ell}$ are radial integrals defined in Eq. (3.6), I are the Gaunt $_f\ell_i$ integrals (3.7) and we have also made use of the identity (3.5).

Eq. (6.16) has a plausible physical interpretation: The internal matrix element is decomposed into a series of atomic transitions at different planes of the crystal. The strength of these single-site transitions and the selection rules imposed by them are given by the radial integrals and Gaunt coefficients, respectively. The phase and amplitude relations between transitions occuring in the different layers are determined by the projections $A^1_{\alpha n L_i}$ and $A^f_{n L_f}$ of ground state and excited state. The phase coherence within planes, finally, leads to the parallel momentum conservation as indicated explicitly in (6.16). The sum over atomic layers is effectively limited by the escape depth of the outgoing electron.

The above described scheme and physically equivalent methods have recently been used with reasonable success to interpret experimental spectra for several systems. These applications are discussed in the remainder of this section.

2. TaS_2

The first theoretical analysis that includes the essential ingredients of the one-electron picture as outlined above, was performed by Liebsch on TaS_2.[2] Systematic measurements[69] of the differential cross section for this system show pronounced angular anisotropies which vary greatly with photon frequency and polarization. $1T$-TaS_2 is a layer-type compound consisting of sandwiches that interact weakly via van der Waals forces. Each sandwich contains a hexagonal plane of Ta atoms between two sulfur planes of the same symmetry. In the $1T$ structure the atoms are arranged in an anti-prismatic configuration, i.e. in one plane, the sulfur atoms lie above one type of the hollows between Ta atoms while in the other plane they lie below the second type.

Band structure calculations[70] show three distinct sets of bands, the lowest corresponding to the occupied sulfur p states, the second to partially filled Ta d bands and at still higher energies the unoccupied anti-bonding p states. Because of the two-dimensional nature of the system, the metallic d bands exhibit very little dispersion in the normal direction, indicating the lack of overlap between wave functions of neighboring sandwiches. Thus, electrons emitted from different sandwiches are detected essentially incoherently. It is therefore appropriate to take the bound states of a single sandwich as initial states. The dispersion of the Ta d bands for such a "thin film"[71] agrees almost exactly with that of the three-dimensional bulk. As surface barrier, an exponential potential

is used on both sides of the sandwich. The final state, on the other
hand, is calculated as a LEED wave function of the semi-infinite
system since at energies above the vacuum, the electron states are
more extendend in the normal direction than the bound states. Their
range is effectively limited by the finite escape depth.

In Fig. 17, theoretical results[2] (solid lines) for the emission
from the Ta-derived d bands are compared with experimental azimuthal
angle distributions[69] (dashed lines) at three separate polarizations,
photon frequencies and polar emission angles. In all three cases,
reasonable qualitative agreement is obtained. In order to analyze
the origin of the observed spectral features, analogous calculations
have been carried out using the plane-wave approximation of the
final state. As pointed out in Sec. II (see Eq. (2.19)), the spectrum
is in this limit characteristic of the Fourier transform of the
ground state wave function. The result for a particular experimental
configuration is shown in Fig. 18 superimposed on the two-dimensional surface Brillouin zone. Three lobes are found along ΓM' which
corresponds to the direction between next nearest neighbors of Ta
and S atoms. Similar lobes with only small variations in width and
magnitude are obtained at the other final energies and polar angles
used in Fig. 17. (All three spectra shown in Fig. 17 are taken by
rotating the crystal about its normal, i.e. $\cos(\underset{\sim}{k}_f, \underset{\sim}{A})$ in Eq. (2.9)
remains constant as the azimuth is varied.)

The comparison of the Figs. 17 and 18 suggests that the experimentally detected lobes near the ΓM direction as well as the
rather pronounced splittings are caused by the non-plane wave nature of the final state. These results clearly demonstrate the necessity of treating the ground state and excited state in a consistent manner by deriving them from the same one-electron potential.
A particularly striking example is the spectrum in Fig. 17(a) for
which the polarization vector and detector direction are perpendicular, i.e. the plane-wave limit would predict zero intensity at
all angles,

The better agreement between the full calculation and the data
for the configuration in Fig. 17(a) is partly due to its special
symmetry. For $\underset{\sim}{k}_\parallel$ along ΓM or ΓM', the detector lies in a mirror
plane of the crystal, i.e. the final state is even with regard to
reflections about these planes.[84] The initial state wave function
turns out to be also even at these angles. Thus, for s polarized
light incident in the plane of emission, the cross section vanishes
along the ΓM and ΓM' directions and varies symmetrically about them.
Moreover, according to the bound state calculation, there are no
occupied states below the Fermi energy outside the pockets around M
and M' that are indicated by the dashed lines in Fig. 18. The photo-
current should, therefore, vanish also for $\underset{\sim}{k}_\parallel$ within narrow cones
about the ΓK symmetry lines. The measured spectrum exhibits minima
along the M,M' and K directions and the relative magnitude of the

(a)   (b)   (c)

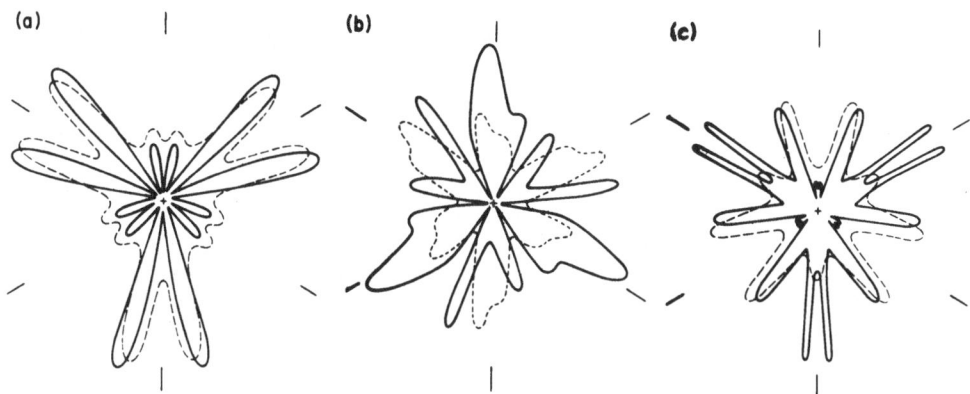

Fig. 17. Comparison of experimental azimuthal spectra for 1T-TaS$_2$
         (dashed lines) with corresponding theoretical results
         (solid curves). (a) E_f = 10 eV, Θ_f = 42^0; s-polarized light
         with $\underset{\sim}{A}$ perpendicular to k_f. (b) $E_f^{\perp}$ = 8 eV, Θ_f = 57^0 (theo-
         retical curve: 48^0); s polarized light with angle of 36^0
         between $\underset{\sim}{A}$ and $k_{f\parallel}$; (c) E_f = 15 eV, Θ_f = 60^0; unpolarized
         light at normal incidence. (Ref. 2)

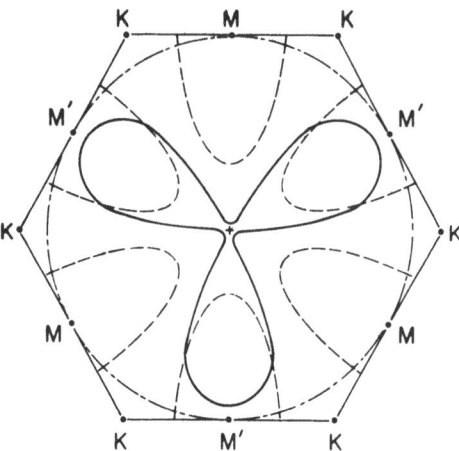

Fig. 18. Theoretical azimuthal distribution for 1T-TaS$_2$,plane wave
         final state approximation (solid curve) for E_f^2 = 8 eV,
         Θ_f = 48^0, and normally incident unpolarized light. The dot-
         dashed circle indicates the position of the parallel mo-
         mentum component within the two-dimensional Brillouin zone.
         The dashed lines enclose the occupied regions. (Ref. 2)

intensity extrema between these angles is correctly reproduced by
the theoretical results.

The fact that the observed cross section in the M,M' and K
directions in Fig. 17(a) is non-zero, although in theory it should
be (it actually amounts to 40% of the maximum signal in the spectrum), may have several explanations. The finite acceptance angle
of the analyzer tends to smooth out structure, especially at angles
where the cross section is rapidly varying as function of emission
angles, e.g. for emission near the ΓK plane. Also, quasi-elastic
scattering of the outgoing electrons via particle-hole excitations
and phonons or diffuse scattering at non-ideal surfaces might lead
to sizeable redistributions of the electron momenta, thereby changing the angular emission pattern. It is to be expected that these
effects are, to a considerable degree, responsible for the discrepancies between experimental and theoretical spectra in Fig.
17(b) and (c). The narrow doublet near ΓM' in (c), for example, is
presumably not resolvable and appears instead as broad peak in the
data. Sizeable fractions of the emitted intensity attributed to
phonon or diffuse scattering have been observed for several materials.[72,73]

3. Cu(001)

So far, we have assumed that the inital state is energetically
well defined and that only the excited electron has a finite life
time. Pendry[74] has emphasized that the created hole interacts with
the surrounding electrons as well. This interaction leads to a
complex self-energy whose imaginary part Γ_1 is a measure of the inverse life time of the hole. In a nearly free electron metal, Γ_1
varies quadratically with the energy relative to the Fermi level.
Typical values lie between 0.1 and 0.5 eV. The main decay mechanism
of the hole state is via Auger excitation of electron-hole pairs.

A surface state, for example, has in the absence of life time
effects a discrete energy position $E_s(\underline{k}_\parallel)$ at a given $\underline{k}_\parallel$. Decay
processes lead to a Lorentzian-type broadening of the infinitely
sharp peak:

$$\frac{d\sigma}{d\Omega} \sim \frac{\Gamma_1}{(E_i - E_s(\underline{k}_\parallel))^2 + \Gamma_1^2} \qquad (6.17)$$

The life time of the outgoing electron, on the other hand, does not
cause any broadening since its energy resolution is limited only by
the spectrometer. Instead, its finite escape depth contributes to
the uncertainty of its normal momentum component which amounts to
an effective width of the final state energy bands of the system.
In the case of emission from bulk valence bands, therefore, the range

of initial states that is sampled close to a particular inter-band transition, depends on both electron and hole life times.

An indication of the effect of hole decay processes can be obtained by comparing photoemission spectra for Cu and Ni. Cu spectra tend to show considerably more fine structure in the range of the 3d bands (on a scale of 0.1 eV) than Ni data which are taken under similar experimental conditions. Also, X-ray photoelectron spectra are nearly identical with the calculated d band density of states in the case of Cu. For Ni, the measurements give a considerably narrower d band than theoretically predicted. Pendry[74] has shown that this narrowing in Ni can be understood by means of an increased broadening. Taking Γ_1 to vary like 0.1 $(E-E_F)^2$ (i.e. 1.6eV at 4 eV below E_F), the lower part of the d band is so strongly smoothed out that it becomes part of the background. Convoluting this life time with the calculated density of states leads to reasonable agreement with the X-ray data.

To incorporate the hole life time into the evaluation of the differential cross section, the expression (2.15) is used. Rather than formulating the Green's function (2.16) as the sum over eigenstates of the semi-infinite system, it is written in terms of the scattering matrices of the solid as in the Dyson equation (2.2). An attractive feature of this method is the fact that both initial and final state can be handled by computationally efficient procedures. Extensive calculations have been performed for the (001) face of clean Cu. The Chodorow potential is used up to the surface layer. The incident light is used as exciting optical field.

Fig. 19 shows a comparison of various theoretical[75] and experimental[76,77] spectra taken at normal incidence. The data in (a) are by Lloyd et al.[76], those in (b) by Ilver et al.[77] The inverse hole life time is 0.05 eV while the inverse electron life times are 4 eV in (a) and 1 eV in (b) to account for the differences in the final state energies. Pendry points out that although many transitions are allowed according to the bulk energy band structure, several do not appear in the spectrum. This is a clear indication of the importance of matrix elements which determine the actual weight of a transition.

This is illustrated in greater detail in Fig. 20 for the $\phi = 40^0$ spectrum of Fig. 19(b). Near -2.7 eV, for example, a transition is possible according to the bulk energy bands while in the spectrum it is not observed. Similarly the transition from the s-p band at -6 eV is energetically possible but not observed. The weak dispersion of the s-p band near -0.5 eV, on the other hand, is clearly seen as a result of the two final states into which electrons can be excited. In the d-band region, a coherent superposition of many transitions takes place whose relative weight depends sensitively on the emission angles as apparent in Fig. 19(b). The general agreement between the measured and calculated spectra suggests that the

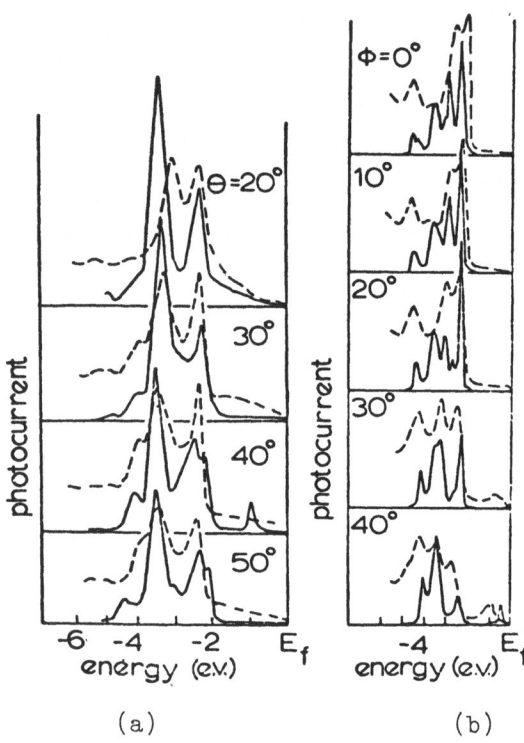

(a)   (b)

Fig. 19. Calculated (solid curves) and measured (dashed curves)
         photoemission spectra for Cu(001) taken at a photon energy
         of (a) 21.2 eV and (b) 16.8 eV. Unpolarized light is nor-
         mally incident. In (a), the polar emission angle is varied
         the azimuth is in the (110) direction; in (b), the polar
         angle is held at 45° and the azimuth is varied. (Ref. 74)

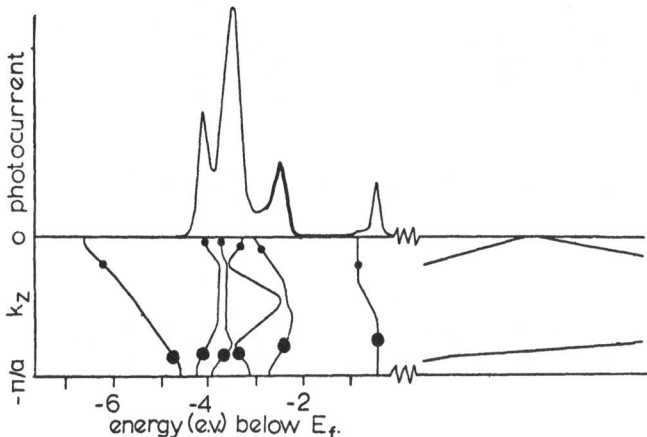

Fig. 20. Calculated spectrum of Fig. 19(b) at $\phi = 40^0$ together
with appropriate energy bands. The dots indicate values
of E and k_z at which inter-band transitions can occur.
(Ref. 74)

analysis of peak shapes and intensities can indeed provide information on the electronic states of the system. It appears likely
that angle resolved photoemission can be used to investigate the
behavior of these states in the vicinity of the surface.

4. Cu(111): Surface State

     Due to its high surface sensitivity, ultra-violet photoemis-
sion is a particularly well suited technique for the investigation
of surface states. The existence of such a state on the (111) face
of copper, at about 0.4 eV below the Fermi energy, has recently
been suggested by several experimental[76-78] and theoretical[79-80]
groups. Additional evidence for such a state is the fact that field
emission energy distributions also show a feature in the same
energy range.

     Danese and Soven[5] have performed calculations of the photo-
emission cross section for this surface state using essentially the
procedure described in Sec. VI.1. The exact energy position of the
state is somewhat uncertain since it depends sensitively on the
details of the surface barrier. Reasonable potential models, however, lead to an energy at the zone center and a dispersion with
parallel momentum that are in close agreement with the experimental

results. Thus, the initial state wave function can be assumed to be
a realistic one. The evaluation of the final state includes the
multiple scattering by the ion cores as well as that by the surface
potential. The inelastic scattering of the outgoing wave is treated
by means of an optical potential.

Particular emphasis is placed upon the influence of the polarization of the photon field on the angular dependence of the cross
section. The vector potential is assumed to be constant in space.
Its value is determined using the macroscopic optical constants of
the bulk solid. The use of the incident light as exciting field is
not considered appropriate in view of the relatively slow decay of
the surface state into the crystal. In order to analize the general
features of the angular dependence, the matrix element is expressed
as

$$| < \Psi_f | \underset{\sim}{A} \cdot \underset{\sim}{p} | \Psi_i > |^2 \sim | \hat{\varepsilon} \cdot \underset{\sim}{M} |^2 \qquad (6.18)$$

where $\underset{\sim}{M} \equiv <\Psi_f | \underset{\sim}{p} | \Psi_i>$. For normal emission, both initial and final
state have $\Delta_1 (\underset{}{s}+p_z)$ symmetry, i.e. M_z is finite whereas $M_{||}$ vanishes.
For small $\underset{\sim}{k}_{||}$, M_z is rougly constant while $M_{||}$ varies proportionally
to $\underset{\sim}{k}_{||}$. (The surface state exists only close to the center of the
zone, so that the azimuthal anisotropy of the matrix element is
negligible.)

Fig. 21 shows the differential cross section at $\hbar\omega = 16.8$ eV
as function of $|\underset{\sim}{k}_{||}|$ or, equivalently, of polar emission angle. In
curve (a), the light is incident in the y-z plane, s polarized, i.e.
$\hat{\varepsilon} = (\varepsilon_x, 0, 0)$, and the detector is in the x-y plane. The emission
intensity is zero at $k_{||} = 0$ and varies symmetrically and quadratically
at finite angles as predicted by the above symmetry arguments. If
unpolarized light is used instead (curve (b)), the cross section has
the form

$$| \varepsilon_x M_x |^2 + | \varepsilon_z M_z |^2 \qquad (6.19)$$

The first term corresponds to the intensity observed in (a) while
the second accounts for the angular variation of M_z which makes the
cross section non-vanishing in the normal direction. Finally, curve
(c) shows the results for p polarized light with emission in the
plane of incidence. The cross section is proportional to

$$| \varepsilon_y M_y + \varepsilon_z M_z |^2 \qquad (6.20)$$

i.e., it is asymmetric about the normal, with the degree of anisotropy depending on the relative size of M_y and M_z.

The rather pronounced asymmetry indicates that the analysis of
measured peak positions in terms of the initial state dispersion can

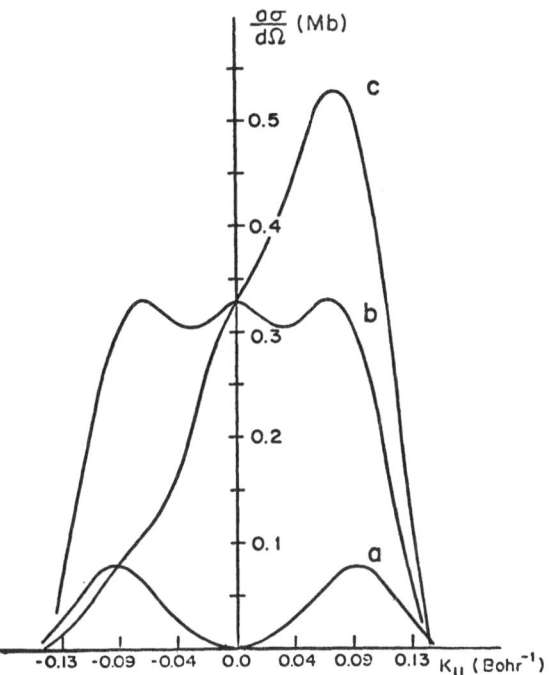

Fig. 21. Calculated differential cross sections for Cu(111) surface
using 16.8 eV radiation. Curve (a) is for s polarized
light and emission in a plane perpendicular to the plane
of incidence; (b) is for unpolarized light and emission
perpendicular to the plane of incidence; (c) is for p
polarized light and emission in the plane of incidence.

be complicated. Due to finite aperture of the analyzer, the intensity
is dominated by contributions from $k_\parallel$ points which do not necessarily
correspond to the wave vector on which the analyzer is centered. It
is evident that the effective averaging over the accepted angles
might lead to rather erroneous predictions of the dispersion relation. Danese and Soven suggest that the rapid angular variation
of the cross section causes the apparent anisotropy in the surface
state energy observed in Ref. 77.

An interesting phenomenon is the decrease of intensity at
larger angles in the three curves of Fig. 21. In the theoretical
model, the gap between surface state and conduction band edge di-
minishes for larger $k_\parallel$, i.e. the decay length of the state increases. This implies that an increasing fraction of the excitation
takes place in regions of the crystal which are beyond the escape
depth of the outgoing electron, thus decreasing the calculated intensity. However, the extent to which this effect is a real one depends sensitively on the precise dispersion relation of the surface
state. The experimental results of Ref. 78 suggest that the separa-

tion of the surface state from the conduction band edge remains
nearly constant as function of $\underset{\sim}{k}_{\parallel}$.

Finally, calculations of the frequency dependence of the photocurrent are performed in order to interpret corresponding experimental spectra.[77] A rapid decrease in intensity is obtained as the
photon energy increases from 10.6 eV to 21.2 eV, in qualitative
agreement with the measured results.

References

1. "Photoemission from Surfaces", Eds. B. Feuerbacher, B. Fitton, and R.F. Willis, Wiley, London, 1977.
2. A. Liebsch, Sol. St. Commun. 19, 1193 (1976).
3. J.B. Pendry, Surf. Sci. 57, 679 (1976).
4. D.J. Spanjaard, D.W. Jepsen, and P.M. Marcus, Phys. Rev. B15, 1728 (1977).
5. J.B. Danese and P. Soven, Phys. Rev. B16, 706 (1977).
6. D. Dill and J.L. Dehmer, J. Chem. Phys. 61, 692 (1974).
7. J.W. Davenport, Phys. Rev. Lett. 36, 945 (1976).
8. A. Liebsch, Phys. Rev. Lett. 38, 248 (1977).
9. M. Scheffler, K. Kambe, and F. Forstmann, Sol. St. Commun., to be published; K. Jacobi, M. Scheffler, K. Kambe, and F. Forstmann, Sol. St. Commun. 22, 17 (1977).
10. C. Caroli, D. Lederer-Rozenblatt, B. Roulet, and D. Saint-James, Phys. Rev. B8, 4552 (1973).
11. P.J. Feibelman, Phys. Rev. Lett. 34, 1092 (1975); K.L. Kliewer, Phys. Rev. Lett. 33, 900 (1974); G. Mukhopadhyay and S. Lundqvist, Sol. St. Commun. 21, 629 (1977).
12. J.W. Gadzuk, Phys. Rev. B14, 2267 (1976); and in Ref. 1.
13. For a list of references, see various chapters in Ref. 1.
14. I. Adawi, Phys. Rev. 134A, 788 (1964).
15. G.D. Mahan, Phys. Rev. B2, 4334 (1970).
16. W.L. Schaich and N.W. Ashcroft, Phys. Rev. B3, 2452 (1971).
17. P.J. Feibelman and D.E. Eastman, Phys. Rev. B10, 4932 (1974).
18. E.O. Kane, Phys. Rev. 127, 131 (1962).
19. G.W. Gobeli and F.G. Allen, Phys. Rev. 127, 141 (1962).
20. C.N. Berglund and W.E. Spicer, Phys. Rev. 136, A1030 (1964).
21. A. Liebsch, Phys. Rev. Lett. 32. 1203 (1974).
22. H. Bethe, L. Maximon, and F. Low, Phys. Rev. 91, 417 (1953); G. Breit and H. Bethe, Phys. Rev. 93, 888 (1954).
23. J.W. Gadzuk, Phys. Rev. B10, 5030 (1974).
24. C.B. Duke and C.W. Tucker, Surf. Sci. 15, 231 (1969).
25. P.J. Feibelman, Surf. Sci. 46, 558 (1974).
26. J. Cooper and R.N. Zare, Lectures in Theoretical Physics, Gordon and Breach, New York, 1969, Vol XLC.
27. E.W. Plummer, in "Interactions on Metal Surfaces", Ed. R. Gomer, Springer, New York, 1975.
28. D.E. Eastman and M. Kuznietz, J. Appl. Phys. 42, 1396 (1971).
29. J.W. Cooper, Phys. Rev. 128, 681 (1962).
30. A. Liebsch, chapter 7 in Ref. 1.
31. J.W. Gadzuk, Phys. Rev. B12, 5608 (1975).
32. J.F. Herbst, Phys. Rev. B15, 3720 (1977).
33. C.N. Yang. Phys. Rev. 74, 764 (1948).
34. J.L. Dehmer and D. Dill, Phys. Rev. Lett. 37, 1049 (1976).
35. J.L. Dehmer and D. Dill, Phys. Rev. Lett. 35, 213 (1975).
36. J.C. Slater and K.H. Johnson, Phys. Rev. B5, 844 (1972).
37. G.R. Wright, C.E. Brion, and M.J. van der Wiel, J. Electr. Spectrosc. Relat. Phenomena 1, 457 (1973).

38. E.A. Stern, Phys. Rev. B10, 3027 (1974); P.A. Lee and J.B. Pendry, Phys. Rev. B11, 2795 (1975).

39. B.M. Kincaid and P. Eisenberger, Phys. Rev. Lett. 34, 1361 (1975).

40. J.W. Gadzuk, Sol. St. Commun. 15, 1011 (1974); Surf. Sci. 53, 132 (1975).

41. T. Gustafsson, E.W. Plummer, W. Gudat, and D.E. Eastman, to be published.

42. C.L. Allyn, T. Gustafsson, and E.W. Plummer, Chem. Phys. Lett. 47, 127 (1977).

43. J.W. Davenport, thesis, University of Pennsylvania, 1976.

44. J.C. Tracy, J. Chem. Phys. 56, 2736 (1972).

45. R.J. Smith, J. Anderson, and G.J. Lapeyre, Phys. Rev. Lett. 37, 1081 (1976).

46. T. Gustafsson, E.W. Plummer, D.E. Eastman, and J.L. Freeouf, Sol. St. Commun. 17, 391 (1975).

47. J.C. Fuggle, T.E. Madey, M. Steinkilberg, and D. Menzel, Phys. Lett. 51A, 163 (1975).

48. J.B. Pendry, J. Phys. C 8, 2413 (1975).

49. B.W. Holland, J. Phys. C 8, 2697 (1975).

50. A. Liebsch and E.W. Plummer, Discuss. Faraday Soc. 58, 19 (1975).

51. A. Liebsch, Phys. Rev. B13, 544 (1976).

52. S.Y. Tong and M. Van Hove, Sol. St. Commun. 19, 543 (1976).

53. S. Anderson, Sol. St. Commun. 21, 75 (1977).

54. N.V. Smith and P.K. Larsen, in ref. 1.

55. B.I. Lundqvist, O. Gunnarsson, and H. Hjelmberg, in Ref. 1.

56. D.E. Eastman and J.K. Cashion, Phys. Rev. Lett. 27, 1520 (1971).

57. I.P. Batra and O. Robaux, Surf. Sci. 49, 653 (1975).

58. G.E. Becker and H.D. Hagstrom, Phys. Rev. Lett. 22, 1054 (1969).

59. N. Nicolaou and A. Modinos, Phys. Rev. B11, 3687 (1975).

60. N. Kar and P. Soven, Sol. St. Commun. 20, 977 (1976); N. Kar, thesis, University of Pennsylvania, 1977, unpublished.

61. A. Liebsch, Phys. Rev. B17, (1978).

62. S. Anderson and C. Nyberg, Surf. Sci. 52, 489 (1975).

63. G.G. Tibbetts, J.M. Burkstrand, and. J.C. Tracy, Phys. Rev. B15, 3652 (1977); K.Y. Yu, W.E. Spicer, I. Lindau, P. Pianetta, and S.F. Lin, Surf. Sci. 57, 157 (1976).

64. G.J. Lapeyre, to be published.

65. V. Heine, Proc. Phys. Soc. 81, 300 (1968).

66. J.W. Jepsen, P.M. Marcus, and F. Jona, Phys. Rev. B5, 3933 (1972); J.B. Pendry, Low Energy Electron Diffraction, Academic, London, 1974.

67. C.B. Duke, Adv. Chem. Phys. 27, 1 (1974).

68. B. Feuerbacher and R.F. Willis, J. Phys. C 9, 169 (1976); G. Hansson and A. Flodström, to be published.

69. N.V. Smith and M.M. Traum, Phys. Rev. B11, 2087 (1975); N.V. Smith, M.M. Traum, J.A. Knapp, J. Anderson, and G.J. Lapeyre, Phys. Rev. B13, 4462 (1976).

70. L.F. Mattheiss, Phys. Rev. B8, 3719 (1973).

71. N. Kar and P. Soven, Phys. Rev. B11, 3761 (1975).

72. T. Gustafsson, P.O. Nilsson, and L. Wallden, Phys. Lett. A37, 121 (1971).
73. B. Feuerbacher and B. Fitton, Phys. Rev. Lett. 30, 923 (1973); H. Becker, E. Dietz, U. Gerhardt, and H. Angermuller, Phys. Rev. B12, 2084 (1975).
74. J.B. Pendry, in Ref. 1.
75. J.B. Pendry and D.J. Titterington, Commun. on Phys. 2, 31 (1977).
76. D.M. Lloyd, C.M. Quinn, and N.V. Richardson, J. Phys. C 8, L371 (1976).
77. L. Ilver and P.O. Nilsson, Sol. St. Commun. 18, 677 (1976).
78. P.O. Gartland and B.J. Slagsvold, Phys. Rev. B12, 4047 (1975),
79. N.Kar and P. Soven, Sol. St. Commun. 19, 1041 (1976).
80. S.J. Gurman, J. Phys. C 9, L609 (1976).
81. S.Y. Tong, C.H. Li, and A.R. Lubinsky, Phys. Rev. Lett. 39, 498 (1977).
82. K. Jacobi, to be published; the originally observed oxygen-induced peak at -8 eV below E_f is now believed to be due to a second binding site, presumably below the first Ni plane. As in the angle-integrated spectra, angle resolved spectra reveal for chemisorbed oxygen a broad feature at -6 eV.
83. M. Van Hove and S.Y. Tong, J. Vac. Sci. Technol. 12, 230 (1975).
84. J. Hermanson, Sol. St. Commun. 22, 9 (1977).

CHEMISORPTION AND CATALYSIS ON METALS: APPLICATIONS OF SURFACE SPECTROSCOPIES

G. Ertl

Institut für Physikalische Chemie

Universität München, W. Germany

Our knowledge on the processes occurring at solid/gas-interphases has tremendously increased during the past years due to the introduction of new surface-sensitive techniques. Since each method probes different aspects evidently only a combined use of various techniques can yield a more or less close picture of the microscopic properties of a surface. There is frequently still a lack of complete theoretical understanding of the available experimental information, but even 'fingerprint' characterization or semiquantitative analysis may lead to valuable insights. This series of lectures concentrates on the experimental information for characterization of the elementary steps in chemisorption and catalysis on metals rather than on the problems encountered with the particular methods themselves. The discussion will be mainly restricted to well-defined single-crystal surfaces since with 'real' surfaces additional complications come into play.

I. CHARACTERIZATION OF CHEMISORBED SYSTEMS

1. The Molecular Nature of the Adsorbate

Elemental analysis is now widely performed on a routine basis e.g. by Auger (AES) or X-ray photoelectron spectroscopy (XPS) so that no need is felt to discuss this aspect in the present context.

Since however particles interacting with surfaces frequently undergo
chemical transformations (like dissociation etc.) probably the most
important problem in characterizing a chemisorbed system concerns
the molecular nature of the adsorbed species. Since bond formation
involves the valence electrons of atoms measurements of the valence
electron ionization energies by means of ultraviolet photoelectron
spectroscopy (UPS) and comparison of the spectra with those of the
free (gaseous) adsorbates or (if available) of corresponding complex
compounds with well-known structure yield probably the most direct
information. It is often sufficient to use UPS in this context only as
a 'fingerprint' technique without taking great care of the occurring
energy shifts.

As an example Fig. 1 shows the photoemission difference
spectrum (i.e. between the adsorbate covered and the clean surfaces)
from an Fe(110) surface which was exposed to PF_3 in comparison
with the spectrum of the compound $Fe(PF_3)_5$ [1]. The energy scales
were shifted with respect to each other by 7.1eV(work function and
relaxation effects) in order to line up most of the peaks with each
other. The different peaks in the spectrum of $Fe(PF_3)_5$ can be assigned to the various valence molecular orbitals of this compound and
evidently quite similar features are found with the chemisorbed

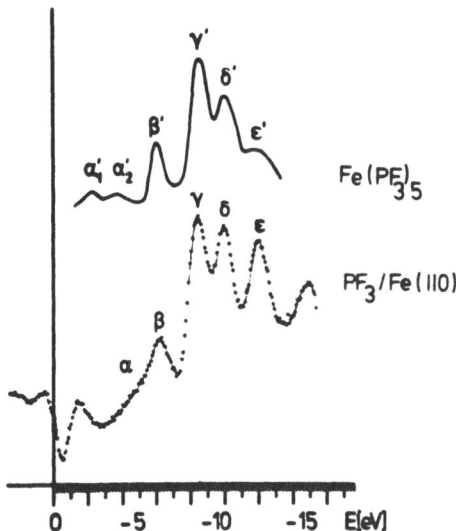

Fig. 1   He-I photoemission difference spectrum from PF_3 chemisorbed on Fe(110) and spectrum from $Fe(PF_3)_5$ [1].

system so that it can be safely concluded that PF_3 is non-dissoci-
atively adsorbed and moreover that the type of bond formation is presumably similar to that within the complex compound.

Another example illustrating stepwise chemical transformation
of the adsorbate is shown in Fig. 2[2]. Spectrum b was recorded from
an Fe(110) surface which was exposed to NH_3 at 150 K. The two
maxima outside the d-band are identified with the $3a_1$- and 1e-levels
of NH_3 (ionization potentials in the gaseous state with respect to the
vacuum level: 11 and 17 eV). If NH_3 interacts with a clean Fe(110)
surface at 350 K spectrum c results which obviously is due to a
different molecular species (presumably NH_2, ad). Above 400 K
spectrum d is observed which is identical to that recorded after dis-
sociative chemisorption of nitrogen and is therefore due to N_{ad}.
(Adsorbed hydrogen causes a weak maximum in about the same
energy range but desorbs at about 400 K [3]).

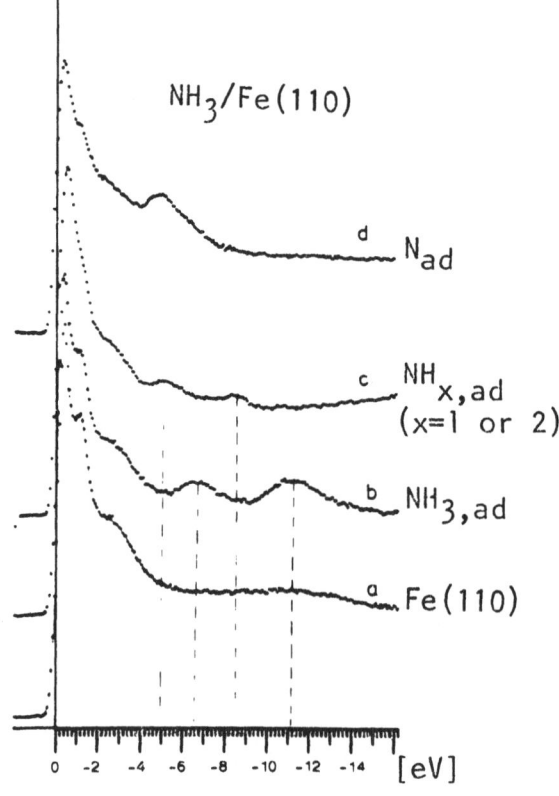

Fig. 2   Photoemission spectra ($h\nu$ = 40.8 eV) of the system NH_3/
Fe(110)

The difficulties encountered with UPS as a method for identify-
ing the molecular nature of surface species are evident: The maxima
are usually relatively broad($\gtrsim 1$ eV) compared with the interesting
energy range (< 15 eV) and their position may be influenced by 'relaxation', 'bonding' and 'chemical' shifts [4] so that identifica-
of more complex species becomes ambiguous. An alternative
source of information is offered by analysis of vibrational spectra as
often applied in 'normal' chemistry: High resolution energy loss
spectroscopy (HRELS) offers promising aspects in this context as
recently demonstrated for example by Ibach [5] in connection with
the problem of reaction intermediates in the catalytic hydrogenation
of ethylene.

Thermal desorption spectroscopy (TDS) can yield some (more
indirect) information on the question whether small molecules (like
H_2) are dissociatively adsorbed or not by determining the reaction
order for desorption in connection with eventual isotope exchange.
Since with this method, however, the temperature of the sample has
to be continuously increased until desorption occurs, activated dissociation on the surface may eventually take place during heating,
i.e. probably not the species adsorbed at low temperatures will be
identified. Such an effect was for example observed with NO interacting with a Ni(111) surface [6].

Auger line shape analysis [7] and Secondary Ion Mass Spectroscopy (SIMS) are other techniques which may be applied in the
present context. The problems and applicabilities of the latter method
are discussed in detail during this course by Werner [8].

2. The Surface Concentration (Coverage)

For adsorption on single crystal surfaces the coverage is most
conveniently defined as

$$\Theta = \frac{\text{number of adsorbed particles}}{\text{number of substrate atoms in topmost layer}}$$

With this definition saturation will only in exceptional cases be associated with $\Theta = 1$ (in contrast to the nomenclature frequently used
with 'real' polycrystalline surfaces).

Whereas determination of the relative coverage may frequently be achieved with a rather high degree of accuracy, the absolute coverage was so far only in rare cases derived more precisely than to within $\overset{+}{-}$ 10 %. The applied methods include:

a) Auger electron spectroscopy. The peak-to-peak height of a (differentiated) Auger signal from a certain element is very often a good measure for its relative surface concentration. Although this technique is widely applied special care has to be taken with respect to disturbing influences of the primary electron beam. The problems connected with absolute concentration determinations are discussed by Gallon [9].

b) X-ray photoelectron spectroscopy. The integrated intensity of a core-electron XPS peak is proportional to the surface concentration of the corresponding element. Quantitative analysis may be achieved according to Yates et al. [10] through the relation

$$ n_a = \frac{Y_a}{Y_s} \frac{N_L \, x_s \, \rho \, \lambda \, \cos \alpha}{x_a M} $$

where n_a = number of adsorbed particles per cm^2

x_s, x_a = substrate and adsorbate X-ray mass absorption coefficients

Y_a, Y_s = photoyields from adsorbate and clean substrate

ρ = density of substrate

α = angle of incidence of the X-rays

N_L = Avogadro's number

M = atomic weight of adsorbate

λ = electron attenuation length

c) Low energy electron diffraction (LEED). Determination of the absolute coverage is frequently achieved from the LEED pattern of highly ordered adsobate layers. The LEED pattern yields usually very simply the unit cell of the adsorbate structure. If this is small enough so that for physical reasons it may contain only a single adsorbed particle the coverage is automatically obtained. This may be used for calibration of other quantities which are continuously varying with Θ .

d) Thermal desorption spectroscopy (TDS). If the pumping speed of the vacuum system is high enough the area $\int p dt$ below a thermal desorption trace is proportional to the surface concentration. Quantitative analysis is achieved through

$$n_a = \frac{S_{eff}}{A k T_g} \int p \, dt \ ,$$

where S_{eff} is the effective pumping speed, A the surface area and T_g the temperature of the desorbing gas.

e) Change of the work function ($\Delta \varphi$). The substrate-adsorbate bond is usually associated with a dipole moment giving rise to a change of the work function. $\Delta \varphi$ is generally not proportional to Θ so that separate calibration (e.g. by means of TDS) is necessary. Once this has been performed, however, $\Delta \varphi$ may be used as a convenient means for continuously monitoring coverage variations without influencing the adsorbed layer.

f) Ultraviolet photoelectron spectroscopy (UPS). Adsorbate valence electron levels as probed by UPS may be used in a similar way as XPS data for the determination of relative coverages. The sensitivity is, however, usually not very high.

g) Vibrational spectroscopy. The intensity of vibrational excitations, either determined by HRELS or by reflection infrared absorption spectroscopy [11], may serve for determining relative coverages, provided that the dipole matrix element is independent on Θ .

h) Molecular beam techniques. If a calibrated flux of molecules strikes a surface and the reflected portion is measured this enables a rather accurate determination of the adsorbed amount [12]. Unfortunately this straightforward technique needs some experimental effort.

i) Radioactive adsorbates. In few cases a radioactive isotope of the adsorbate may serve for deriving its surface concentration [13].

j) SIMS . There are still considerable problems involved with the quantitative aspects of this technique so that general application for surface concentration determinations is not yet possible [8].

It has to be pointed out that determination of the coverage is only meaningful with systems where the surface does not reconstruct under the influence of the adsorbates which may be associated with partial incorporation of the adsorbates into the second or even deeper atomic layers. In this case different techniques will - depending on their probing depth - yield different results. The system N_2/Fe(111) is an example of this type [14].

3. Electronic Properties

Fig. 3 shows the well-known orbital scheme for chemisorption
of CO on a transition metal surface [15]. Thereafter the highest
filled CO-level (5σ) couples to the metal and some back-donation
of metallic d-electrons into the lowest empty CO state ($2\pi^*$) takes
place. The result will be some redistribution of the electronic density
of states within the d-band region and a lowering ('bonding shift') of
the energy of the 5σ-level with respect to that of the 1π- and 4σ-
orbital. The general validity of this picture has been proved by means
of UPS [16]. A photoemission spectrum from a CO covered Ni- surface is reproduced in Fig. 4 which exhibits two maxima below the d-
band region which are associated with the ($5\sigma + 1\pi$)- and 4σ-levels
[16]. Analysis of angular-resolved photoelectron spectroscopy revealed that the CO molecule is in fact oriented to the surface as
drawn in Fig. 3 [17].

Of course UPS does not yield the orbital energies (this would
imply the validity of Koopman's theorem). The problems encountered
with the so-called 'relaxation' shifts have often been discussed [4]
and can only be tackled by proper theoretical treatment. The assump-

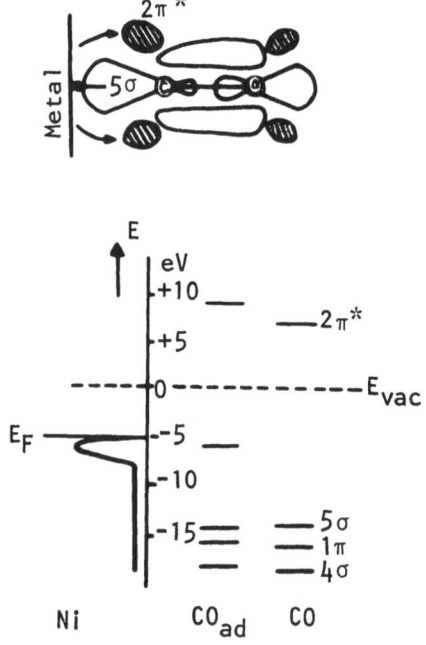

Fig. 3   Mechanism of CO chemisorption on transition metals [15].

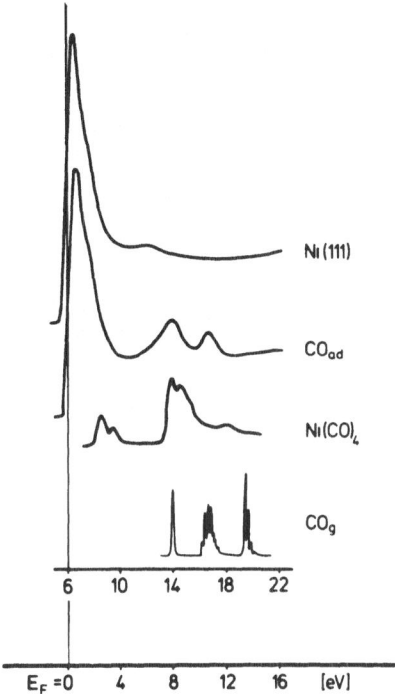

Fig. 4 Photoemission spectra from clean and CO covered Ni(111) surfaces compared with data for Ni(CO)$_4$ and gaseous CO. In the latter two cases the energy scale is referred to the vacuum level which is displaced with respect to the Fermi level E_F by the work function ($\sim$ 6 eV)

tion of uniform relaxation shifts for all adsorbate levels is certainly not generally justified and so far a satisfactory solution to this problem is still lacking.

Electronic excitations into empty levels at the surface may be recorded by means of electron energy loss spectroscopy (ELS). Fig. 5 shows the energy loss spectrum from a clean and a CO covered nickel surface [18]. The latter exhibits two distinct losses at 8 and 15 eV. Quantitative analysis of such data can of course again only be performed with a proper theoretical treatment of the chemisorbed system although a plausible assignment of these date was proposed [19].

The mean dipole moment of an adsorbate complex is another

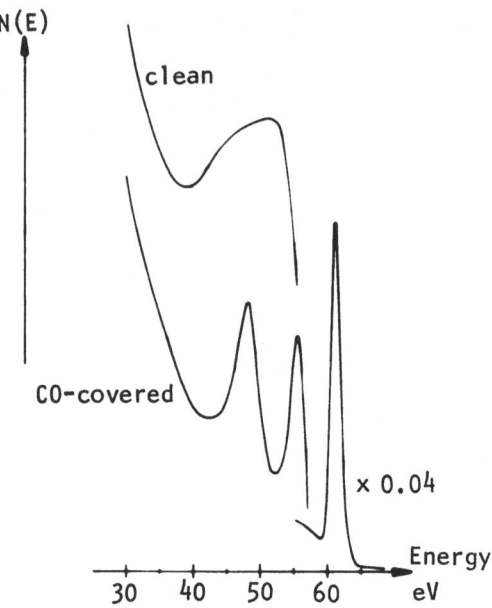

Fig. 5   Electron energy loss spectra from clean and CO covered
Ni(110) surfaces [18].

electronic property of adsorbate systems which is easily accessible
by experiment. The dipole moment μ_a is related with the change of
the work function $\Delta\varphi$ through

$$\mu_a = \varepsilon_0 \Delta\varphi/n_a \quad [\text{Cb} \cdot \text{cm}] \; ,$$

where $\varepsilon_0 = 8.85 \times 10^{-14}$ Cb/V $\cdot$ cm. (Note that 1 Cb$\cdot$cm = $3 \cdot 10^{27}$
Debye). Several methods for determining work function changes are
reviewed for example in ref. [20]. Usually the dipole moment is not
constant over the whole range of coverages, variations being mainly
caused by mutual depolarization or by the occupation of different adsorption sites. A particularly striking example of the latter effect is
offered by the system CO/Pt(111) [21]: As can be seen from Fig. 6
with increasing exposure $\Delta\varphi$ passes at first through a pronounced
minimum (at $\Theta = 1/3$), then through a maximum (at $\Theta = 1/2$) and
finally through a second (shallow) minimum (at about $\Theta = 0.65$).

Saturated adsorbed layers exhibit usually work function changes
of less than 1 eV (with the exception of adsorbed alkali metals) indicating relatively small dipole moments. Therefore it has to be

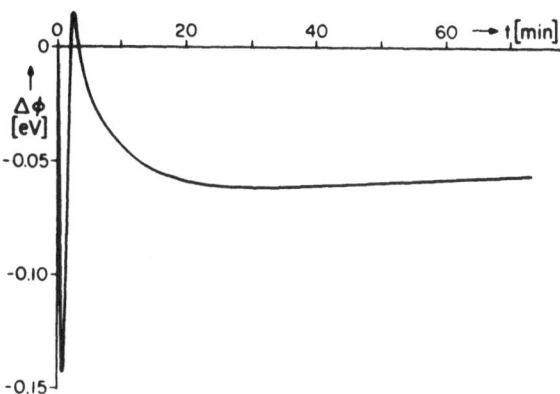

Fig. 6   CO/Pt(111): Variation of the work function with time at
$p_{CO} = 10^{-7}$ torr and T = 190 K [21].

concluded that usually only small charge transfer is associated with
chemisorption and the bond is predominantly of a covalent character.

4. Structure of Chemisorbed Layers

The desorption of the structure of an adsorbed overlayer may be
divided into two parts, namely the configuration of the adsorbed particles with respect to each other and the local arrangement of the ad-
particle with respect to the neighboring substrate atoms (adsorption
site). The formation of long-range order within adsorbed layers revealed to be more the rule than the exception. In these cases the unit
cell may easily be derived from the geometry of the LEED pattern
thus yielding frequently the mutual configuration of the adsorbed
particles. The different structures may be divided into the following
types [21]:
a) Lattice gas structures. The adsorbed particles are located in
identical adsorption sites as is frequently observed with strongly held
species such as O. The LEED patterns usually exhibit simple period-
icities such as 2x2, c2x2 etc.
b) Incoherent structures and coincidence lattices. The vectors
describing the unit cell of the adsorbate structure are no longer composed of integral multiples of the substrate lattice vectors. As a consequence the adsorbed particles are no longer located in sites with

identical local symmetry but rather show a tendency for the formation of close-packed layers. Such a behaviour is found with relatively weakly held species such as physisorbed particles or adsorbed CO at high coverages. As an example, Fig. 7 shows the proposed arrangement of CO on a Ni(111) surface at various coverages as derived from the LEED patterns [22].

c) Surface reconstruction. If the interaction between adsorbate and substrate atoms is strong enough the latter may be displaced from their original locations. This effect is easily observed in cases where the clean substrate surface itself exhibits a surface structure which differs from that of a corresponding plane within the bulk (e.g. Pt(100), Ir(100), Si(111)) but may also occur with non-reconstructed clean surface like with the system N/Fe(111) [14]. In the latter case there is strong evidence for the incorporation of N atoms not only in the topmost atomic layer, a situation which may be regarded as an intermediate between chemisorption and the formation of a true bulk compound.

The determination of the actual geometry of the adsorption site from an analysis of LEED intensity data is still a very elaborate task. However it appears as if LEED theory is now in a stage where at least for simple systems a complete structural analysis may be performed with a satisfactory degree of confidence [23, 24]. This seems to be possible even for weakly scattering H atoms [25].

Sometimes safe conclusions on the geometry of the adsorption site may be drawn even without LEED intensity analysis: Park and

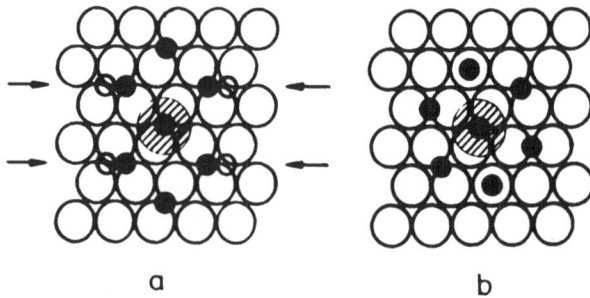

a   b

Fig. 7   Structure models for CO/Ni(111). a) $0.33 \leqslant \theta \leqslant 0.5$; b) $\theta =$ = 0.57 (saturation) [22].

Madden [26] proposed the structure model for CO/Pd(100) at $\Theta =$
= 0.5 as reproduced in Fig. 8 since this is the only configuration
where all CO molecules are located in identical sites which would be
compatible with the observed LEED pattern. Quite recently Bradshaw
and Hoffmann [11] demonstrated that in fact only a single infrared
absorption band is associated with this structure thus confirming the
earlier conclusions. Vibrational spectroscopy in general offers a
promising tool for determining the local geometry of the adsorption
sites [27, 28], without of course being able to deduce the actual
distances.

Angular resolved photoelectron spectroscopy offers promising
aspects for determining the orientation of adsorbed molecules with
respect to the surface plane [17]. Plummer et al. [16] demonstrated recently that the molecular axis of CO is oriented perpendicular
to a Ni(100) surface. This result, however, appears to be in contradiction to the results of a LEED intensity study where a skewed configuration was concluded [29].

5. The Strength of the Substrate-Adsorbate Bond

Information on the strength of the chemisorbed bond can most

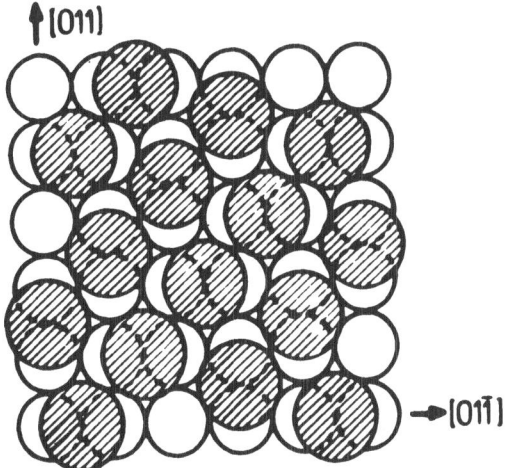

Fig. 8   Structure model for CO/Pd(100) at Θ = 0.5

conveniently be obtained from the analysis of thermal desorption spectra. The rate of desorption of a surface species i may be written as

$$- dn_i/dt = v_i n_i^x \exp(-E_i^*/RT),$$

where v_i is a frequency factor ('preexponential'), x the reaction order and E_i^* the activation energy for desorption. Since adsorption is frequently a non-activated process the latter should be equal to the adsorption energy. The main difficulty lies in the fact that v_i and x are usually unknown so that a simple determination of E_i^* from the desorption peak temperature [30] has to rely on reasonable assumptions on these quantities. Nevertheless this procedure yields often reasonable estimates of the adsorption energy. A more reliable derivation has to be based on an analysis of TDS line shapes [31] which may become a much more elaborate task.

In cases where a true adsorption-desorption equilibrium is established a more straightforward determination of the adsorption energy is possible via measurements of adsorption isotherms and by application of the Clausius-Clapeyron equation

$$\left. \frac{d \ln p}{d(1/T)} \right|_{\theta = const} = - \frac{E_{ad}}{R}$$

That means a plot of $\ln p$ over $1/T$ at constant coverage (or a quantity which is a unique function of coverage such as the intensity of an IR absorption band, the work function change $\Delta \varphi$ or sometimes the LEED pattern) yields the isosteric heat of adsorption at the respective coverage. As an example Fig. 9 shows a series of adsorption isotherms($\Delta \varphi = f(p)$ at various temperatures) for the system CO/Pd(111) from which E_{ad} as a function of Θ as reproduced in Fig. 10 was evaluated. The absolute coverage was in this case calibrated by means of the LEED pattern [32].

In cases of dissociatively adsorbed diatomic molecules A_2 the substrate-adsorbate bond strength E_{M-A} is related with the adsorption energy E_{ad} and the dissociation energy of the free molecule E_d through $E_{M-A} = 1/2 (E_{ad} + E_d)$.

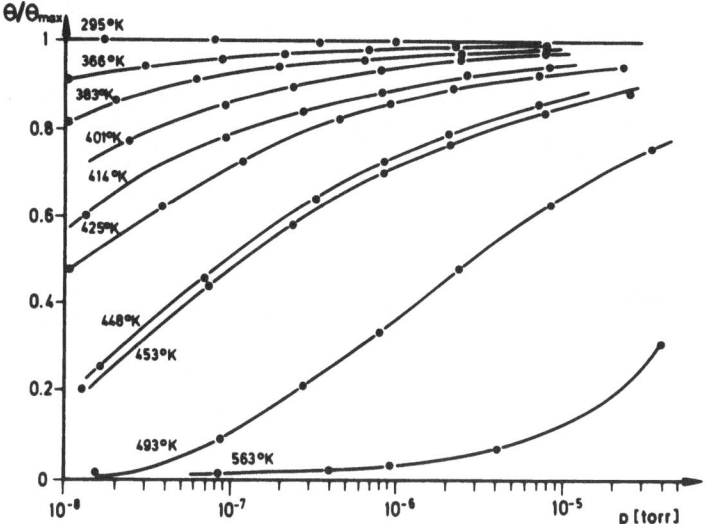

Fig. 9   Adsorption isotherms for CO/Pd(111) [32]

6. Vibrations

Vibrational frequencies of particles adsorbed on single crystal planes can be studied by means of infrared reflection spectroscopy [33] as well as by high resolution electron energy loss spectroscopy (HRELS) [34]. Both techniques are based on the dipole moment

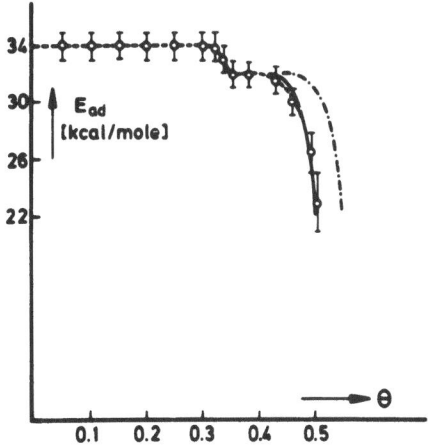

Fig. 10   Isosteric heat of adsorption for CO/Pd(111) [32]

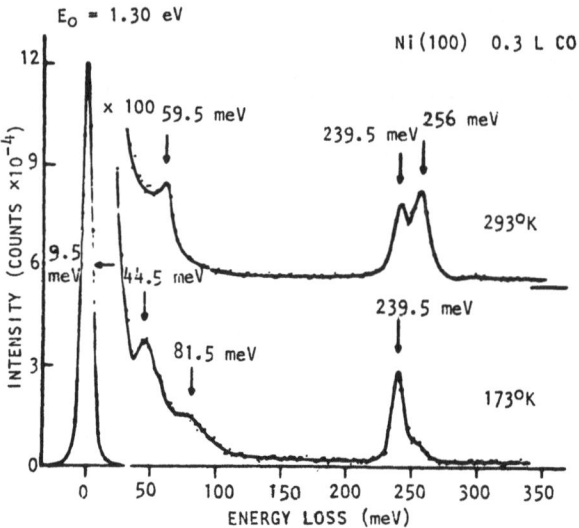

Fig. 11   High resolution energy loss spectra for CO/Ni(100). The
band at 239.5 meV is associated with the C-O stretching vibration
of a molecule in 'bridge' position, whereas that at 256 meV belongs
to a 'linear' species. [28].

of surface vibration and should therefore yield essentially the same
physical information [35]. Infrared spectroscopy is probably experimentally simpler and may also be applied at higher pressures which
makes this technique attractive for applications under more 'real' conditions. On the other hand electron loss spectroscopy has presumably
a higher sensitivity and also a larger accessible spectral range. In
particular it allows for example also the determination of the M-CO
vibration frequency, whereas infrared spectroscopy is restricted to the
C-O stretching vibration. Therefore HRELS yields direct information
on the shape of the potential well of the substrate-adsorbate bond.
Analysis of the symmetry of vibrations and eventual splitting of vibrational bands may be related with the symmetry of the adsorption
site.

As an example Fig. 11 shows electron loss spectra taken from
a Ni(100) surface with low coverage of CO at two different temperatures indicating a thermal equilibrium of the occupation of two adsorption sites with different local symmetry [28].

II. FACTORS INFLUENCING THE CHEMISORPTION BOND

1. The Nature of the Substrate and Adsorbate

The chemical nature of the adsorbed molecules as well as of the surface atoms are of course the most important factors determining the properties of the chemisorption bond. There are certainly no simple rules connecting a single property of the metal (e.g. its percentage d-character) with its chemisorptive behaviour. Table 1 lists some selected values for the substrate-adsorbate bond energy of different particles on the most densely packed planes of a series of transition metals. For the bond strength a general trend appears to be valid with respect to the nature of the adsorbate, viz. $N > O > N > CO \gtrsim NO$, which is obvious for simple chemical reasons. However it is still very difficult to make any predictions on the behaviour of a particular metal. Since chemisorption theories are not yet able to yield accurate quantitative data for the adsorption energies, careful experimental determinations will be necessary also in the future.

2. The Local Character of the Chemisorption Bond

Cluster models where a chemisorbed species is coupled to a small number of metal atoms offer an attractive approach for a description of the chemisorption bond [36]. This implies that this bond is rather localized and involves essentially only the neighboring surface atoms instead of exhibiting 'bandlike' features of an infinitely periodic structure. Experimental support for this idea is obtained from a comparison of the properties of chemisorbed systems with those of

	N	O	H	CO	NO
Fe(110)	140		64		
Ni(111)	135		63	27	25
Cu(111)			56	12	
Pd(111)	130	87	62	34	31
Ag(111)		80		7	25
W(110)	155		68		27
Ir(111)	127	93	63	34	20
Pt(111)	127		57	32	27

Table 1   Substrate-adsorbate bond energies [kcal/mole] [42]

	Cr(CO)$_6$	Mo(CO)$_6$	Fe(CO)$_5$	Ni(CO)$_4$	Mn$_2$(CO)$_{10}$
E	26	36	29	35	24 [kcal/mole]

Table 2   Average M-CO bond energies in carbonyls

corresponding complex compounds and from studies with alloy surfaces.

Fig. 1 already indicates some close relation of the electronic
properties for bonding a molecule to a single metal atom and to an
extended single crystal surface. Table 2 lists some values for the
average bond energies in metal carbonyls which are of about the same
order of magnitude as the heats of chemisorption of CO on metal surfaces. A closer inspection, however, reveals that no identical trends
are observed with both types of systems: Whereas the adsorption
energies of CO on Pd and Pt are higher than on Ni, on the other hand
Ni(CO)$_4$ is considerably more stable than the congeners Pd(CO)$_4$ and
Pt(CO)$_4$. Since the ionisation energy of a single metal atom is typically
about 2 eV larger than that of a metal surface (= work function) this finding

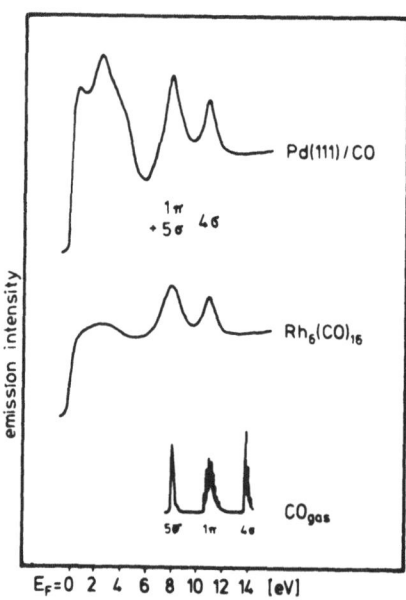

Fig. 12   Photoemission spectra from Pd(111)/CO and from Rh$_6$(CO)$_{16}$
          [37].

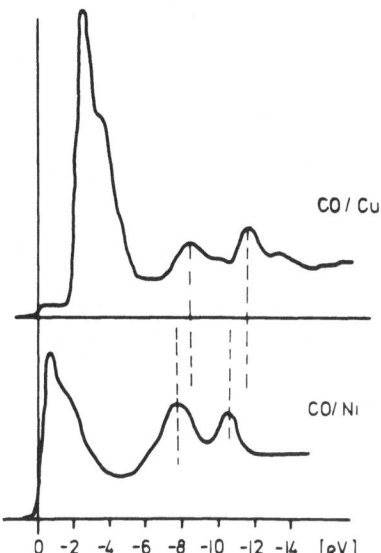

Fig. 13   UPS data for CO adsorbed on Cu and Ni surfaces [38].

is not so surprising. However a cluster compound such as $Rh_6(CO)_{16}$ exhibits already very similar properties to a corresponding chemi-sorption system. Fig. 12 shows photoelectron spectra from $Rh_6(CO)_{16}$ and from CO/Pd(111) where the widths of the d-states are practically identical and the CO-derived peaks exactly line up with each other [37]. Also the vibrational and thermal properties of both systems are very similar so that it was concluded that a small number of metal atoms may in fact be regarded as a good model for chemisorption.

Experiments with alloy surfaces may yield additional informa-tion: Fig. 13 shows photoemission spectra from CO covered Ni and Cu surfaces. The energies of the CO derived peaks differ by about 1 eV between these two metals. Moreover with Cu the 4σ-maximum exhibits a tail towards higher ionisation energies which is presum-ably caused by a shake-up process [38]. Thus these spectra may be regarded as fingerprints for chemisorption on Ni or Cu atoms. The photoemission spectrum of a Cu/Ni alloy surface (note the super-position of the two d-bands !) saturated with CO at low temperature is reproduced in Fig. 14. The CO derived peaks are relatively broad and also exhibit a tail beyond the 4σ-level. If the sample is warmed

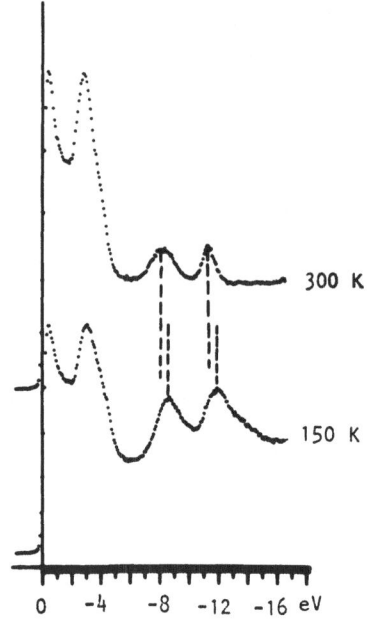

Fig. 14   Photoemission spectra from a Cu/Ni alloy surface exposed
to CO at 150 and 300 K

to room temperature part of the CO desorbs and the spectrum is now
quite similar to that for CO adsorbed on a pure Ni surface [39]. It is
thus concluded that the alloy surface is energetically heterogeneous
and that CO is more strongly bonded on Ni than on Cu sites which is
in agreement with the behaviour of the pure constituents. In fact the
CO molecules are not coupled to single metal atoms but instead the
chemisorption bond involves a small ensemble [40] of neighboring
surface atoms. This becomes nicely evident from thermal desorption
spectra recorded by Spicer et al. [41] which indicate the existence
of 'mixed' adsorption sites besides those which may be attributed to
adsorption on pure Ni or Cu ensembles.

3. The Crystallographic Orientation of the Surface

Table 3 contains values for the initial (i.e. at $\Theta = 0$) adsorption energies for CO on different planes of the fcc metals Ni, Cu and

	(111)	(100)	(110)	(210)	(211)	(311)
Ni	26.5	30	30			
Cu	12	13.5			14.5	
Pd	34	36.5	40	35		35.5

Table 3   Adsorption energies [kcal/mole] for CO on different crystal planes of Ni, Cu and Pd [42]

Pd. Data for H_2 chemisorption on various W surfaces are reproduced in table 4. Similar results have been found for a whole series of different chemisorption systems so that the general conclusion is allowed, that the variation of the adsorption energy with the surface orientation is about one order of magnitude smaller than the bond strength itself. This is a somewhat surprising result if one thinks in terms of strongly directed free valencies of the surface atoms which of course should be markedly influenced by the surface orientation [43]. Such a picture is probably more appropriate for solids where also the bond formation in the bulk is governed by highly oriented orbitals as for example with the elemental semiconductors.

There are other properties of chemisorbed systems which are more strongly influenced by the surface structure, e.g. the work function change (where even the sign of the dipole moment may vary), vibrational frequencies, and in particular kinetic phenomena. The latter effect becomes plausible if it is kept in mind that rate constants are exponentially dependent on activation energies which therefore may strongly be influenced by small energetic variations. The rate of adsorption may typically vary by about a factor of 10 (sometimes even more) between different crystal planes. As an example Fig. 15 shows the variation of the surface concentrations of adsorbed nitrogen atoms (as followed by AES) on three different Fe planes as a function of N_2 exposure exhibiting a variation of the rate by about a factor 50

Plane	(110)	(100)	(211)	(111)	(013)	(122)	(123)	(144)
E_{ad}(kcal/mole)	33	35	40	36.5	33	36.5	39	34

Table 4   Adsorption energies for H_2 on different W planes [42]

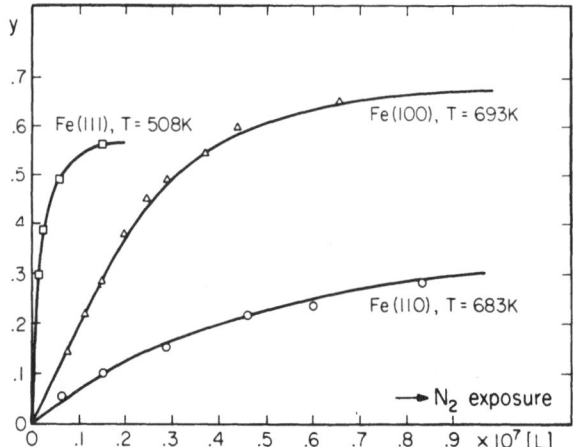

Fig. 15   Relative surface concentration of N$_{ad}$ on three Fe planes as
a function of N$_2$ exposure   (F. Boszo et al., J. Catalysis, in
press).

between Fe(111) and Fe(110). Since this step is regarded to be rate-
determining in the catalytic sythesis of ammonia it becomes evident
that this is a 'structure-sensitive' ('demanding') reaction [44] .
There are on the other hand however many examples of catalytic reactions whose rate is only slightly influenced by the surface structure of the catalyst ('structure insensitive' or 'facile' reaction)[44].

4. The Role of Monoatomic Steps

It is an old idea that structural imperfections on surfaces might
probably play the role of 'active sites' in surface reactions. It was
recognized that periodic arrays of steps may sometimes be easily
formed on single crystal surfaces and may also be rather stable so
that such systems are convenient models for studying such effects
under well-defined conditions. The (average) step orientation, height
and terrace width may conveniently monitored by LEED [45].

The general conclusion which can be drawn from the results of
experiments with stepped surfaces is that the influence of surface
imperfections is similar to the effect of the surface orientation of
single crystal planes on the chemisorption and reaction properties:
Whereas the energetics are only slightly altered this may well be the

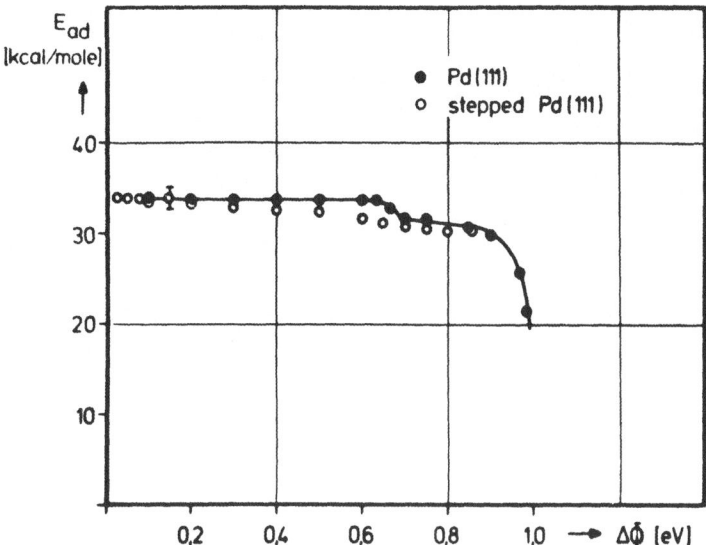

Fig. 16   Variation of the adsorption energy of CO with relative coverage on a flat and stepped Pd(111) surface [46].

case with the kinetics.

     Measurements of the isosteric heat of CO adsorption on flat and
stepped Pd(111) surfaces (the latter consisting of 9 atoms wide
terraces followed by <111> oriented monoatomic steps) revealed in
both cases practically identical results (Fig. 16) [46]. With H_2 adsorption on the other hand a difference in the initial heats of adsorption by about 3 kcal/mole was observed which is caused by a corresponding difference in bond energy between step and terrace sites
[47]. A similar effect was observed with the system H/Pt(111) [48].
In this case also an interesting effect with respect to the work function
change $\Delta\varphi$ was observed: As can be seen from Fig. 17 with the flat
surface $\Delta\varphi$ decreases continuously with increasing coverage, whereas with the stepped surface it at first increases, exhibits a first break
at $\Theta \approx 0.12$ and a maximum at $\Theta \approx 0.25$ from where on the dipole
moments of additionally adsorbed H atoms change their sign. Since
with the latter system about 11% of the surface atoms are located at
steps obviously two H atoms (with slightly different positive dipole
moments) have to be associated with each step atom. Since the
metal-H bond strength on the flat portions of the surface is slightly
smaller (56.3 kcal/mole compared to 57.6 kcal/mole at the steps)
the step sites are occupied first.

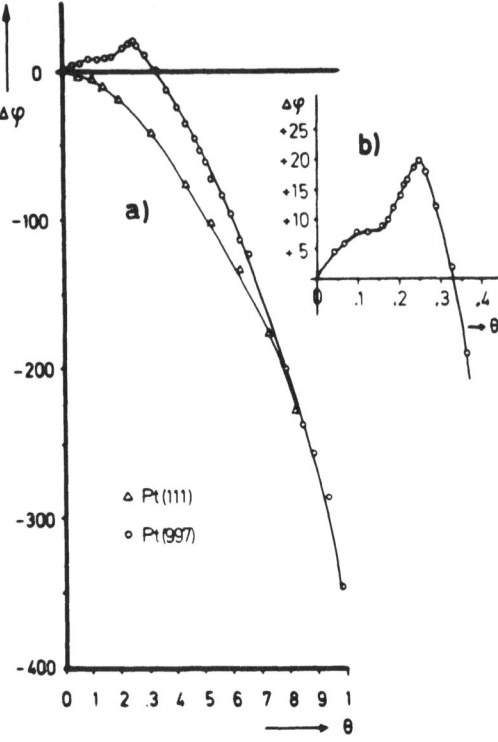

Fig. 17   Variation of the work function with hydrogen coverage on a flat and a stepped Pt surface [48].

In spite of this very small energetic differences the rate of H_2 chemisorption as well as that of the catalytic H_2/D_2 isotopic exchange reaction are increased by about one order of magnitude by the presence of $\sim 10\%$ step atoms [48]. That this effect is really associated with the interaction of H_2 with the atomic steps has recently been nicely demonstrated by molecular beam experiments [49].

With more complex surface reactions where different alternative reaction paths are offered such effects may of course be of pronounced influence on the overall reactivity and selectivity. This becomes particularly evident from Somorjai's studies on catalytic reactions involving hydrocarbon compounds [45].

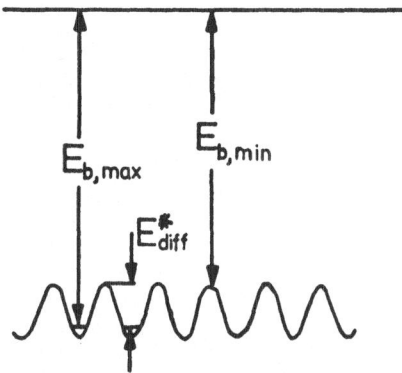

Fig. 18   Periodic potential of an adsorbed particle in a particular
direction on a single crystal surface

5. The Adsorption Site

Fig. 18 illustrates schematically the potential energy variation
of an adsorbed particle along a certain crystallographic orientation
of the surface. Complete description needs a two-dimensional energy
profile over the substrate unit cell. The energy difference E_{diff}^{*} is equal
to the activation energy for surface diffusion in that particular direction. Two cases may be distinguished: i) If $E^{*} \lesssim kT$ the adsorbed
particle will be highly mobile on the surface without preferential
coupling to a certain adsorption site (delocalized adsorption). Restrictions of the mobility may however occur at higher coverages or
if the mutual interactions between adsorbed particles are sufficiently
strong. ii) For $E^{*} > kT$ the adsorbed particle will remain preferentially at a site associated with a potential minimum and the duration
of a place exchange process will be fast compared to the mean residence time at these sites (localized adsorption).

Within a substrate unit cell there may exist sites of different
geometry with nearly equal adsorption energy. The energy difference
ΔE between two sites then governs their relative occupations in
thermal equilibrium at a given temperature. For the system CO/Pt(111)

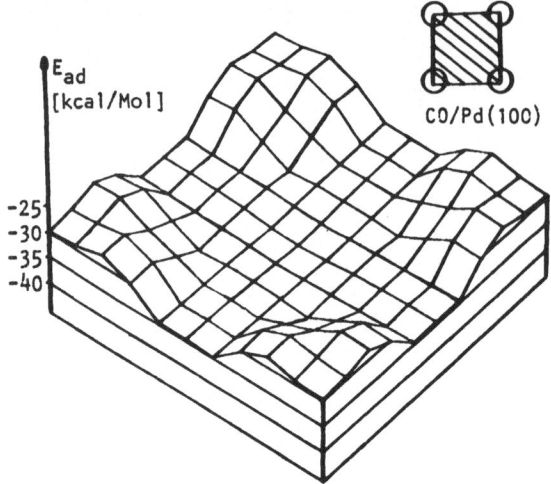

Fig. 19   Energy profile for CO on Pd(100) [50].

it was for example concluded that two sites (presumably twofold and threefold coordinated) differ only by 0.5 kcal/mole in their adsorption energy but may be distinguished by different dipole moments [21]. Adsorption of CO on Ni(100) is another example where the variation of the population of two sites (bridge and linear) with temperature becomes evident from the vibrational spectra (as can be seen from Fig. 11)[28]. It is estimated that also in this case the energy difference is only of the order of about 1 kcal/mole.

Fig. 19 shows a theoretical energy profile for CO/Pd(100) whereafter the adsorption energy varies only by a few kcal/mole across the surface, bridge and fourfold-coordinated sites being energetically nearly equivalent [50]. This agrees with the structure model of Fig. 8 as well as with the observation that the adsorption energy drops by about 6 kcal/mole above $\Theta = 0.5$ where the molecules are displaced from their symmetric 'bridge' positions [51]

Little quantitative experimental data are available for other adsorbates, however the general conclusion is that transition metal surfaces are usually energetically rather 'smooth'. This enables high mobilities of adsorbed particles (at least into distinct directions) which is a necessary prerequisite for surface reactions involving bimolecular recombination steps. So far, specific interactions between adsorbed particles, however, were not taken into account which may modify the effective potential considerably.

III. INTERACTIONS BETWEEN ADSORBED PARTICLES

Fig. 20 illustrates how superposition of the periodic single-particle potential with a pairwise interaction potential yields the effective potential of an adsorbed particle as a function of its distance from a second particle which is placed at the origin of the coordinate system. Obviously two different situations may occur: i) The effective potential minima are still associated with adsorption sites determined by the lattice periodicity. This will give rise to equilibrium configurations described by the 'lattice gas' model, i.e. the adsorbed particles will be located on sites with identical local geometry but varying energy. ii) The positions of adsorbed particles with respect to each other are primarily determined by the minima of mutual interaction potentials, giving rise to incoherent structures where the actual locations are no longer determined by the lattice periodicity. Sometimes a compromise between both limiting cases is reached leading to 'coincidence' lattices [20].

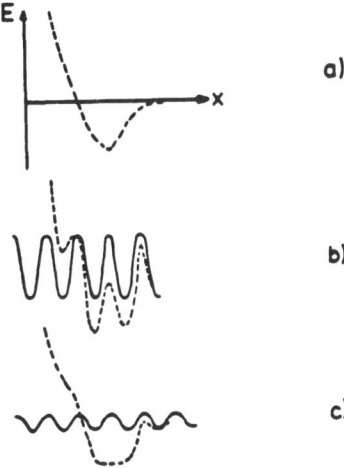

Fig. 20   A pairwise interaction potential between two adsorbed particles (a) superposed over the periodic single-particle potential (full lines in b) and c)) gives rise to effective potentials as drawn by dashed lines in b) and c)

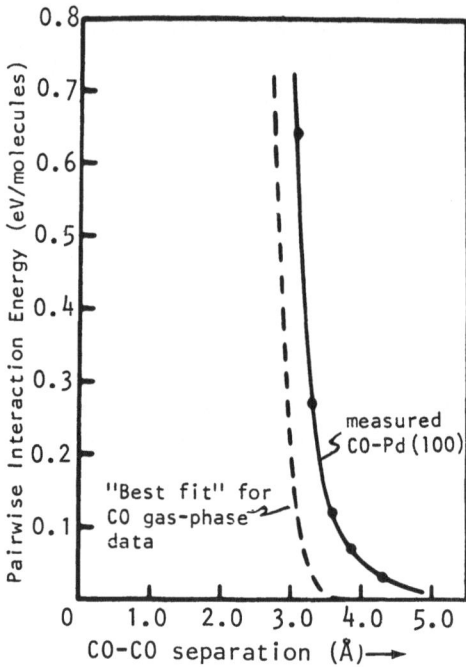

Fig. 21 Experimentally derived interaction potential for CO on
Pd(100) (full line) compared with gas-phase data (dashed line) [51]

The origin of interactions may be threefold:

a) Dipole-dipole interactions. The chemisorption bond is usually associated with a dipole moment μ which can easily be derived
from the work function change $\Delta\varphi$ through $\Delta\varphi = 4\pi n_s u$, where n_s
is the density of adsorbed particles per cm . The pairwise interaction
energy between two dipoles separated by a distance r is given by U =
μ^2/r^3 and summation over all interacting dipoles per area yields
$W = 1/2\, n_s^{5/2}\, \mu^2 \cdot \mathcal{X}$, where $\mathcal{X}$ is a factor (≈ 9) which is slightly dependent on the geometry of the arrangement [52]. The repulsive
interaction energy per adsorbed particle thus results to be $w \approx$
$\approx 4.5\, n_s^{3/2} \cdot \mu^2$. Numerical estimates show that for normal gaseous
adsorbates the effect on the effective adsorption energy is practically
negligible and attains appreciable values ($\sim$ 1 kcal/mole) only with
strongly ionic adsorbates like the alkali metals.

b) Orbital overlap. The 'size' of the adsorbed particle is expected to determine their closest mutual approach (if no other interactions come into play) and therefore also their maximum coverage.
The interaction potential may presumably be approximated   by

(empirical) relations for free molecules such as the Lennard–Jones
potential etc. There is strong evidence that this type of interaction
dominates chemisorption of CO on transition metals at high coverages
and reflects itself in the decrease of the differential heat of adsorption
with coverage near saturation.

With the system CO/Pd(100) Tracy and Palmberg [51] derived
the pairwise repulsive interaction potential as a function of the CO-
CO distance from the variation of the adsorption energy with coverage.
Their data (Fig. 21) are in fair   agreement with the interaction potential between gaseous CO molecules and indicate the dominant contribution from orbital overlap in the high coverage range.
c) Indirect interactions. Fig. 22 shows the variation of the adsorption energy with coverage for the just discussed system where
orbital overlap comes into play above Θ = 0.5 . However E_{ad} decreases continuously already at rather low coverages indicating the
operation of another (long-range) type of interaction. The continuous
change of the effective bond strength is also reflected in the variation
of frequency of the C-O stretching vibration with coverage (Fig. 23)
[11]. The operating interactions are of an indirect ('through bond')

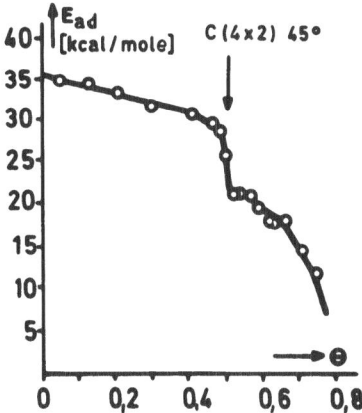

Fig. 22 CO/Pd(100). Variation of the adsorption energy with coverage [51]

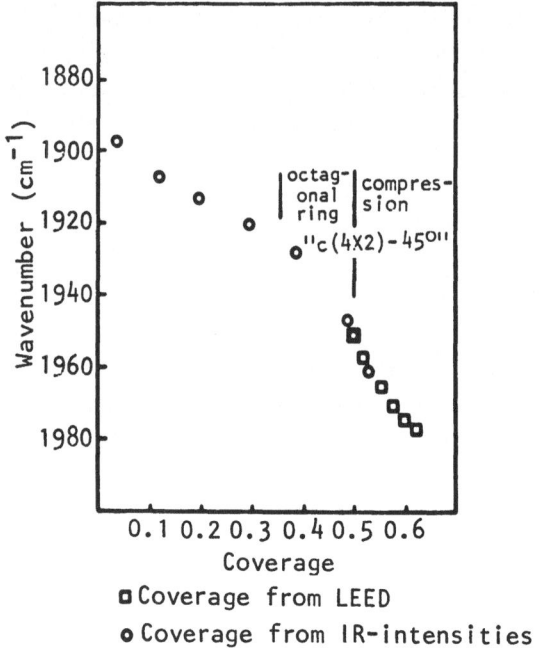

Fig. 23   CO/Pd(100). Variation of the C-O stretching frequency with coverage [11].

character, i.e. they are acting via the valence electrons of the metal-lic substrate as treated theoretically for several model systems [53]. The essential features of these interactions are: i) They exhibit oscil-latory character, i.e. they may be repulsive as well as attractive. ii) Their values are about one order of magnitude smaller than the strength of the chemisorption bond. iii) They are decaying within a few lattice constants to values below kT.

Indirect interactions are certainly very important with strongly held adsorbates(like O and N) and are also responsible for the forma-tion of ordered overlayers of the 'lattice-gas' type. On the other hand the observation of LEED patterns of partially disordered adsorbate structures andtheir computer simulation by means of the Monte Carlo technique [54,55] may be used to evaluate numerical values for the interaction energies in various directions and distances. As an example Fig. 24 shows a model for the perfectly ordered 2x1-structure of O on W(211) at θ = 0.5 and the schematic LEED pattern for θ < 1/2. Computer simulation of the intensity and shape of the half-order LEED

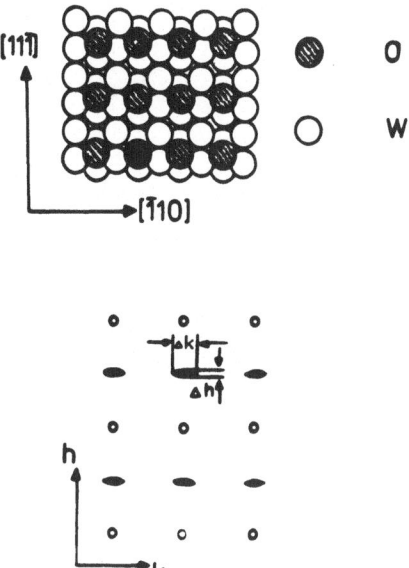

Fig. 24   O/W(211). Structure model at   Θ = 0.5 and LEED pattern
for Θ < 0.5 [56].

spots at various coverages yielded best agreement with experimental
data for the interaction energies denoted in Fig. 25 by circles. The
dashed curves are thus reasonable estimates for the mutual interaction

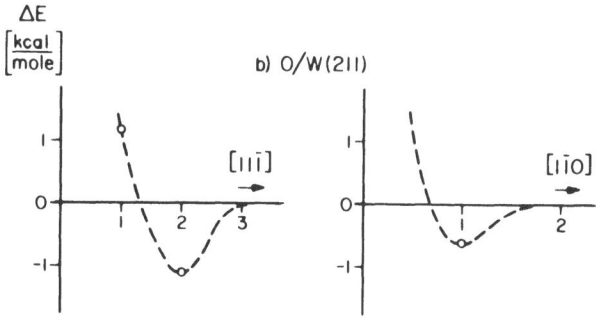

Fig. 25   O/W(211). Interaction potentials between two neighboring
oxigen atoms in [11$\bar{1}$] and [$\bar{1}$10]-directions as derived from a 'best-
fit' computer simulation of the LEED data. (from G. Ertl and
D. Schillinger, J. Chem. Phys. 66, 2569 (1977)).

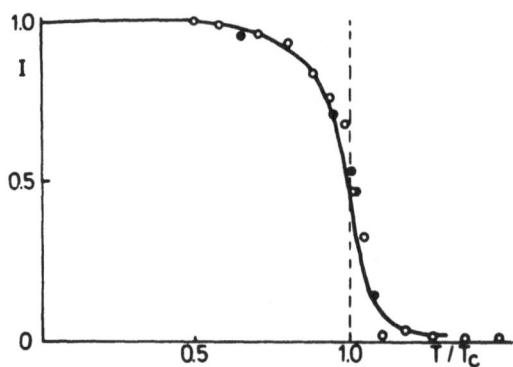

Fig. 26   H/W(100). Order-disorder transition of the c2x2-structure.
Full circles: experimental data. Open circles and full line: theoretical
data. T_c = 500 K [55]

potentials as a function of distance (in units of the lattice constants)
in the $[11\bar{1}]$-and $[\bar{1}10]$-directions on the surface [56] which exhibit
the qualitative features of indirect interactions as outlined above.

The occurence of order-disorder transitions in adsorbed layers
may experimentally be followed by recording the intensity of fraction-
al-order LEED spots as a function of temperature which represents another source of information on mutual interactions. The simplest system is offered by a c2x2-structure on a square substrate lattice which
may be modelled by a single repulsive interaction energy ε between
nearest neighbors and which is adequately described by the two-dimensional Ising model [55, 57]. Fig. 26 shows experimental data
for the system H/W(100) [58], - i.e. the variation of the intensity
of the half-order LEED spots at Θ = 1/2 with temperature, corrected
by background and thermal vibration contributions - together with
theoretical results for a square lattice of limited size ($\triangleq$ coherence
width of the LEED beam) [55]. The transition temperature T_c = 500 K
is related with ε through ε = 1.76 kT_c leading to ε = 1.75 kcal/
mole.

Another general prediction from theory is that the order-disorder
transition temperature should vary with coverage [57]. Such effects
were for example observed with the systems O/W(110) [59] and
H/Ni(111) [60]. In the latter case a 2x1-structure is formed and the
variation of the intensities of half-order LEED spots with temperature

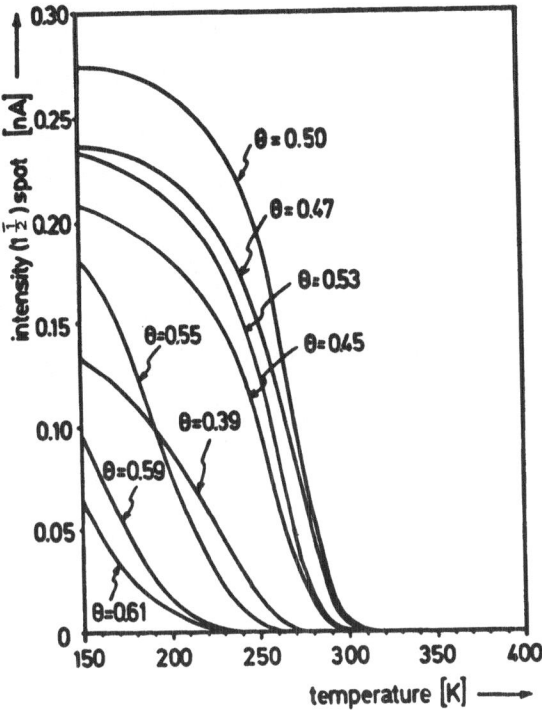

Fig. 27   H/Ni(111). Variation of the intensities of half-order spots with temperature at various coverages (corrected by contributions from the Debye-Waller factor) [60]

at various coverages is reproduced in Fig. 27. Obviously the highest transition temperature is reached for Θ = 1/2. The phase diagram (Fig. 28), however, is not symmetric about this coverage which would be the case for strictly pairwise and independent interactions between adsorbed atoms. This result suggests that in this case the interactions are in fact more complicated, i.e. involve presumably also many-body effects.

The operation of interactions between adsorbed particles influences also thermodynamic properties of the overlayers such as the adsorption isotherms. In the case of attratice interactions condensation phenomeny may occur, i.e. a sudden increase of the (equilibrium) coverage about a critical gas pressure (which depends on

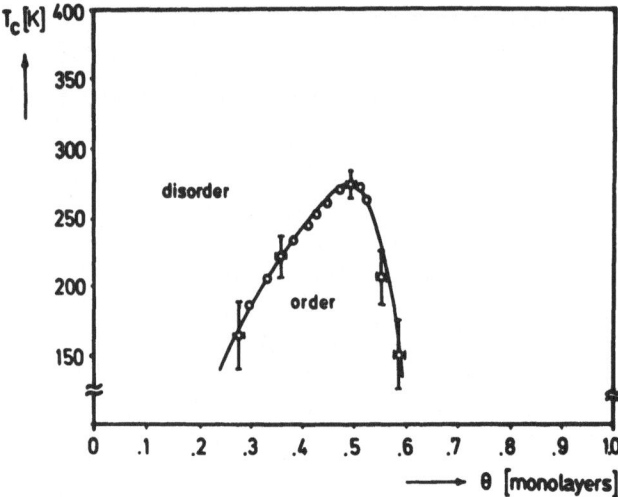

Fig. 28   Phase diagram for H/Ni(111) [60]

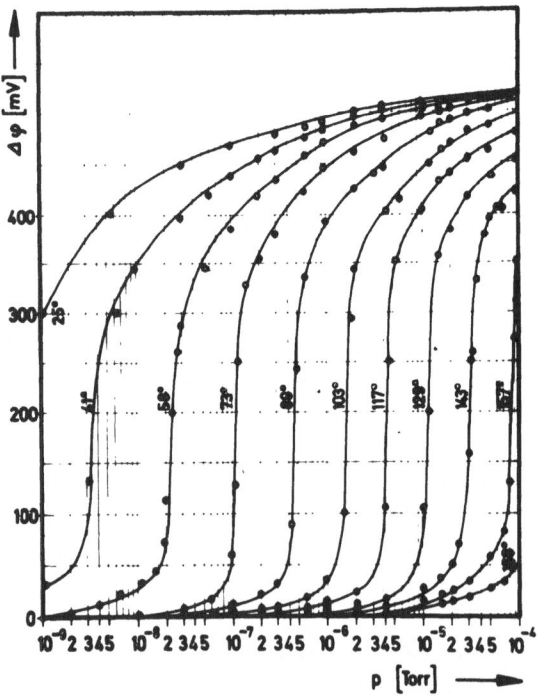

Fig. 29   Adsorption isotherms for H/Ni(110). The work function
change $\Delta\varphi$ serves as a measure for the relative coverage [62].

temperature). Whereas effects of this kind were frequently discussed
for physically adsorbed overlayers [61] the system H/Ni(110) is so
far the only case of chemisorption where this phenomenon was clearly
observed [62]. The corresponding adsorption isotherms are reproduced
in Fig. 29. Interestingly, the adsorption energy increases at low
coverages by about 2 kcal/mole which also demonstrates the operation
of attractive forces between chemisorbed hydrogen atoms.

A final point in this connection concerns the effect of mutual
interactions with systems consisting of two different types of adsor-
bates, A and B. Roughly speaking, the value of $\Delta \varepsilon = \varepsilon_{AA} + \varepsilon_{BB} -
- 2 \varepsilon_{AB}$ will now determine the qualitative features of the surface
phase diagram, where the following two limiting cases may be distinguished:
       a) Cooperative adsorption. If $\Delta \varepsilon \gg kT$ a regular mixed phase
will be formed which is usually characterized by a new surface periodicity as sketched schematically in Fig. 30. As a consequence a new
LEED pattern may appear, as well as the effective adsorption energies,
dipole moments or orbital ionization energies may be altered. An example of this type is offered by the system O + CO/Ni(111)[22], where UPS
shows that the ionization energy of the CO-4σ-level is increased by
about 0.3 eV by the presence of coadsorbed oxygen, whereas the other
levels are essentially not affected (Fig. 31). This effect was tentative-
ly interpreted in terms of a partial electron transfer from the 4σ-orbital
(lone electron pair at the O atom of CO) to neighboring chemisorbed
oxygen atoms.
       b) Competitive adsorption. If $\Delta \varepsilon \ll kT$ both species will not

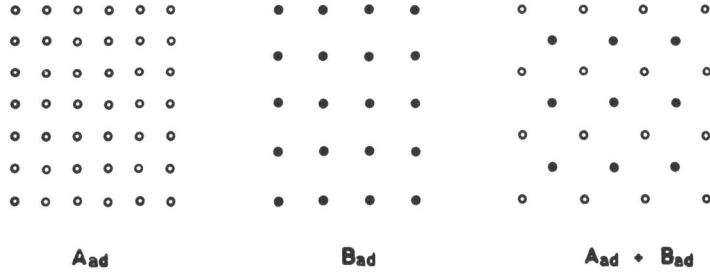

A_{ad} B_{ad} $A_{ad} + B_{ad}$

Fig. 30   Cooperative adsorption of two species (schematic)

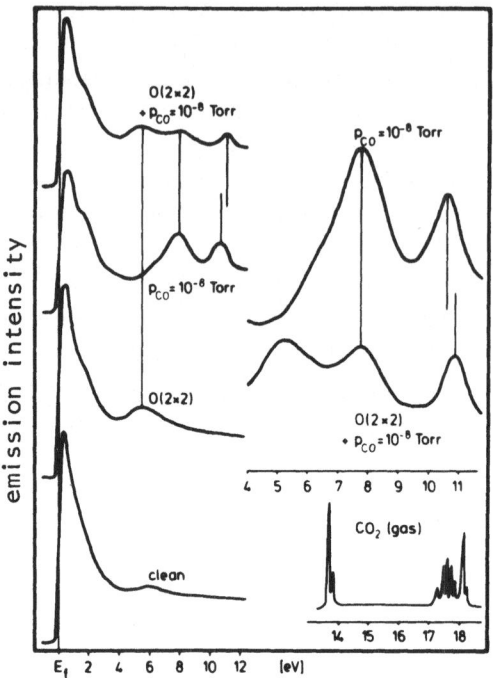

Fig. 31   Photoemission spectra of the coadsorption of O + CO on
Ni(111) [22]

be miscible but coexist on the surface in separate domains (Fig. 32).
Frequently the mean domain size exceeds the coherence width of the
LEED electron beam so that the diffraction pattern consists of a superposition of spots arising from either of the adsorbed species alone. An
example may be found with the system O + CO/Pd(111) [63].

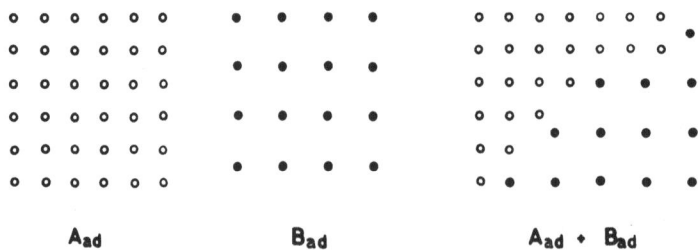

Fig. 32   Competitive adsorption of two species (schematic)

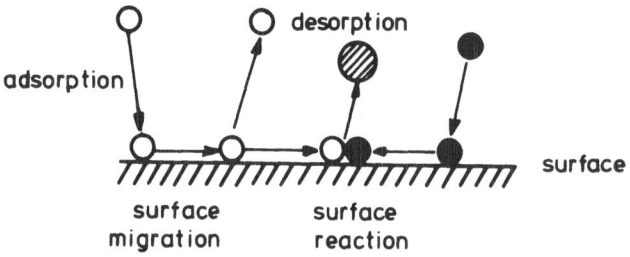

Fig. 33   Elementary steps in heterogeneous catalysis

IV. DYNAMICS OF SURFACE PROCESSES

So far mainly the static aspects of gas/metal interactions were discussed. However in order to understand the mechanism of a complex surface reaction investigation of the individual rate processes is of vital importance . It is felt that this field will concentrate major efforts in the near future.

Some of the elementary steps involved in a heterogeneously catalyzed reaction are illustrated schematically in Fig. 33: It is evident that the individual rate constants for adsorption, desorption and surface reaction are determining the surface concentrations and geometric distributions of reactants and products under steady-state conditions and thereby the overall rate of product formation. Surface diffusion is usually believed to occur rapidly enough, but quantitative data on this step are so far still rather scarce.

The rate of adsorption is characterized by the sticking coefficient s, i.e. the probability that a particle striking the surface from the gas phase becomes adsorbed. s will be dependent on the coverage Θ . The simplest case whereafter s is proportional to the empty fraction of the surface, i.e. $s = s_0 (1 - \Theta / \Theta_{max})$, (Langmuir model) is however only rarely realized. Frequently adsorption kinetics may be described in terms of a 'precursor' model [64] which is illustrated

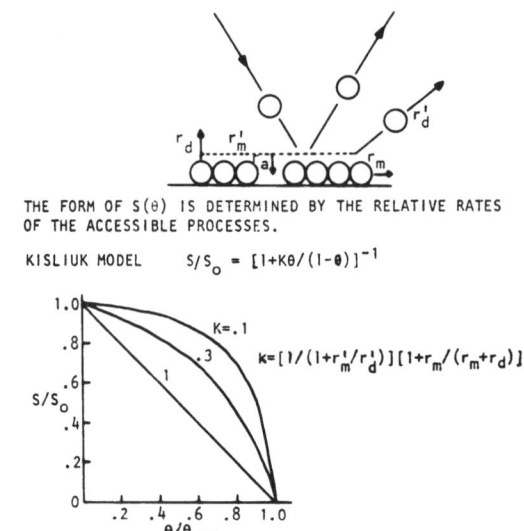

Fig. 34   Kisliuk model for adsorption kinetics

by Fig. 34. This model assumes a certain lifetime of particles in a
second layer (precursor state) during which they may reach unoccupied chemisorption sites or otherwise desorb. As a consequence the
decrease of s with θ becomes flatter at low coverages than that
according to the Langmuir model. The system CO/Pd(111) offers an
example of this type as can be seen from Fig. 35 [65]. This figure

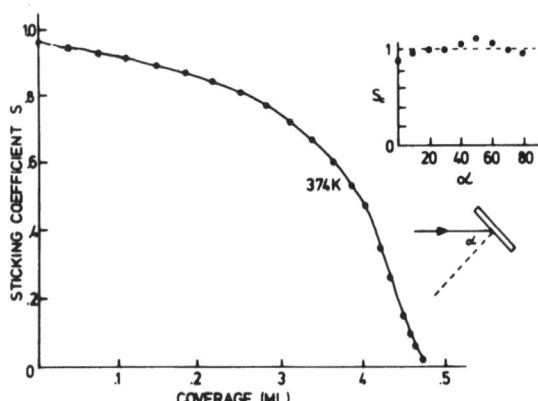

Fig. 35   CO/Pd(111). Variation of the sticking coefficient s with
coverage and of the initial sticking coefficient s_o with the angle of
incidence α. [65].

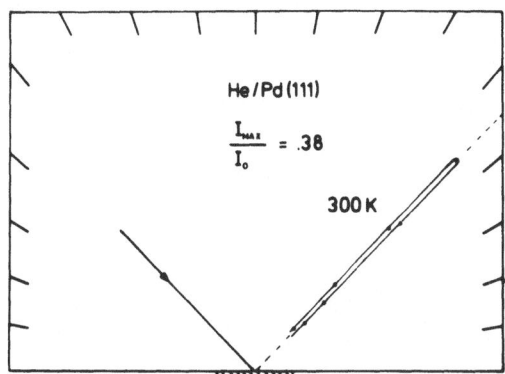

Fig. 36   Angular distribution of He atoms scattered from Pd(111)
(specular reflection). The arrow indicates the direction of the primary
beam.

also shows that the initial sticking coefficient, s_0 = 0.96, is in this
case independent of the angle of incidence of the gaseous molecules.

These data were obtained by measuring the portion of molecules
reflected from the surface by means of a molecular beam apparatus.
This technique yields certainly the most accurate values and the most
detailed insight into the dynamics of surface reactions [66]. More
frequently sticking coefficients are derived by determining the adsorbed amount (e.g. by thermal desorption) as a function of gas exposure. In the case of activated adsorption the sticking coefficient
will increase with increasing temperature, whereby in principle the
influence of gas and surface temperature should be studied separately [67].

Information on the dynamics of gas/solid interactions may
further be obtained by studying the angular distribution of scattered
particles. As can be seen from Fig. 36 a Pd(111) surface acts for
He atoms at room temperature like a hard wall since a high fraction
is simply specularly reflected which has to be attributed to a rather
shallow minimum of the interaction potential and to the mass ratio.
The degree of perfection of the surface has a marked influence on this
result so that He scattering is a very sensitive means (far better
than LEED) for checking the quality of a surface [68]. By contrast
CO molecules come off a Pd(111) with an almost perfect cosine

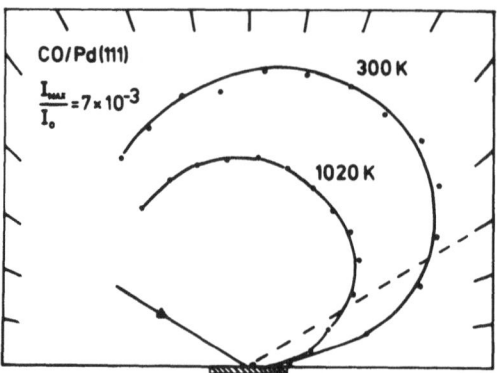

Fig. 37   Angular distribution of He atoms scattered from a Pd(111) surface at 300 and 1020 K (cosine distribution)

distribution (Fig. 37) which indicates that the molecules stay even at 1000 K at the surface for a long enough time to loose completely the 'memory' of their initial momentum.

This question as well as that for the rate of desorption is closely related with the mean surface residence time τ of a particle on the surface. For a first-order desorption process $\tau = 1/k_d$, where k_d is the rate constant for desorption. τ may be written in terms of the

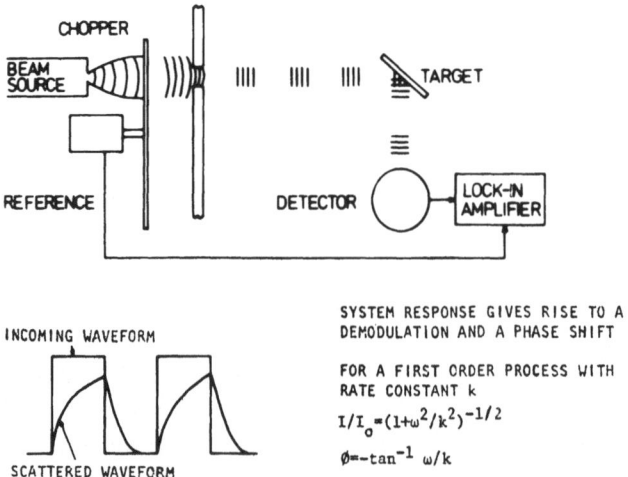

Fig. 38   Principle of the modulated molecular beam technique

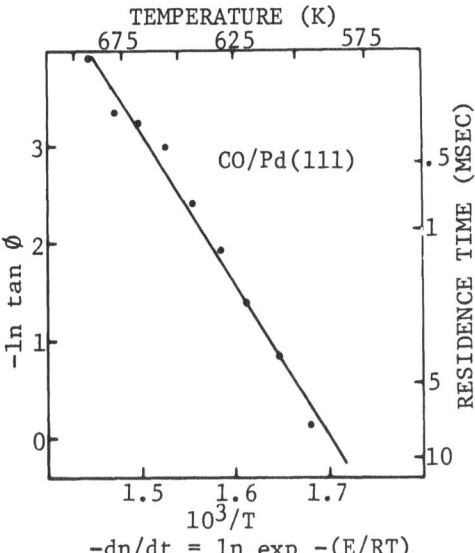

Fig. 39   Phase shift tan ϕ and mean surface residence time for CO/Pd(111) as a function of 1/T [65]

so-called Frenkel equation $\tau = \tau_0 \cdot e^{-E^*/kT}$, where E^* is the activation energy for desorption and τ_0 is expected to be of the order of magnitude of a vibrational period normal to the surface ($\sim 10^{-13}$sec), eventually modified by contributions from entropy effects.

Surface residence times may directly be determined down to about 10^{-6} sec by means of modulated molecular beam experiments the principle of which is illustrated by Fig. 38 [69]. Most conveniently the phase shift ϕ (of the first Fourier component) between the primary and the scattered beam is recorded, which for a first-order rate-process is related with its rate constant k_r through tan ϕ = $= \omega/k$, where ω is the chopping frequency. As an example Fig. 39 shows an Arrhenius plot (In tan ϕ resp. In τ vs. 1/T) for the system CO/Pd(111) [65]. From the slope the activation energy for desorption E^*_d = 33 kcal/mole results which is in good agreement with the isosteric heat of adsorption, as expected since the adsorption process is non-activated in this case. τ_0 = 3×10^{-13} sec is also of the expected order of magnitude.

The oxidation of CO on a Pd(111) surface will serve as an

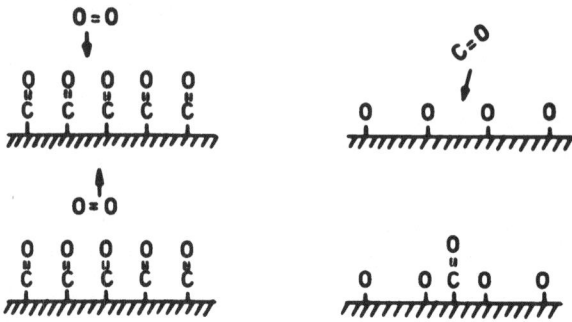

Fig. 40   Interactions of CO and oxygen on Pd (schematic)

example for a catalytic reaction whose mechanism appears to be fairly
well established now. This reaction has been studied extensively in
the past with single crystals [63, 70] as well as with polycrystalline
material [71], whereby the surface structure revealed to be only of
minor importance. The involved adsorption steps may be formulated
schematically as follows:

$$CO + * \rightleftharpoons CO_{ad}$$
$$O_2 + * \rightleftharpoons 2\,O_{ad}$$

* denotes a free adsorption site on the surface. The adsorption energy
of O_2 is considerably higher ($\gtrsim$ 60 kcal/mole) than that of CO
(34 kcal/mole). As a consequence desorption of O_2 starts only above
400 °C and can therefore be neglected in the present discussion. Coadsorption of CO and O is competitive at lower coverages (island
formation) [63]. If, however, a surface already saturated with O_{ad}
is exposed to CO also cooperative adsorption may take place [72].
On the other hand preadsorption of CO to $\theta_{CO} \geq$ 0.33 (i.e. about
2/3 of the CO saturation coverage) completely inhibits the dissocia-
tive adsorption of O_2. This situation is illustrated schematically by
Fig. 40. Since the catalytic reaction proceeds only via adsorbed O
atoms obvioulsy CO acts as an inhibitor for the steady-state reaction
and desorption of CO is the rate-limiting step below 200 °C [70, 71].

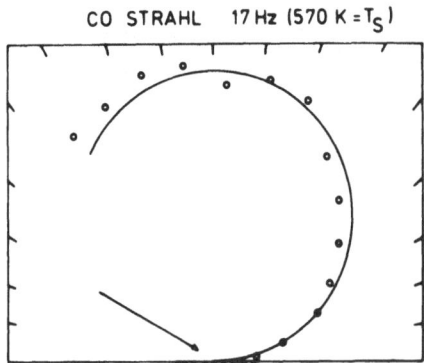

CO STRAHL   17 Hz (570 K = T_S)

Fig. 41   Angular distribution of CO_2 formed by the catalytic oxidation
of CO on a Pd(111) surface. The arrow indicates the direction of the
CO molecular beam.

As can bee seen from Fig. 41 CO_2 formed during the reaction comes
off the surface with a complete cosine distribution. CO_2 itself is only
rather weakly adsorbed on the surface so that its mean residence time
is negligibly small above room temperature.

The surface reaction itself may proceed in principle via two
different mechanisms which are illustrated by Fig. 42, namely

$$CO + O_{ad} \longrightarrow CO_2 \quad \text{(Eley-Rideal mechanism)}$$
$$\text{or} \quad CO_{ad} + O_{ad} \longrightarrow CO_2 \quad \text{(Langmuir-Hinshelwood mechanism)}$$

With the ER mechanism the reaction should take place by direct
collision of CO from the gas phase or via a weakly held (physisorbed)
species whereas the LH- mechanism proceeds between both reactants
in their chemisorbed state. A clear distinction between these two
mechanisms on the basis of stationary kinetic measurements appears
to be rather questionable. However analysis of the results from modulated molecular beam experiments demonstrated that in the present
case there is no evidence for the occurrance of the ER mechanism.
The surface residence times for CO associated with the catalytic oxidation of this molecule were evaluated to lie in the range of 10^{-2} to

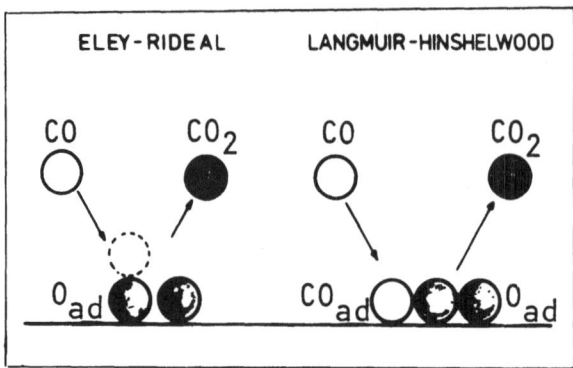

Fig. 42   Reaction mechanisms for the catalytic oxidation of CO

10^{-5} sec for temperatures between about 200 and 400 $^{\circ}C$; the activation energy for the LH reaction at low coverages was derived to be
about 24 kcal/mole [73].

REFERENCES

1.   G. Ertl, J. Küppers, F. Nitschke and M. Weiss, Chem. Phys.
     Lett. (in press)

2.   G. Ertl, F. Nitschke and M. Weiss, to be published

3.   F. Bozso, G. Ertl, M. Grunze and M. Weiss, Appl. Surf. Sci.
     1 (1977), 103

4.   P. S. Bagus and K. Herrmann, Solid State comm. 20 (1976), 5

5.   H. Ibach, H. Hopster and B. Sexton, Appl. Surface Sci. 1
     (1977), 1

6.   H. Conrad, G. Ertl, J. Küppers and E. E. Latta, Surface Sci.
     50 (1975), 296

7.   T. Kawai, K. Kunimori, T. Kondow, T. Onishi and K. Tamaru,

Jap. J. Appl. Phys. Suppl. 2, Pt. 2 (1974), 513

8.   H. M. Werner, this volume

9.   T. E. Gallon, this volume

10. J. T. Yates, T. E. Madey and N. E. Erickson, Chem. Phys. Lett. 19 (1973), 487

11. A. M. Bradshaw and F. M. Hoffmann, Surface Sci. (in press)

12. C. Wang and R. Gomer, Proc. 7th Int. Vacuum Congr., Vienna 1977, p. 1155

13. M. Kostelitz and J. Oudar, Surface Sci. 27 (1971), 176

14. F. Bozso, G. Ertl, M. Grunze and M. Weiss, J. Catalysis 49 (1977), 18

15. G. Blyholder, J. Phys. Chem. 68 (1964), 2772
     G. Doyen and G. Ertl, Surface Sci. 43 (1974), 197

16. C. L. Allyn, T. Gustafsson and E. W. Plummer, Chem. Phys. Lett. 47 (1977), 127

17. A. Liebsch, this volume

18. J. Küppers, Surface Sci. 36 (1973), 53

19. F. P. Netzer, R. A. Wille and J. A. D. Matthew, Solid State Comm. 21 (1977), 97

20. G. Ertl and J. Küppers: "Low energy electrons and surface chemistry", Verlag Chemie, Weinheim 1974

21. G. Ertl, M. Neumann and K. M. Streit, Surface Sci. 64 (1977), 393

22. H. Conrad, G. Ertl, J. Küppers and E. E. Latta, Surface Sci. 57 (1976), 475

23. J. B. Pendry: "Low energy electron diffraction", Academic

Press, New York 1974

24. M. A. van Hove, in "The nature of the surface chemical bond"
     (Eds. T. N. Rhodin and G. Ertl), North Holland, Amsterdam
     1978

25. M. A. van Hove, G. Ertl, W. H. Weinberg, K. Christmann
     and R. J. Behm, Proc. 7th Int. Vacuum Congr., Vienna 1977,
     p. 2415

26. R. L. Park and H. H. Madden, Surface Sci. 11 (1968), 188

27. H. Froitzheim, H. Ibach and S. Lehwald, Phys. Rev. B14
     (1976), 1362

28. S. Andersson, Solid State Comm. 21 (1977), 75

29. J. B. Pendry and S. Andersson, to be published

30. P. A. Redhead, Vacuum 12 (1962), 203

31. D. A. King, Surface Sci. 47 (1975), 384

32. G. Ertl and J. Koch, Z. Naturforsch. 25a (1970), 1906

33. R. G. Greenler, J. Chem. Phys. 44 (1966), 310; 50 (1969),
     1963
     R. A. Shigeishi and D. A. King, Surface Sci. 58 (1976), 379

34. H. Ibach, J. Vac. Sci. Techn. 9 (1972), 713

35. H. Ibach, Surface Sci. 66 (1977), 56

36. R. P. Messmer, in "The nature of the surface chemical bond",
     (T. N. Rhodin and G. Ertl, eds.), North Holland, Amsterdam
     1978

37. H. Conrad, G. Ertl, J. Küppers, H. Knözinger and E. E. Latta,
     Chem. Phys. Lett. 42 (1976), 115

38. H. Conrad, G. Ertl, J. Küppers and E. E. Latta, Solid State
     Comm. 17 (1975), 613

C. R. Brundle and K. Wandelt, Proc. 7th Int. Vacuum Congress, Vienna 1977, p. 1171
C.L. Allyn, T. Gustafsson and E. W. Plummer, to be published

39.   J. Küppers, K. Wandelt and G. Ertl, to be published

40.   W. M. H. Sachtler and R. A. van Santen, Adv. Catalysis 26 (1977), 69

41.   K. Y. Yu, D. T. Ling and W. E. Spicer, J. Catalysis 44 (1976), 373

42.   G. Ertl, in "The nature of the surface chemical bond", (T. N. Rhodin and G. Ertl, eds.), North Holland, Amsterdam 1978

43.   G. C. Bond, Disc. Faraday Soc. 41 (1966), 200

44.   M. Boudart, Adv. Catalysis 20 (1969), 153

45.   G. A. Somorjai, Adv. Catalysis 26 (1977), 2

46.   H. Conrad, G. Ertl, J. Koch and E. E. Latta, Surface Sci. 43 (1974), 462

47.   H. Conrad, G. Ertl and E. E. Latta, Surface Sci. 41 (1974), 435

48.   K. Christmann and G. Ertl, Surface Sci. 60 (1976), 365

49.   R. J. Gale, M. Salmeron and G. A. Somorjai, Phys. Rev. Lett. 38 (1977), 1027

50.   G. Doyen and G. Ertl, Surface Sci. 69 (1977), 157

51.   J. C. Tracy and P. W. Palmberg, J. Chem. Phys. 51 (1969), 4852

52.   J. Topping, Proc. Roy. Soc. A114 (1927), 67

53.   T. B. Grimley and M. Torrini, J. Phys. C6 (1973), 868
      T. E. Einstein and J. R. Schrieffer, Phys. Rev. B7 (1973), 3629

54.   G. Ertl and J. Küppers, Surface Sci. 21 (1970), 61

55.   G. Doyen, G. Ertl and M. Plancher, J. Chem. Phys. 62 (1975), 2957

56.   G. Ertl and M. Plancher, Surface Sci. 48 (1975), 364

57.   K. Binder and D. P. Landau, Surface Sci. 61 (1976), 577

58.   P. J. Estrup, in "The structure and chemistry of solid surfaces" (G. A. Somorjai, ed.), Wiley, New York 1969, p. 19 - 1

59.   T. M. Lu, G. C. Wang and M. G. Lagally, Phys. Rev. Lett. 39 (1977), 411

60.   R. J. Behm, K. Christmann and G. Ertl, to be published

61.   J. H. de Boer, "The dynamical character of adsorption" (Clarendon, Oxford 1953)
      J. M. Honig, "The solid-gas interface" (Dekker, New York 1967), Vol. 1, p. 371

62.   K. Christmann, O. Schober, G. Ertl and M. Neumann, J. Chem. Phys. 60 (1974), 4528

63.   G. Ertl and J. Koch, in "Adsorption-Desorption Phenomena", (F. Ricca, ed) Academic Press 1972, p. 345

64.   P. J. Kisliuk, J. Phys. Chem. Solids 3 (1957), 95; 5 (1958), 78
      D. A. King and M. G. Wells, Surface Sci. 29 (1972), 454

65.   T. Engel and G. Ertl, Proc. 7th Int. Vacuum Congr., Vienna 1977, p. 1365

66.   W. H. Weinberg, Adv. Colloid Interf. Sci. 4 (1975), 301

67.   M. Balooch, M. J. Cardillo, D. R. Miller and R. E. Stickney, Surface Sci. 46 (1974), 358

68.   S. L. Bernasek, G. A. Somorjai and R. P. Merrill, J. Vac. Sci. Techn. 12 (1975), 655

69.   R. H. Jones, D. R. Olander, W. J. Siekhaus and J. A.
      Schwarz, J. Vac. Sci. Techn. 9 (1972), 1429

70.   G. Ertl and P. Rau, Surface Sci. 15 (1969), 443;
      G. Ertl and J. Koch, Proc. 5th Int. Congr. on Catalysis, North
      Holland, Amsterdam 1973, p. 969
      G. Ertl and M. Neumann, Z. phys. Chem. N.F. 90 (1974), 127

71.   T. Masushima and J. M. White, J. Catalysis 39 (1975), 265
      J. S. Close and J. M. White, J. Catalysis 36 (1975), 185

72.   H. Conrad, G. Ertl and J. Küppers, to be published

73.   T. Engel and G. Ertl, to be published

X-RAY PHOTOELECTRON SPECTROSCOPY OF SOLIDS

G. K. Wertheim

Bell Laboratories

Murray Hill, New Jersey   07974

I.   INTRODUCTION

X-ray photoelectron spectroscopy (XPS) is a deceptively simple technique. The basic physics, namely the photoelectric effect, has been well understood since the beginning of this century. Ultra violet photoelectron spectroscopy has been practiced with a high degree of sophistication for decades, and even X-ray induced photoemission was studied over 50 years ago (see the work of P. Auger [1]. We owe to Kai Siegbahn [2] the current activity in XPS. He realized that useful work could be done by X-ray induced photoemission in spite of the obvious problems due to lack of X-ray intensity and monochromaticity. He and his collaborators had developed this technique to a point, 10 years ago, where the applications were sufficiently promising to warrant the manufacture of commercial XPS units. Part of this enthusiasm was undoubtedly generated by the acronym ESCA (Electron Spectroscopy for Chemical Analysis) that was coined by Siegbahn for this technique [2]. It is a mistake, however, to take this appellation literally; the applications of the technique transcend chemical analysis, and presentation of even a selected subset of its contributions will require a number of lectures.

Historically, the instrumental techniques are derived from those of beta-ray spectroscopy. In fact, some of the first ESCA experiments were carried out with such machines, modified to admit X-rays to the sample position. It was soon realized that the best resolution and intensity is obtained with Kα X-rays of low-Z elements [3]. Practical considerations require the use of a metal as the X-ray tube anode, narrowing the choice to Na, Mg, or Al.

The latter two have been almost universally adopted. The low
energy of the Kα radiation of Mg and Al, 1253 and 1486.6 eV, makes
magnetic deflection analysis of the photoelectron cumbersome, but
is ideally suited for electrostatic analyzers. Two types have
been found to offer a suitable combination of resolution and
aperture, cylindrical and hemispherical mirror analyzers.

The use of the complete spectrum from an X-ray tube for the
excitation of photoelectrons, of course, presents serious limitations on resolution and spectral purity. In addition to the
desired characteristic $K\alpha_{1,2}$ lines there are the bremsstrahlung
from the incident electron beam, the other characteristic lines,
e.g., Kβ, as well as the so-called nondiagram lines, e.g., $K\alpha_{3,4}$
The bremsstrahlung reduces the signal-to-background, and the
discrete lines produce satellites which can interfere with the
observation of other lines. The obvious solution is to disperse
the X-rays by Bragg reflection in order to select the desired
wavelength. In practice, this requires a much more sophisticated
instrument because the loss of intensity (10^{-3}) due to the mono-
chromatization of the X-ray must be made up by using a multi-
element detector and a high-power X-ray tube. The improvement in
resolution which has been achieved so far is about a factor of
four, i.e., ∿0.25 eV FWHM. A second direction is to use synchrotron radiation for excitation. In principle a continuous spectrum
from the UV to 100 keV is available, but only the spectrum up to
∿100 eV has so far been used to any extent. Much interest lies in
the 100–1000 eV region where instrumentation remains a problem.
This region includes the K edges of B, C, N, O and F so that a
variety of threshold effects become amenable to study.

Although core electron binding energies go up to 100 keV in
the region of the actinides, AℓKα photons make accessible most of
electronic states of the elements, see Fig. 1. For any particular
Z, the binding energies have characteristic values which serve to
identify the element and make the technique suitable for elemental
analysis, motivating the term ESCA. Identification is often aided
by the fact that core hole states with orbital angular momentum
have characteristic spin-orbit splitting. The concept of ESCA,
however, goes well beyond elemental analysis. The technique is
used to reveal more subtle details of the electronic structure.
Best known are the chemical shifts due to change of valence.
These are normally measured for core electrons and represent the
opposing effects of an increase in binding energy due to the
removal of an outer electron and a decrease due to the Madelung
energy.

The detectability of the elements is largely determined by
the cross sections for photoemission, which have been calculated
by Scofield [4] for photon energies between 1 keV and 1.5 MeV.

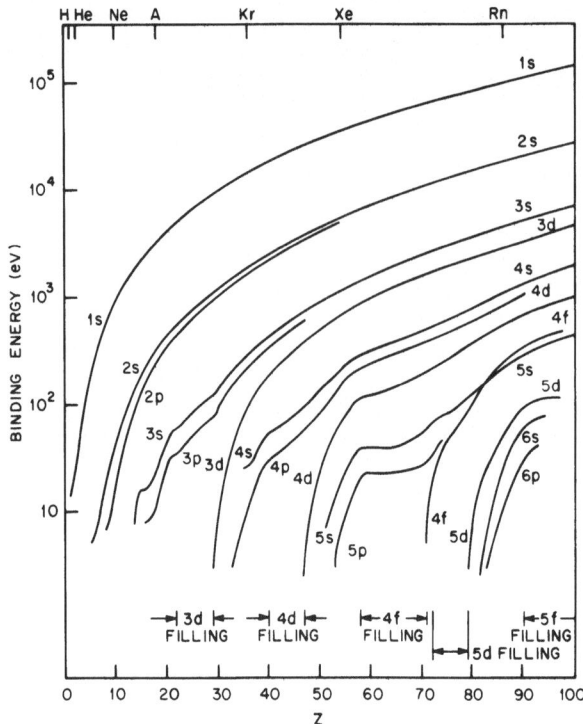

Figure 1. Core electron binding energies of the elements. Note
that the majority of levels is accessible with AℓKα radiation.
Plotted from Siegbahn et al., Ref. 2.

A plot of the 1.5 keV values, appropriate for work with AℓKα
radiation are shown in Fig. 2. It is clear that except for some
low-Z elements, shells with large, 10^5 barn, cross sections are
available for the detection of most elements. It is equally
apparent that the weakly bound valence electrons have very much
smaller cross sections. (One can understand this intuitively in
terms of the well-known fact that a free electron cannot undergo
photoelectric effect because momentum conservation cannot be satisfied.) Because of the small cross section of valence band
electrons, ESCA is often considered to be inherently a core-level
spectroscopy. We shall see, however, that the cross section is not
a serious limitation, and that quite valuable information can be
obtained about valence and conduction bands. A number of attempts
have also been made to determine empirical cross section ratios
which could be useful for chemical analysis [5]. The agreement
among them is, so far, not encouraging, and factor of two discrepancies with theory are not uncommon.

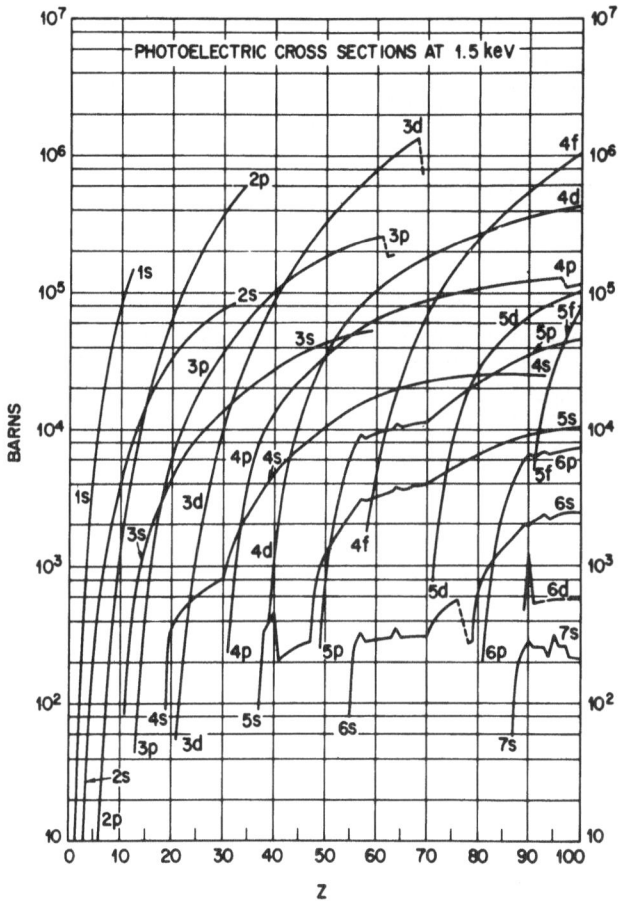

Figure 2. Cross sections for photoelectric effect at 1.5 keV.
Plotted from Scofield, Ref. 4.

We now turn to examine the photoemission process in greater
detail.   In a solid the effect of photoelectric absorption is to
raise a bound electron into the empty density of state.   A qualitative but significant difference between UPS and XPS is that the
empty density of states at energies more than 100 eV above the
vacuum level has little, if any, resolved structure or characteristic orbital character.   The electron wave functions resemble
those of plane waves.   The transition probability in XPS will
therefore depend largely on the occupied density of states.   In
fact one can visualize the photoexcitation process as generating
a replica of the occupied density of state in kinetic energy space,
albeit with an intensity factor which depends on the orbital
angular momentum of the occupied state.

Energy conservation leads to the expression

$$h\nu = E_{kin} + e\Phi + E_b \qquad\qquad (1)$$

where Φ is the work function and E_b the "binding energy" of the electron, or more accurately the energy of the hole state produced by the removal of the electron (more on this below).

The kinetic energy is generally measured only after the electron has emerged from the solid into vacuum. The mean-free path of electrons with energy in 100 to 1000 eV range is known to be in the 5 to 25 Å range in most solids [6], Fig. 3. Consequently, information about the kinetic energy of Eq. (1) can only be obtained for electrons produced close to the surface. The absorption length of AℓKα X-rays is, however, orders of magnitude greater, typically microns. As a result the kinetic energy distribution in the vacuum space outside the sample is dominated by degraded electrons coming from deep within the sample. The reason that the undegraded part can be observed in XPS is that it is

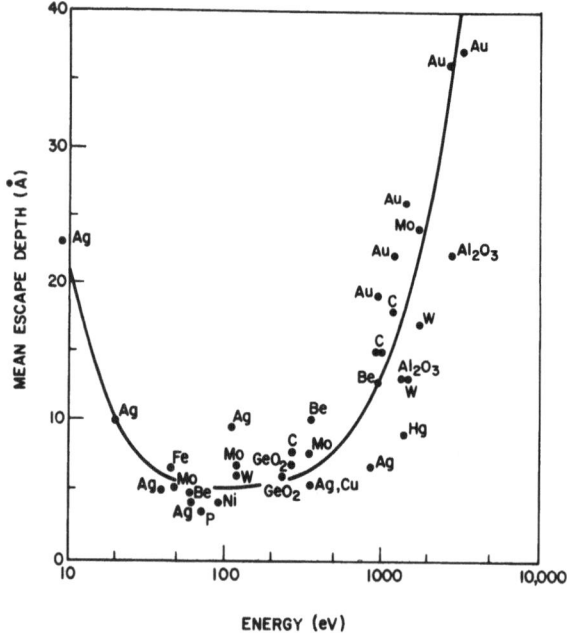

Figure 3. Mean free path of hot electrons in solids. From Rivière, Ref. 6.

narrow while the degraded part is spread out over an energy typically representing the excitation of one or more plasmons [7]. In favorable cases the individual and multiple plasmon losses are resolved in the XPS spectrums, Fig. 4, but more generally a broad featureless loss tail is associated with each line. This extrinsic energy loss tail constitutes one of the major obstacles to a detailed analysis of XPS spectra.

The small mean-free-path makes XPS a surface-sensitive technique, but not to the exclusion of the bulk response. If we consider a 15 Å escape depth from an fcc solid with a lattice constant of 4 Å we find that 1/8 of the signal comes from surface atoms. Two methods are available to enhance the surface contribution relative to that of the bulk. Data can be taken with the photoelectron emerging at a glancing angle, or the photon energy can be changes so that the kinetic energy is reduced to the 50-100 eV range where the escape depth is comparable to the lattice constant. These techniques have been used to accentuate the contribution from chemically distinct surface (oxide) layers, but the core levels of surface atoms of a clean surface don't appear to be distinguishable from those of the bulk. On the other hand, surface states have been repeatedly identified in UPS valence band spectroscopy.

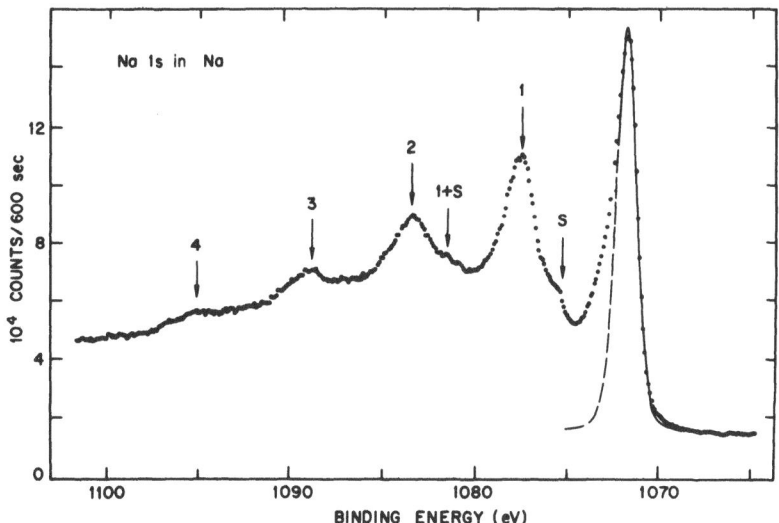

Figure 4. Plasmon loss tail accompanying the Na 1s photoemission line in metallic Na. From Citrin, Phys. Rev. B8, 5545 (1973).

Quite naturally major interest has always focused on the
binding energies of core and valence electrons in XPS spectra.
The early finding of shifts related to valence and bonding gave a
clear indication of the power of ESCA as a technique for studies
beyond elemental analysis. The interpretation of binding energies
starts with the realization that experimental values differ by
large amounts from the Hartree-Fock eigen-energy of the corresponding electron in the initial state atom. The difference
represents the energy liberated by the atom as it relaxes to
accommodate the core hole [8]. The binding energy calculated
with the orbital frozen is called the Koopmans energy. Although
it does not correspond to a realizable state it does have a
physical significance and is, at least in principle, measurable.
To see how this comes about we describe the photoemission process
in two steps: (1) excitation of the Koopmans state, and (2)
projection into final eigen-state of the hole-state atom. The
lowest of these is separated from the Koopmans state by the
relaxation energy and is normally identified with the binding
energy, but there are additional higher-lying excited states which
are also populated. A simple derivation shows that the centroid
of the final state population must coincide with the Koopmans
state. The higher-lying states are usually called the satellites
of the XPS spectrum which, while understood in principle, have
been convincingly identified in only a few cases. The problem is
that satellites represent excited states of a hole-state atom
which are not accessible by normal spectroscopic means.

This does not exhaust the spectrum of complexity, however,
because in a solid there are additional modes of relaxation and
excitation. For example, the electronic relaxation is not confined to photo-ionized atom itself. Neighboring atoms will contribute because they are polarized by the excess unit charge.
The vibrational modes of the lattice can be excited by the
resultant forces. In a metal the conduction electrons will react
to screen the photo-hole. These phenomena all manifest themselves
in E_b. It thus turns out that a full understanding of XPS requires
consideration of all the degrees of freedom of the solid.

II. VALENCE BANDS

A. Metals and Alloys

The valence band XPS data of metals are in many respects the
easiest to interpret. The reason is simple: both relaxation and
screening are relatively unimportant for a delocalized band
electron. We therefore expect the data to reflect the structure
of the occupied density of states. The favorite examples are

the noble metals where excellent agreement between band structure
calculations and XPS data has been repeatedly demonstrated, at
least with respect to the location of prominant density of states
peaks [9]. Intensities remain more problematical because of s-d
mixing which increases toward the bottom of the d-band, and also
because the lifetime of the hole in the conduction band will be
a function of energy, producing an increasingly significant
broadening toward the bottom of the band. The results obtained
by XPS are quite similar to those obtained by UPS at higher photon
energies. One interesting recent result is the finding that there
are clearcut changes in the intensities of the band components
with crystallographic direction. These have been related to the
composition of the wave functions in the particular part of the
band [10].

Good agreement between band structure calculations and XPS
data has now been reported for all types of metals including
transition metals and the alkalis. The major anomaly is Ni whose
XPS d-band width appears to be comparable to that of Cu, i.e.,
much narrower than the calculated width. Recent angle-resolved
measurements suggest that this problem arises simply from the
dispersion of a weak peak at the bottom of the d-band, which disappeared into the energy-loss background in angle-integrated
measurements [11]. This obviates more fascinating explanations,
like those predicated on band narrowing at the surface and a small
escape depth, or on some many-body property of a strongly correlated metal.

The study of alloys, especially those involving transition
metals, makes it possible to obtain answers to more detailed
questions. We can test the applicability of the rigid band model
and obtain information about charge transfer. An instructive
example is provided by a comparison of the XPS band structure of
Cu, Zu and β-brass [12], the ordered equiatomic alloy, see Fig. 5.
The Cu and Zn d-bands remain entirely distinct in the alloy, and
become slightly narrower. Since the width is dominantly due to
band structure, i.e., the overlap of the d-bands in the periodic
lattice, it is not surprising that "dilution" with a different
atom makes the bands narrower. The second observation is that the
bands shift by small amounts, especially that of Cu which shifts
toward greater binding energy. (The Fermi level is the common
reference level.) At first sight one is tempted to interpret this
as due to electron transfer from Cu to Zn, although it goes counter
to expectations based on the electronegativities. The problem is
that the Fermi level is not the proper energy reference, although
it is the obvious and convenient one in XPS. Binding energies are
more meaningfully measured from the vacuum level. Using work
functions from the literature we find that those of Zn and β-brass
are similar while that of Cu is larger by an amount very close to

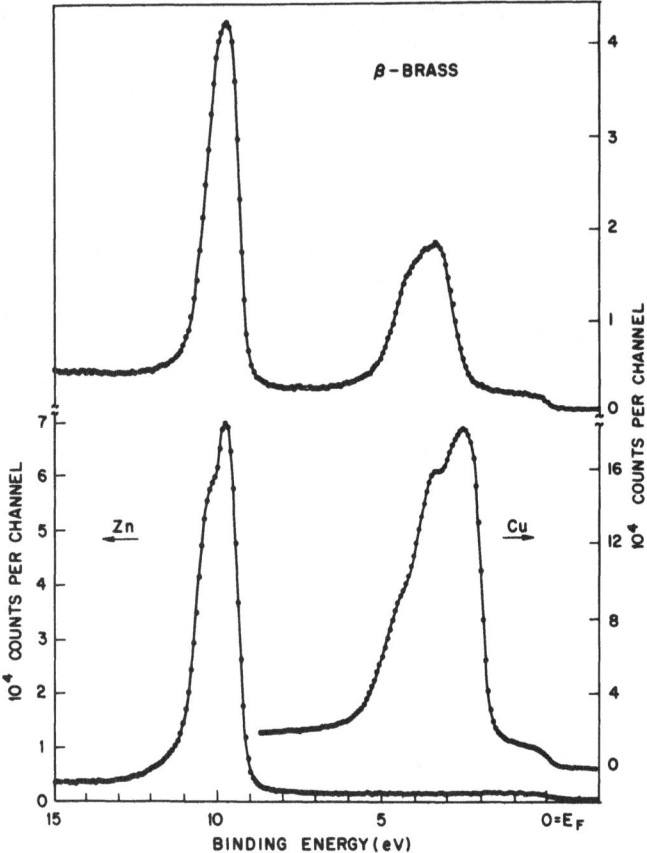

Figure 5. Comparison of the band structures of copper, zinc and
β-brass. Note the shift of the Cu 3d band in the alloy. From
Wertheim, Campagna and Hüfner, Ref. **12**.

the observed shift of the Cu d-band. As a result we find that
there is at most a small charge transfer to the Cu and very little
change in binding energy. It is unfortunately more difficult to
measure charge transfer from core level shifts because composition
dependent relaxation and screening must be taken into account.

   The Ag-Pd alloy system forms an interesting subject because
it exists as an fcc alloy over the entire range of composition.
Moreover, the d-bands of the pure metals overlap in energy from
4 to 6 eV. The properties of this alloy system have often been
discussed from the rigid band point of view in which the addition
of Ag to Pd fills the fractional d-band hole of the Pd. A decrease
in the DOS at the Fermi energy near mid-composition is cited as
support for this model. The XPS data [13] Fig. 6, lead to

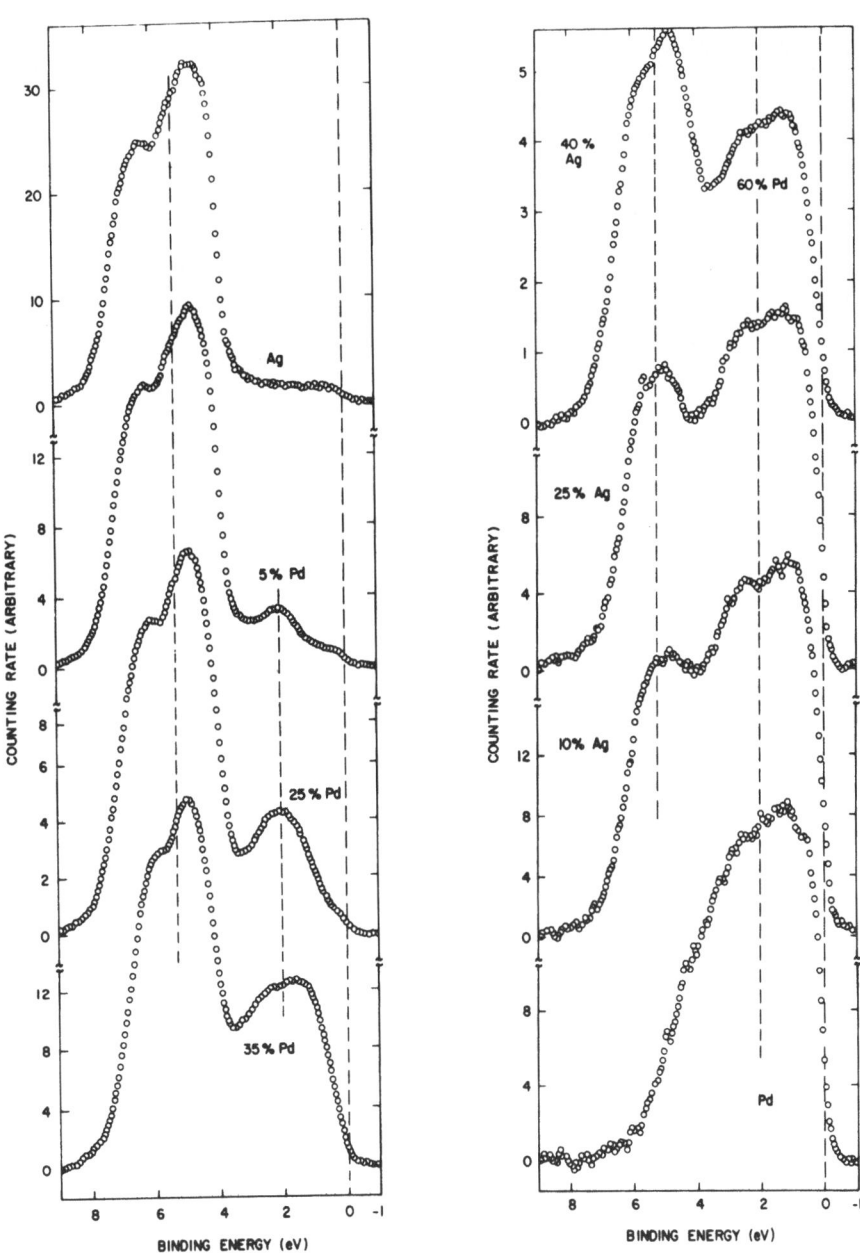

Figure 6. Band structure of the silver palladium alloys. Note the virtual bound state of dilute palladium 2 eV below the Fermi energy. From Hüfner, Wertheim and Wernick, Ref. 13.

a different interpretation. The data show that the d-bands of the
two components remain distinct, i.e., there is no common band as
assumed in the rigid band model. Furthermore, the band of the
minor constituent narrows but remains at the centroid of the full
band. In the dilute limit, the Pd d-band has been reduced to a
narrow resonance, 2 eV below E_F, superimposed on the Ag 5s band,
i.e., it is the virtual bound state described by Friedel and
Anderson. The behavior over the entire composition range is in
accord with the coherent potential approximation.

Similar results have also been obtained for the Cu-Ni [14]
and other [15] alloy systems.

B. Compounds with Metallic Conductivity

Metallic behavior is of course not confined to the elemental
metals and their alloys; there are many stoichiometric compounds
in which the Fermi level passes through a band resulting in
metallic conductivity. Well-known examples are found among the
oxides, sulfides and selenides of the transition metals. These
materials are characterized by an anion p-derived valence band and
a cation d-conduction band. There is, however, substantial p-d
hybridization so that anion p character is found at the Fermi level
and d character in the valence band.

Rhenium trioxide, ReO_3 provides an instructive example. It
has a simple cubic structure, which can be visualized as that of
the ABO_3 perovskite with the A site empty. It had been thoroughly
studied by the optical and de Haas-Van Alphen techniques, and the
band structure had been calculated and adjusted to fit the experimental data. At first sight XPS data [16] on a vacuum-cleaved
surface of a single crystal appeared to contradict the band
structure calculation. The p-d gap appeared to be much larger
than in the calculations. It turns out, however, that there is
no justification for a direct comparison of the total density of
states and the XPS data because the photoelectric cross sections
for oxygen 2p and rhenium 5d electron differ by more than an order
of magnitude, the 5d cross section being much larger. Comparison
with 5d density of states alone yields much better agreement,
Fig. 7, because the strong DOS peak at 3 eV has dominantly O 2p
character and is barely seen in the XPS data. Agreement between
theory and experiment is very satisfactory with regard to such
parameters as the width and shape of the 5d conduction band and
the width of the 2p valence band. There is, however, a 0.45 eV
discrepancy in the location of the valence band relative to the

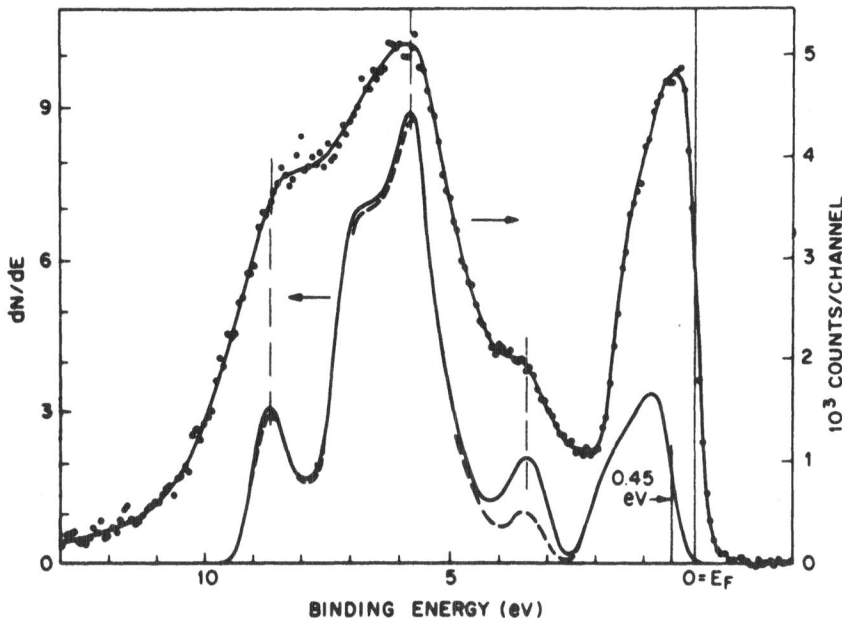

Figure 7. Comparison of the rhenium 5d component of the calculated density of states of ReO_3 with the XPS spectrum. The oxygen 2p component has a strong peak 3 eV below the Fermi energy. From Wertheim et al., Ref. 16.

Fermi level. This raises an interesting issue which is more fully discussed elsewhere. It is important to remember that the "binding energy" represents a final state property, i.e., the energy of the screened and relaxed hole-states solid. It is reasonably clear that the screening and relaxation around a hole dominantly localized on the oxygen atoms will be different from that of a delocalized metal d-band hole. Unfortunately, no estimates of the magnitude of this effect exist for this case even though it is fundamental to the interpretation of data from the various electron spectroscopies.

A second example of a metallic oxide is provided by VO_2, $3d^1$, which has the interesting property of undergoing a metal insulator transition at 65°C. The high temperature metallic phase has the rutile structure. The low temperature phase is greatly distorted. The XPS data [17] clearly show that the states at E_F move toward greater binding energy, in effect creating a gap.

Other examples of metallic compounds are $NbSe_2$ and the various modifications of TaS_2 and $TaSe_2$ which have one electron in a d-band just like VO_2. These layer structure materials also have an instability at lower temperature, but the result is a charge density wave in the d-conduction band without loss of the metallic conductivity [18,19].

C.   Insulators

The valence bands of ionic insulators, e.g., the alkali halides, yield simple looking XPS spectra. To first approximation, they can be interpreted in terms of the free-ion spectra with all energies shifted by the Madelung energy. The finer details depend on relaxation and polarization corrections which reflect the properties of the lattice [20,21]. The resolution of XPS is generally not sufficient to observe the details of the anion p-electron derived valence band, although the band width is generally measurable.

It has long been recognized that insulators present a special problem in XPS because the conductivity may be so low that the charge lost from the surface by photoemission is not replaced [22]. The insulator thus acquires a positive surface potential which reduces the kinetic energy of the photoelectrons. This is usually called "the charging problem". A number of solutions have been proposed, ranging from grinding up the insulator with graphite, to independently measuring the surface potential. The former is not compatible with work on well-characterized surfaces and the latter does not necessarily solve the problem since the potential may be a strong function of the distance into the sample. Two approaches have been found to be useful. In one the surface is flooded with low-energy electrons to replace the photoemitted electrons. This is often sufficient to obtain data on insulators, but there is generally considerable ambiguity with regard to the binding energy. The second method relies on an evaporated metallic overlayer, typically gold, which is taken to define the Fermi level for the sample below. While this does seem to give <u>reproducible</u> results it is based on dubious assumptions. The very fact that the Fermi level in a insulator may be anywhere in the gap and is defined only in thermal equilibrium should suffice to discourage its use as a reference level under X-ray excitation. The best approach is to reference all data to the edge of the valence band, which can be related to the vacuum level through optical measurements.

For further details on valence band spectra of insulators the the reader is referred to the literature.

III. PHOTOEMISSION FROM OPEN SHELLS

A. The Rare Earths

The 4f electrons of the rare earths (RE) provide an interesting contrast to the behavior of the band states of the metals
(including the 5d, 6s conduction bands of the RE's). Although the
binding energies are only a few eV, the 4f's remain localized
because their radial extent is very limited, typically 0.5 Å.
Photoemission from the valence band region of one of the heavier
rare earths yields a complex spectrum which cannot be interpreted
as a replica of a 4f band since the 4f states are known to be
very narrow. How then do we understand such spectra?

The approach is the following [23]: We recognize that the
$4f^n$ shell of the unperturbed ion in the initial state is in its
ground state. Photoemission of an f electron produces an f shell
with n-1 electrons, not necessarily in its ground state. The
probability with which the ground and excited states of $4f^{n-1}$ are
occupied may be calculated by projecting the initial configuration
onto the combination of the outgoing photoelectron plus the manifold of final states. The appropriate final state spectrum is
very similar to that of the element with next lower atomic number,
and can be taken from optical studies of the trivalent rare earth
ions [24] because the crystalline environment has only a small
influence on the 4f shell. In other words, the final state
spectrum of the trivalent element Z, $4f^n$, has the same components
as the optical spectrum of element Z-1, $4f^{n-1}$. The only correction is a ~10% increase in all energies due to the unit difference in Z.

The 4f spectrum of Gd^{3+} then should consist of a single broad
line. The simplicity of the spectrum traces back to the fact that
the seven 4f electrons in the half-filled shell are coupled
according to Hund's rule to produce a $^8S_{7/2}$ configuration. Upon
photoionization it becomes a ^7F state. The j states of ^7F cover
an interval of about 0.6 eV sufficient to account for the observed
broadening. In this case, the simplest encountered in the rare
earths, photoemission does not lead to resolved structure with the
resolution achievable in insulators.

Additional complexity should be observed beyond the half-
filled shell. In Tb^{3+} ($4f^8$ with $^7F_6$ ground state), the main peak
develops discernible structure and a new narrow peak appears with
smaller binding energy, see Fig. 8. From the final state analysis
it is clear that the structure must represent a large number of
final state configurations which can be populated by the removal

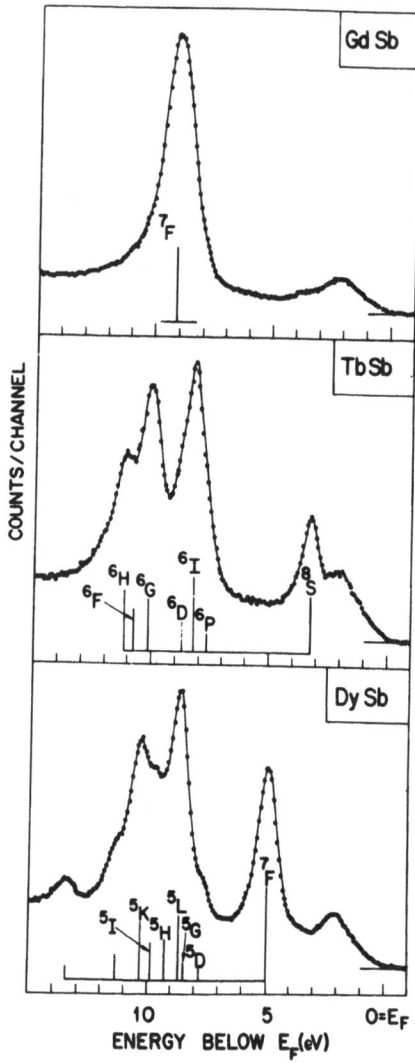

Figure 8. The 4f and conduction band region of GdSb, TbSb and
DySb. The final state multiplets of the 4f shell are indicated.
From Campagna et al. Refs. 25 and 26.

of one electron. In Tb^{3+}, removal of the spin-down electron leads
to the unique ^8S state, the ground state of the $4f^7$ configuration
in Hund's rule coupling. This is the lowest energy state of $4f^7$
and therefore corresponds to the sharp weak line with smallest
binding energy. In the rest of the spectrum, we can identify the
sextet states populated by removal of a spin-up electron. They
are ^6P, ^6I, ^6D, ^6G, ^6H, ^6F and have energies which can be estimated
from the optical spectrum of the excited states of Gd^{3+}. The
separation between ^8S and the sextet states in the optical spectrum
of Gd^{3+} is in reasonable agreement with the separation between the
two features in the XPS data for Tb^{3+}. The final state analysis
thus readily accounts for the observed structure in every detail
in terms of configuration which do not exist in the initial state.
Since the observed structure is associated with states of the $4f^{n-1}$
configuration, it clearly cannot be ascribed to the $4f^n$ initial
state. The detailed analysis, using the method of fractional
parentage, was carried out independently by Cox and by Zabolotskii,
et al. The calculations of Cox have been published in detail and
have general applicability to the analysis of rare earth spectra.

A significant test of the theory has been provided by recent
high resolution data for the RE metals, chalcogenides, and antimonides [25]. The interesting, but not unexpected finding is that
the $4f$ spectra of the metals and the semiconducting antimonides
(and even those of the insulating halides) are identical except for
an overall shift in absolute binding energy. This is a result of
the well-known fact that the $4f$ shell is in many respects core-like,
being well-shielded from the crystal field by the $5s$ and $5p$
electrons.

A surprising result is that these characteristics are not
lost in the final state even when the $4f$ electrons are initially
in an intermediate valence state [26]. This type of behavior has
been variously described by representing the $4f$ electrons by a
virtual bound state at the Fermi energy, or through the use of the
interconfiguration fluctuation concept. We illustrate work in
this area by recent results for samarium hexaboride, which has long
been recognized as an intermediate valence compound [27]. The XPS
data, Fig. 9 show the structure associated with the final states
corresponding to the 2+ and 3+ initial states. Agreement with
theory is very satisfactory, and the relative areas of the two
parts of the spectrum accurately reflect the intermediate valence.
The $4d$ multiplet spectrum of SmB_6 can also be represented as a
linear combination of 2+ and 3+ spectra. TmSe shows similar
behavior, although the valence ratio as well as the lattice
constant and other physical properties are very sensitive to
stoichiometry.

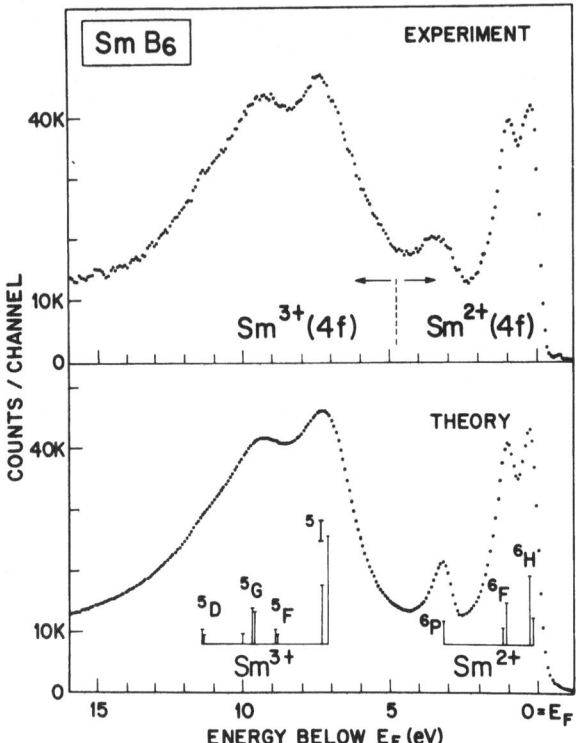

Figure 9.   Final state 4f structure in the intermediate valence
compound SmB_6. The coulomb correlation energy can be estimated
from the separation of the ground states of the two multiplet
groups.   From Chazalviel et al. Ref. 27.

B.   Transition Metal Compounds

The great universality that typifies the rare earth elements
is lost in the d-group transition metal compounds largely because
the crystal field splitting may be of the same order of magnitude
as the exchange coupling. Moreover, the d-electrons are hybridized
with the ligand wave functions and contribute to bonding and anti-
bonding orbitals. In the metals they overlap sufficiently to form
bands of considerable width. The question of final state structure is therefore much more problematical and the behavior subject
to wide variation, even for a given ion.

The conditions under which a transition metal ion could
exhibit final state structure should be encountered in ionic

compounds with weak crystal fields. Although weak, the crystal field cannot be ignored because both the initial and final states are crystal field states. The initial state wavefunction is needed only to estimate the matrix element for transition to final excited states of d^{n-1}. The observed structure is defined by the final states. Interpretation of data depends on the ability to make a reasonable estimate of the magnitude of the crystal field in the final state. The discussion is here greatly facilitated by reference to the Tanabe–Sugano (TS) diagrams [28], which show the electronic states as a function of the strength of the cubic field. (Note that the ground state is always shown as the abscissa, accounting for the discontinuity in slope at the high-spin to low-spin crossover.) The experimental data here are unfortunately quite limited, largely because it is difficult, if not impossible, to obtain high resolution spectra on insulating compounds due to phonon broadening. Nevertheless, there are a few suggestive examples which illustrate the concepts [29].

The trivalent Cr ion has three d electrons in $^4A_2$ configuration, i.e., coupled spin parallel in a weak octahedral crystal field. Upon photoionization the final state is $^3T_1$, a unique state. Cr_2O_3 does indeed give a single narrow 3d line. The comparison with $K_2Cr_2O_7$ in which the hexavalent chromium has formally lost all of its d electrons serves to identify the largely oxygen 2p-derived valence band. The case of divalent Fe in FeF_2 is in many ways analogous to that of trivalent Tb, i.e., there is one electron in addition to a Hund's rule coupled half-filled shell. The initial state is $^5T_2$, the final states are $^6A_1$ and a set of quartet states lying about 3 eV higher according to the TS diagram, see Fig. 10. Divalent iron in a strong crystal field, for example in $K_4Fe(CN)_6$, is in a $^1A_1$ state, since the t_{2g} states are completely filled. Photoionization leads to a unique $^2T_2$ state in good accord with the data in Fig. 11. In the case of divalent Ni, $3d^8$, with $^3A_2$ configuration, it appears likely that the final state ion will be near the high-spin to low-spin crossover. Data for $NiCl_2$ can be interpreted on this basis. These results strongly suggest that such data give final state information and should not be interpreted in terms of initial state hybridized orbitals.

Between the ionic compounds with localized d electrons and the metals with delocalized electrons there are compounds with intermediate properties, with narrow bands and modest conductivities. Well-known examples may be found among the oxides and sulphides of the transition metals. The interpretation of XPS data on these materials has so far not been without ambiguity. An example should serve to illustrate the current status. Nickel oxide, NiO, is a $3d^8$ insulator. The 3d portion of the XPS spectrum has been shown to be compatible with the final state interpretation.

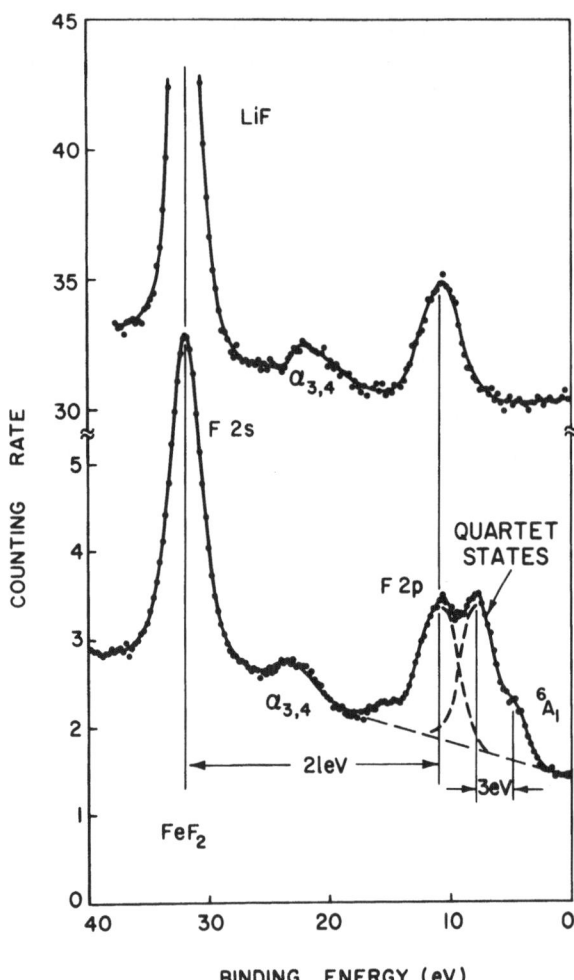

Figure 10. Final state 3d structure of high spin divalent iron in FeF_2. The LiF comparison spectrum identifies the fluorine components. From Wertheim, Guggenheim and Hüfner, Ref. 29.

Good agreement has also been reported between the entire valence band in an Xα molecular cluster calculation. APW band structure calculation are similar to the cluster calculation. However, data taken over a range of energy with synchrotron radiation suggest that part of the XPS spectrum should be classified as a multi-electron satellite rather than band structure. This indicates that neither the Xα molecular cluster nor the band structure

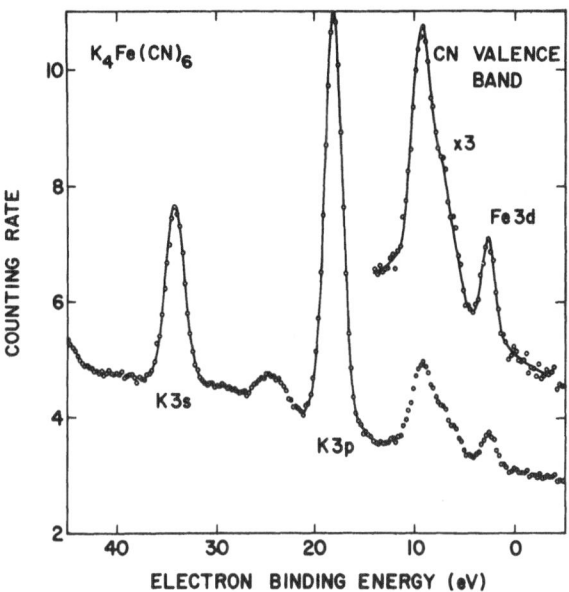

Figure 11.   Final state structure of low spin divalent iron in
$K_4Fe(CN)_6$. There is only a single component.

calculation should be directly compared with the XPS data, and
strongly favors the $3d^{n-1}$ final state interpretation for NiO and
other transition metal oxides [30].

IV.   CORE LEVEL SPECTRA

    It is a common finding in X-ray photoelectron spectroscopy
that core-level spectra are not simple lifetime broadened Lorentzian
lines. (I exclude from this discussion such essentially extrane-
ous phenomena as those arising from excitation with nonmonochromatic
radiation, energy loss, differential charging, or from the instrumental resolution function.) Lines in metals which during the
early days of this subject had the reputation of being simple
because they appeared to be narrow turn out to have a most complex
shape due to the collective response of the conduction electrons
to the creation of a core hole [31]. Lines in insulators are
broadened by the Franck-Condon excitation of the lattice [32].
Beyond the observation that photoemission lines do not have the
simple shape that we might have expected, comes the fact that they
may be split by multiplet effects and be accompanied by additional

resolved structure which has no counterpart in the initial state.
In common parlance such structure is referred to as satellites [33].

A. Shake-up Satellites

Satellites occur at greater binding energy than the fundamental line and therefore must involve processes in which additional energy is used up. It would seem that these must of necessity
be electronic processes involving filled and empty states near the
Fermi energy. However, it is fundamentally wrong to assume that
such structure can be described in terms of electronic transition
of the atom or solid in the initial state. The photoelectric
transition is between a definite initial state of the solid and
final states with one electron removed [34].

Properties of the final state are deduced from the conservation of energy, Eq. (1), in which E_b represents the energy of the
final state. It is customary to refer to this energy as the
"binding energy" of the electron which has been photo-excited.
From an operational point of view this is entirely appropriate,
even though the energy is not equal to the eigen energy of the
electron in the original, unperturbed atom. The difference between
these two energies is not negligible, and contains contributions
from the (electronic) relaxation of the atom and lattice around
the newly created hole. This relaxation does not, however, always
go to the ground state of the ionized system. When sufficient
energy is available to cause an electronic transition, the resulting excited final state appears as a satellite at greater binding
energy in the spectrum. These are called "shake-up" satellites.
For sufficiently deep levels the spectrum of final states will not
depend on which core electron is removed. This is true provided
multiplet coupling and the overlap between valence electrons and
the photoexcited core orbital remain small. The extent to which
the excited states are populated then depends dominantly on the
magnitude of the relaxation energy. The latter decreases the
closer the photohole is to the valence shell. This suggests that
valence shell electrons should have very small shake-up satellite
intensities, but other mechanisms may become important in the
outer shells. The interpretation of shake-up satellite structure
remains a troublesome problem largely because the initial state
electronic structure is a poor guide to that of a core-ionized
atom in the final state. A single-ion atomic description is
likely to be inadequate. Electron transfer from the ligands to
the ionized atom may make important contributions to the final
state. In general a full scale Hartree-Fock calculation for a
representative cluster of atoms of the core-ionized solid may be
required to gain an understanding of the final state core electron
satellite spectrum, especially for transition metal compounds.

B. Configuration Interaction

There are cases of satellite structure in which a single-ion, final state description is useful. These provide clear cut examples of structure which has no counterpart in the initial state of the ion, and is thus clearly final state structure. In these cases the structure is associated with only a single valence shell core level. This is an immediate proof that they are not due to the shake-up process, as usually conceived.

Such structure was first identified in XPS studies of alkali metal ions [35] in which an unusually strong satellite, which had no counterpart on other lines of those atoms, was found accompanying the K 3s and Rb 4s lines, see Fig. 12. This was quite unexpected because s-shells of closed shell ions were expected to give single, broad lines. (They are, however, split by multiplet coupling to the incomplete shells of transition metal, lanthanide and actinide ions, see below.)

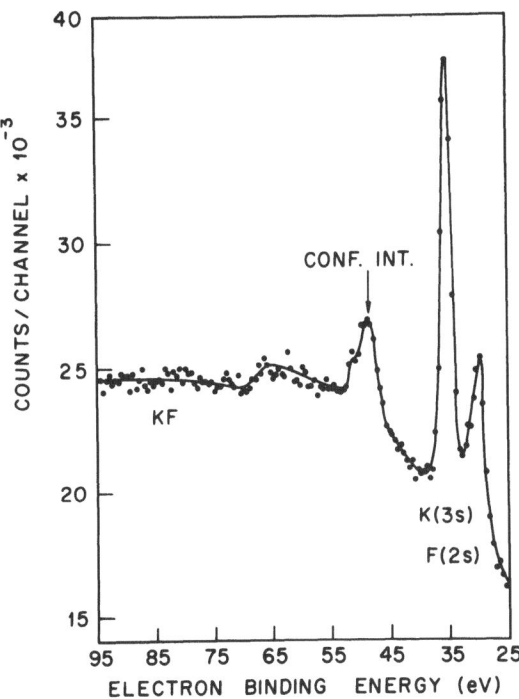

Figure 12. Configuration interaction satellite of the potassium 3s shell in KF. The satellite corresponds to the configuration $3s^2 3p^4 3d$, ^2S. Its separation from $3s^1 3p^6$ is in good accord with atomic spectra.

These satellites are due to a two-hole final state in which
one electron is promoted into a normally empty d orbital. The
unique feature here is that the core hole corresponding to the
fundamental line is filled in the satellite configuration. To be
more specific, consider the case of K^+. The 3s line corresponds
to the final state $3s^13p^6$ with symmetry ^2S; the satellite to
$3s^23p^43d$, also coupled to give ^2S symmetry. The identity of the
two symmetries is an essential condition for the occurrence of
such satellites, because the two-hole state cannot be directly
excited by the photoelectric interaction. The process may be
characterized as due to final state configuration interaction.
The energy of the two-hole state must be close to that of the one
hole parent if the satellite is to have appreciable intensity.
These conditions facilitate a quick scan of the table of binding
energies to identify those levels which should exhibit this type
of satellite. They can occur for 3s electrons from argon to the
3d group transition metals, the 4s electrons from krypton to the
4d group transition metals, for 4p electrons in the vicinity of
the rare earths, etc.

Theoretical calculations [36] account quite well for the
large intensities of these satellites which are often comparable
to that of the main line. They indicate further that intensities
from final states involving excitation from one shell into the
next higher one will be very small, i.e., final states like
$3s^23p^44s$, ^2S, although satisfying the symmetry and energy requirements, will be very weak. This, incidentally, provides a simple
explanation of the fact that a thorough search for the corresponding satellite of the Na 2s electrons gave a negative result. We
can now ascribe it simply to the fact that there are no 2d electronic states.

It should be emphasized that these satellites are entirely
compatible with a two-step description of the photoemission
process, i.e., excitation of the Koopmans state and projection
into final eigenstates. The symmetry of the Koopmans state determines the symmetry of the final states. The condition which has
been used in the discussion of core level satellites, namely,
monopole selection rules in the outer shell excitation, is too
restrictive because it ignores the multiplet coupling between the
photo-produced core hole and the outer-shell hole and electron.
The monopole selection rule may be appropriate when this coupling
is extremely weak, but it must fail when the coupling is strong,
as is the case of configuration interation. The state corresponding to the dominant configuration interaction satellite cannot be
generated under the monopole restrictions. In order to subsume
configuration interaction satellites under the point of view
characterized by the phrase "shake-up", it is necessary to include
the possibility that the original hole is itself shaken up.

C. Multiplet Splitting

The extra complexity which appears when core levels of ions with incomplete outer shells are examined by XPS becomes apparent from the following simple consideration. Imagine an ion with a net spin, S (and L=0) in an outer d or f shell, and remove one core s-electron. Even in the absence of satellites there will be two distinct final states depending on whether the core spin is parallel or antiparallel to the spin of the outer shell. The energy difference is given by an exchange integral. This phenomenon is, of course, not limited to core s-shells but applies to all core levels of such ions. The presence of both spin and orbital angular momentum in both the core and outer shell can clearly result in very complex spectra.

The description in terms of exchange is misleading in that it suggests that a core s-electron spectrum will be split into two lines of equal intensity. A proper description requires that we consider the multiplicity of the final states of the multiplet coupled shells [37]. Some examples should serve to clarify these concepts. (Configuration interaction is left out of these examples although it modifies actual spectra significantly, see below.) (1) Core hole in s-shell; spin-only outer shell. The spin of the core shell is coupled to that of outer shell (denoted by S), yielding two final states with spin $S \pm 1/2$. Mn^{2+}, $3d^5$, ^6S, according to this theory, should have a 3s-spectrum corresponding to ^7S and ^5S final states with intensity ratio 7/5. Gd^{3+}, $4f^7$, ^8S, has ^9S and ^7S final states. (2) Core hole in p-shell; spin-only outer shell. All final states will be P-states. The net spin is again $S \pm 1/2$. The 3p-spectrum of Mn^{2+} will consist of ^7P and ^5P final states. Neglecting the spin-orbit interaction in the p-shell, the ^7P state can be formed only by coupling the core spin to the ^6S state of the $3d^5$ shell. On the other hand, ^5P-states can be obtained by coupling s = 1/2 and ℓ = 1 to the ^6S, ^4P, and ^4D-states of $3d^5$. As a result there are four energetically distinct final states ^7P, $^5P_{(1)}$, $^5P_{(2)}$ and $^5P_{(3)}$. (3) Core hole in s-shell; outer shell with S and L. All final states will have the L of the outer shell. Tb^{3+}, $4f^8$, $^7F_6$ will have ^6F and ^8F final states. (4) When there is orbital angular momentum in both shells, the number of final states tends to become large [38].

The important concept that emerges from case (2) is that the outer incomplete shell may change its net S and L during the core electron photoemission process, i.e., it may be shaken up without obeying the monopole selection rule. The coupling responsible for these transitions is electrostatic and falls into the general concept of electron-electron correlations. This is another manifestation of the multi-electron aspects of photoemission already encountered in closed shell systems in the discussion of configuration interaction.

A critical comparison of the 3s-multiplet structure in Fig. 13 with the theoretical description given above clearly shows that the predicted 7/5 intensity ratio for Mn^{2+} is not realized in the experiment. The measured splitting also turns out to be smaller than the Hartree-Fock value by about a factor of two. We

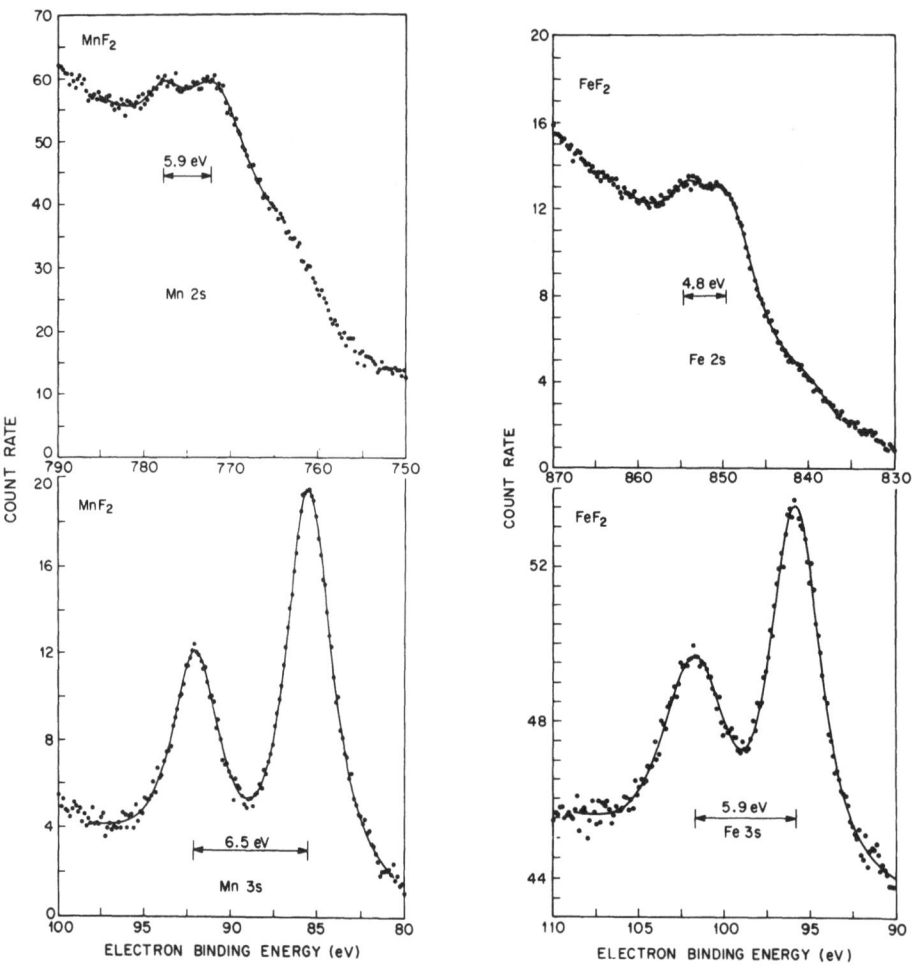

Figure 13. Multiplet structure of core 2s and 3s electrons of divalent ion in MnF_2 and FeF_2. Note that the orbital angular momentum in FeF_2 does not produce additional structure. From Wertheim, Hüfner and Guggenheim, Ref. 37, and Hüfner and Wertheim, Ref. 40.

will return to this problem after a look at multiplet effects in the rare earths.

Data for rare earth compounds are more extensive, and show a similar problem. Figure 14 exhibits the 4s and 5s-multiplet splitting in some rare earth metals and trifluorides. (For the rare earths covalency has negligible effect on the multiplet splitting.) In fact, quite similar splittings have been obtained in the metals, the LnSb, the Ln_2O_3, and the insulating LnF_3. This is no surprise since the 4f-electrons are well shielded by the 5s- and 5p-electrons. Comparison with Hartree-Fock values shows excellent agreement between theory and experiment in the case of the 5s-splittings, but a constant factor of two discrepancy between calculated and measured 4s-splittings. The fact that the discrepancy in 4f-ions appears in the 4s-splitting (in 3d-ions it was in the 3s-splitting) points strongly toward electron correlations as the cause because these are most important within a shell of given principal quantum number. (Recall the discussion of configuration interaction.) The detailed analysis required to evaluate the effects of electron correlation on the 4s splitting in the rare earths has not yet been carried out.

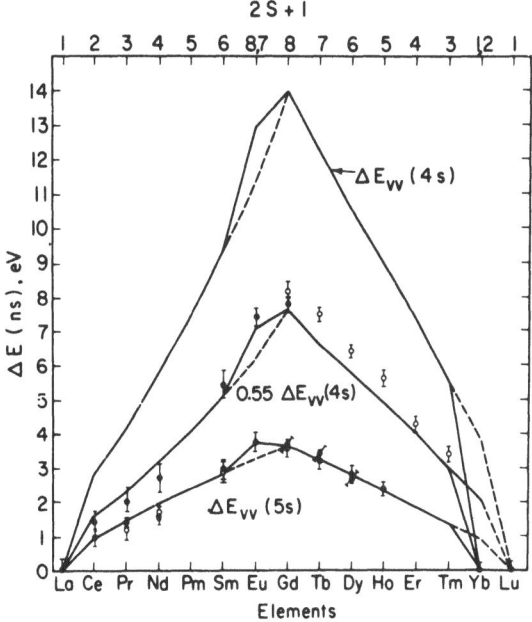

Figure 14. Multiplet splitting of the 4 and 5s electrons in the rare earth metals and trifluorides. Solid dots represent the metals, open circles the fluorides. From McFeely et al., Ref. 37.

The large discrepancy between theory and experiment was most disturbing considering the success of the Hartree-Fock theory in accounting for the related core-polarization hyperfine field. The difficulty arose from the fact that what is measured is not strictly comparable to what had been computed. The calculations were made for the initial state and did not include the effects of electron correlations, while the experiments are greatly affected by them. The effect of electron correlations are analogous to those encountered in the discussion of configuration interaction, involving the promotion of an electron into the empty d states in the d group metals and empty f states in the lanthanides. In the case of the 3s line of Mn^{2+}, satellite structure analogous to that in K^+ is formed, i.e., the $3s3p^63d^5$, ^5S, final state is accompanied by $3s^23p^43d^6$, ^5S structure. A little thought will show that there is no analogous satellite for the ^7S state. As a result the ^5S line is split while the ^7S line remains unaffected. This has the effect of reducing the intensity of the dominant ^5S line, and changing the apparent multiplet splitting. In general, the accessible final states include all ^5S and ^7S states which can be formed by distributing the 12 final state electrons within the M shell (n=3). Some of these may be energetically unfavorable and will not be observed. Excited states involving other shells, e.g., 4s, will be quite weak [39].

The same analysis applied to the 2s multiplet structure shows quite clearly why correlation effects are essentially negligible there. Final states of the type $2s^22p^43d^{n+1}$ are energetically far removed from the primitive $2s^12p^63d^n$ state and require intershell transitions. Final states involving single electron M shell excitations are forbidden by orbital angular momentum considerations. We would therefore expect to find good agreement between calculated and experimental 2s splittings for 3d group metal ions.

A more detailed quantitative comparison of both 2s- and 3s-splittings, Table I, confirms the expectations. Note first of all that the 2s-splittings are in quite good agreement with theory. Note further that when correlation effects are included in the calculation of the 3s splitting, the agreement with experiment is greatly improved. The data in Fig. 14 and the table show the surprising fact that the effect of electron correlations on the splitting results in proportionate reductions of both the 3s splitting of the 3d group elements and the 4s splitting in all of the rare earths. A theoretical explanation of this observation has not yet been given.

The connection between the core s-shell multiplet splitting and the nuclear hyperfine interaction is worthy of further exploration. The latter is due to the interaction between the nuclear magnetic moment and the unbalanced s-electron spin density at the

TABLE I

Comparison of theoretical and experimental multiplet
splittings for 2s and 3s electrons of 3d-group
transition metal ions and compounds

Splitting	2s (eV)	3s (eV)
Mn^{2+} theory[a]	6.10	14.32[b]
MnF_2	5.9	6.5
MnO	5.6	6.1
Fe^{2+} theory[a]	5.40	12.40
FeF_2	4.8	5.9
Co^{2+}	–	10.34
CoF^2	–	5.1

[a] Calculated for multiplet hole theory using optimized
orbital. Freeman, et al. Ref. 37.

[b] Reduced to 8.2 eV by correlation effects, Bagus et al.,
Ref. 37.

nucleus produced by the exchange coupling to the outer shell. The
radial wavefunction of the spin-up s-electron is attracted toward
the partially filled shell while the spin-down electron is not
affected. This process is called core polarization and can lead
to either an increase or a decrease in spin-up density at the
nucleus depending on whether the main lobe of the s-wave function
lies outside or inside the incomplete outer shell.

The same core polarization process is, of course, also
responsible for the s-electron multiplet splitting seen in XPS.
It would be rash, however, to assume that multiplet splittings and
hyperfine interactions must be proportional, because there are
large cancellations between the contributions of the various inner
shells to the contact hyperfine interaction. Moreover, the multi-
electron aspects of photoemission strongly affect the multiplet
splitting, but are not relevant to hyperfine interactions. The
two types of measurement are, therefore, not strictly comparable.

Nevertheless, empirical comparisons show that there is, in
fact, a reasonably good correlation between these two parameters

for Fe^{3+} and Mn^{2+}, which are both ^6S ions [40]. The data suggest
that covalency must change the effective number of d-electrons.
Even metallic iron falls close to the straight line through the
origin, as it should, if the significant parameter in the plot is
the d-shell moment.

D. Many-Body Effects in Metals

The discussion above has emphasized that in insulators the
interpretation of photoemission spectra must take cognizance of
the fact that one or more outer electrons can make transitions
during the photoemission process. Although there is a strong
tendency to classify the resulting final states as shake-up, configuration interaction, charge transfer, etc., there is an underlying unity in that these phenomena are all the result of the
electrostatic interaction of the electrons in the solid, often
termed electron-electron correlations.

The absence of shake-up satellites in metals was early interpreted to show that outer shell excitations were prevented by rapid
screening of the core hole by the conduction electrons. This
screening involves excitation of the electrons at the Fermi surface, and is thus, in a sense, itself analogous to the shake up
process. The energy which goes into these electron-hole pairs is
taken from the photoelectron and should be measurable in XPS. The
important question is the shape of e-h pair excitation spectrum.

Historically the importance of this phenomenon was first
appreciated in connection with X-ray absorption edges in metals [41]
where anomalous edge shapes had been observed many years ago. The
theory developed for the edge problem was then extended for XPS
even though no experimental evidence of the corresponding anomaly
was available [42]. In essence the theory predicts that the
conduction electron response is so strong that the primary line at
ω_0 is replaced by a one-sided singularity of the form

$$I(\omega) = 1/(\omega-\omega_0)^{1-\alpha} \qquad (2)$$

where

$$\alpha = \sum_{\ell=0}^{\infty} 2(2\ell+1)(\delta_\ell/\pi)^2 \qquad (3)$$

is called the singularity index. The δ_ℓ are the Friedel phase
shifts which satisfy the sum rule

$$Z = \sum_{\ell=0}^{\infty} 2(2\ell+1)\delta_{\ell}/\pi \qquad (4)$$

where Z is the charge to be screened which is 1 in the case of photoemission.

Initially this form was known to be valid only in the immediate vicinity of ω_0, but it has now been shown theoretically that it remains a good approximation over a range up to the bulk plasmon energy [43]. In practice this singularity is always modified by the finite lifetime of the core hole, which adds a Lorentzian width γ, yielding

$$I(\omega) \propto \frac{\cos[\pi\alpha/2+(1-\alpha)\arctan \omega/\gamma]}{(\omega^2+\gamma^2)^{(1-\alpha)/2}}.$$

This formula has been shown to give an excellent representation of the shape of core level XPS lines in a variety of metals [44], and to yield values of α which are fully compatible with both theory and other experimental evidence. (In making the comparison between theory and experiment it is important to fold an accurate representation of the instrumental resolution function into the theory.)

The results obtained by this process are illustrated in Fig. 15. Note first of all that the data (represented by the points) have not been modified or corrected in any way; no background subtraction has been done, no smoothing, and no deconvolution. The theoretical lineshape (represented by the solid line) is determined by two parameters, γ and α. In practice γ is determined by the shape of the right-hand part of the line, and by α by the shape of the left-hand part. Two other parameters, which determine the height and position, are essentially trivial since they do not influence the shape. The same is true of a constant background which is entirely determined by the level at the right-hand end of the spectrum. The instrumental resolution function is treated as a known constant.

Detailed analyses of the core level line shapes of Li, Na, Mg and Al have recently been reported [45]. In the case of Li it was found that there is an appreciable phonon broadening, 0.35 eV at 300°K, which is also responsible for the broadening of K-absorption edge. Phonon broadening of lesser magnitude has now also been identified in Na and Mg. The theoretical expectation that the singularity index is the same for all core levels of

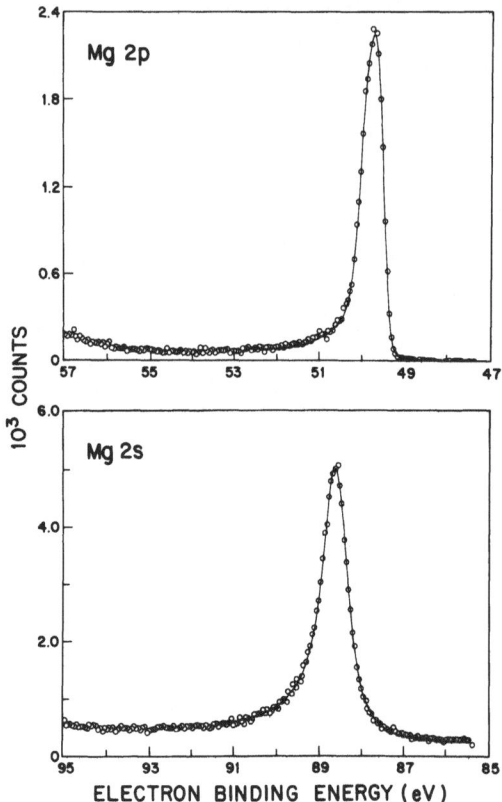

Figure 15. Many-body effects in the 2s and 2p electron spectra of metallic magnesium. From Citrin, Wertheim and Baer (to be published).

given atom was confirmed. The values of the singularity indices are in good accord with recent theoretical calculations. It thus seems safe to conclude that the many-body theory provides an accurate description of the response of the conduction electrons to the sudden creation of a core hole.

The good agreement between theoretical and experimental line shapes generally extends over an energy range of about one-half of the width of the filled conduction band. Eventually two opposing factors become important, the joint density of states and the transition probability cause the response to drop below that of the simple theory, while the intrinsic plasmon response tends to raise it. Data can confirm only that the equation remains a good approximation over an interval smaller than the plasmon energy because the data eventually become obscured by the extrinsic plasmon tail which accompanies the photoemission line. A second point concerns this energy loss tail which accompanies all XPS lines. It is often assumed that at any point this tail is proportional to the integral of the spectrum taken from that point to the Fermi energy. Detailed analysis of XPS data of simple metals gives no indication that such a tail is present; only the well-known plasmon contribution is found. This indicates that application of this process of background subtraction to an asymmetric line will destroy the information residing in its shape, and explains why attempts to fit XPS lines with the Voigt function after background subtraction are never wholly satisfactory. However, since the area of the plasmon peak is proportional to the area of the parent line, wide lines will have more prominent plasmons, and may well overlap the plasmon tail. Under these circumstances a line may appear to be superimposed on a rising background, but it is still incorrect to assume that the background at any point is proportional to the integral of the spectrum taken toward smaller binding energy.

It is intuitively clear that the many-body screening process cannot be well approximated over a large energy interval if the density of states is highly structured, e.g., in transition metals. Some progress toward interpretation of such cases based on a formalism due to Hopfield and Mahan [46] has been made, but a full theoretical treatment is not available. Even nominally simple cases like the noble metals can be understood only when the properties of their band structure are taken into account [47]. Free-electron theory predicts a much larger singularity index than is experimentally realized. de Haas-van Alphen studies show that there is appreciably more screening by higher orbital angular momentum states, reducing α to the range of observed values.

V. SUMMARY

X-ray photoelectron spectroscopy extends photoemission to the deeper core levels, but is not exclusively a core level spectroscopy. In the valence band it provides a clear picture of the occupied density of states. In conjunction with other measurements (or theory) the various contribution to the density of states can be separated.

Core level XPS may be characterized as a final state spectroscopy. The data are relevant to the final, relaxed, and screened state of the ion in the solid. The binding energy reflects this relaxation and screening. The line shape contains the Lorentzian lifetime width of the hole state, a Gaussian width due to the excitation of phonons, and, in metals, a many-body asymmetry.

Open shell ions exhibit multiplet splitting of core levels and final state structure of the open shell itself. In every case the electrostatic coupling between electrons makes the photo-emission process a many-electron phenomenon in which outer shell transitions accompany photoemission.

XPS is consequently a rich discipline which transcends its early promise as a tool for chemical analysis.

REFERENCES

1. P. Auger, Compt. rend 180, 65 (1925); J. Phys. Radium 6, 205 (1925); An. Phys. Paris 6, 183 (1926).

2. K. Siegbahn, et al., ESCA; Atomic, Molecular and Solid State Structure Studied by means of Electron Spectroscopy. (Almqvist and Wiksells, Uppsala, 1967).

3. More recently successful attempts have been made to utilize yttrium $M\zeta$ radiation, M. O. Krause, Chem. Phys. Letts. 10, 65 (1971); M. S. Banna and D. A. Shirley, Chem. Phys. Letts. 33, 441 (1975).

4. J. H. Scofield, J. Electron Spectrosc. 8, 129 (1976).

5. C. D. Wagner, Anal. Chem. 44, 1050 (1972); C. K. Jørgensen and H. Berthou, Faraday Discussions 54, 269 (1972), Anal. Chem. 47, 482 (1975); V. I. Nefedov, et al., J. Electron Spectr. 2, 383 (1973); ibid. 7, 175 (1975); P. C. Kemeny, et al., Phys. Lett. 49A, 171 (1974); J-T. J. Huang and F. O. Ellison, Chem. Phys. Letts. 25, 43 (1974); A. Calabrese and R. G. Hayes, Chem. Phys. Letts. 27, 376 (1974); M. Janghorbandi, M. Vulli and K. Starke, Anal. Chem. 47, 2200 (1975); R. C. G. Leckey, Phys. Rev. A13, 1043 (1976).

6. J. C. Riviere, Contemp. Phys. 14, 513 (1973); C. C. Chang in Characterization of Solid Surfaces; P. F. Kane and G. B. Larrabee, Eds. (Plenum Press, N. Y., 1974) p. 509. J. C. Powell, Surf. Sci. 44, 29 (1974).

7. O. Klemperer and J. P. G. Shepherd, Adv. Phys. 12, 355 (1963). J. Daniels et al. in Springer Tracts in Modern Physics 54, 77 (1970).

8. T. Koopmans, Physica 1, 104 (1933); P. S. Bagus, Phys. Rev. 139A, 619 (1965); L. Hedin and G. Johansson, J. Phys. B2,

1336 (1969); M. E. Schwartz, Chem. Phys. Letts. $\underline{5}$, 50 (1970); D. A. Shirley, Chem. Phys. Letts. 16, 220 (1972); P. H. Citrin, R. W. Shaw, and T. D. Thomas in Electron Spectroscopy, D. A. Shirley, Ed. (North Holland, Amsterdam, 1972) p. 105; L. Ley et al., Phys. Rev. $\underline{B8}$, 2392 (1973); S. P. Kowalczyk, et al., Faraday Discussions $\underline{60}$, 7 (1975); R. E. Watson, M. L. Perlman, and J. F. Herbst, Phys. Rev. $\underline{B13}$, 2358 (1976).

9. D. A. Shirley, Phys. Rev. $\underline{B5}$, 603 (1972); S. Hüfner, et al., Solid State Commun. $\underline{11}$, 323 (1972); G. K. Wertheim, et al., Phys. Letts. $\underline{49A}$, 191 (1974); N. V. Smith, et al., Phys. Rev. $\underline{B10}$, 3197 (1974).

10. R. J. Baird, L. F. Wagner, and C. S. Fadley, Phys. Rev. Lett. $\underline{37}$, 111 (1976); F. R. McFeely, et al., Phys. Rev. $\underline{B14}$, 3273 (1976); P. S. Wehner, et al., Phys. Rev. Lett. $\underline{38}$, 169 (1977).

11. R. J. Smith, et al., Solid State Commun. $\underline{21}$, 459 (1977).

12. G. K. Wertheim, M. Campagna and S. Hüfner, Phys. Cond. Matter $\underline{18}$, 133 (1974); P. T. Andrews and L. A. Hisscott, J. Phys. F: Metal Phys. $\underline{5}$, 1568 (1975); see P. O. Nilsson and I. Lindau, ibid. $\underline{1}$, 854 (1971) for UPS data.

13. S. Hüfner, et al., Solid State Commun. $\underline{11}$, 259 (1972), Phys. Rev. $\underline{B8}$, 4511 (1973); J. Hedman, et al., Physica Scripta $\underline{4}$, 195 (1972); V. V. Nemoshkalenko, et al., Physica Scripta $\underline{11}$, 387 (1975).

14. S. Hüfner, et al., Phys. Rev. Letters $\underline{28}$, 488 (1972).

15. R. E. Watson, J. Hudis and M. L. Perlman, Phys. Rev. $\underline{B4}$, 4139 (1971); J. C. Fuggle, et al., Solid State Commun. $\underline{13}$, 507 (1973); R. T. Poole, et al., J. Phys. F: Metal Phys. $\underline{3}$, L46 (1973); C. R. Helms and D. Collins, Solid State Commun. $\underline{17}$, 459 (1975); S. Hüfner, G. K. Wertheim and J. H. Wernick, ibid. $\underline{17}$, 1585 (1975); N. J. Shevchik and C. M. Penchina, J. Phys. F; Metal Phys. $\underline{5}$, 2008 (1975); J. D. Riley, et al., ibid. $\underline{6}$, 293 (1976).

16. G. K. Wertheim, et. al., Phys. Rev. Letts. $\underline{32}$, 997 (1974).

17. G. K. Wertheim, J. Franklin Inst. $\underline{298}$, 289 (1974); G. K. Wertheim, et al., AIP Conf. Proc. $\underline{24}$, 235 (1975); C. Blaauw, et al., J. Phys. C: Solid State $\underline{8}$, 459 (1975).

18. J. A. Wilson, F. J. DiSalvo and S. Mahajan, Phys. Rev. Lett. $\underline{32}$, 882 (1974); Adv. Phys. $\underline{24}$, 117 (1975).

19. For XPS studies of the CDW instability see G. K. Wertheim,
     F. J. DiSalvo and S. Chiang, Phys. Lett. 54A, 304 (1975);
     Phys. Rev. B13, 5476 (1976).

20. P. H. Citrin and T. D. Thomas, J. Chem. Phys. 57, 4446 (1972);
     R. T. Poole, et al., Chem. Phys. Lett. 22, 101 (1973); ibid.
     23, 194 (1973), ibid. 31, 308 (1975).

21. For studies of cuprous halides see S. Kono, et al., Phys.
     Rev. Lett. 28, 1385 (1972); Phys. Rev. B8, 795 (1973); D. R.
     Williams, et al., Phys. Lett. 49A, 141 (1974); A. Goldman,
     et al., Phys. Rev. B10, 4388 (1974).

22. See for example: D. J. Hnatowich, et al., J. Appl. Phys. 42,
     4883 (1971); M. F. Ebel and H. Ebel, J. Electron Spectrosc. 3,
     169 (1974); C. K. Jørgensen and H. Berthou, Chem. Phys. Lett.
     31, 416 (1975).

23. P. O. Heden, H. Löfgren and S. B. M. Hagstrom, Phys. Rev.
     Lett. 26, 432 (1971); Phys. Stat. Sol. b49, 721 (1972); G. K.
     Wertheim, et al., Phys. Rev. Lett. 27, 505 (1971); P. A. Cox,
     Y. Baer and C. K. Jørgensen, Chem. Phys. Letts. 22, 433 (1973);
     C. Bonnelle, R. C. Karnatak, and C. K. Jørgensen, Chem. Phys.
     Lett. 14, 145 (1972); E. I. Zabolotskii, Yu. P. Irkhin, L. D.
     Finkelshtein, Sov. Phys., Solid State 16, 733 (1974); C. K.
     Jørgensen, Struct. Bonding 22, 49 (1975); P. A. Cox, Struct.
     Bonding 24, 59 (Springer, Berlin 1975).

24. W. T. Carnall, P. R. Fields and K. Rajnak, J. Chem. Phys. 49,
     4412, 4424, 4443, 4447 and 4450 (1968).

25. For spectra of the RE metals see Y. Baer and G. Busch, Phys.
     Rev. Lett. 31, 35 (1973); F. R. McFeely, et al., Phys. Lett.
     45A, 227 (1973); Y. Baer and G. Busch, J. Electron Spectrosc.
     5, 611 (1974); for the antimonides see M. Campagna, et al.,
     Proc. 11th Rare Earth Research Conf., Oct. 1974, Traverse City,
     Mich. (USAEC, TIC, Oak Ridge, Tenn.), and for the insulating
     fluorides see G. K. Wertheim in "Electron Spectroscopy: Theory
     Techniques and Application" Brundle and Baker, Eds. (to be
     published).

26. C. M. Varma, Rev. Mod. Phys. 48, 219 (1976); D. K. Wohlleben
     and B. R. Coles in "Magnetism" Vol. V, G. T. Rado and H. Suhl,
     Eds. (Academic Press, New York, 1973); M. Campagna, G. K.
     Wertheim and E. Bucher, Struct. Bonding 30, 99 (Springer,
     Berlin, 1976) and references therein.

27. A Menth, E. Buehler and T. H. Geballe, Phys. Rev. Lett. 22,
     295 (1969); R. L. Cohen, M. Eibschütz and K. W. West, Phys.

Rev. Lett. 24, 383 (1970); J.-N. Chazalviel et al., Solid State Commun. 19, 725 (1976), Phys. Rev. B14, 4586 (1976).

28. S. Sugano, Y. Tanabe and H. Kamimura, "Multiplets of Transition-Metal Ions in Crystals" (Academic Press, New York, 1970).

29. G. K. Wertheim, H. J. Guggenheim and S. Hüfner, Phys. Rev. Lett. 30, 1050 (1973); S. Hüfner and G. K. Wertheim, Phys. Rev. B8, 4857 (1973); K. Ishii, et al., Chem. Phys. Lett. 27, 126 (1974); R. T. Poole, et al., Phys. Rev. B13, 2620 (1976).

30. Compare the points of view in D. Adler and J. Feinleib, Phys. Rev. B2, 3112 (1970); K. H. Johnson, R. P. Messmer and J. W. D. Connolly, Solid State Commun. 12, 213 (1973); B. Koiller and L. M. Falicov, J. Phys: Solid State C7, 299 (1974), with the data of G. K. Wertheim et al., Ref. 29, K. S. Kim, Chem. Phys. Lett. 26, 234 (1974) and D. E. Eastman, and J. L. Freeouf, Phys. Rev. Lett. 34, 395 (1975).

31. See G. K. Wertheim and P. H. Citrin in "Photoemission in Solids," Cardona and Ley, Eds. (to be published).

32. P. H. Citrin, P. Eisenberger and D. R. Hamann, Phys. Rev. Lett. 33, 965 (1974); U. Gelius et al., Chem. Phys. Lett. 28, 1 (1974); J. A. D. Matthew and M. G. Devey, J. Phys. C: Solid State 7, L335 (1974).

33. T. A. Carlson, M. O. Krause and W. E. Moddeman, J. Physique, Colloque C4, 32, 74 (1971); T. Robert, Chemical Physics 8, 123 (1975); M. A. Brisk and A. D. Baker, J. Electron Spectrosc. 7, 197 (1975).

34. The literature on satellites is extensive: C. S. Fadley and D. A. Shirley, Phys. Rev. A2, 1109 (1970); A. Rosencwaig, G. K. Wertheim and H. J. Guggenheim, Phys. Rev. Lett. 27, 479 (1971); D. C. Frost, A. Ishitani and C. A. McDowell, Mol. Phys. 24, 861 (1972); L. J. Aarons, M. F. Guest and I. H. Hillier, J. Chem. Soc. F2, 68, 1866 (1972); B. Wallbank, C. E. Johnson and I. G. Main, J. Phys. C: Solid State 6, L340 and L493 (1973); T. Robert and G. Offergeld, Chem. Phys. Lett. 29, 606 (1974); K. S. Kim, J. Electron Spectrosc. 3, 217 (1974), Phys. Rev. B11, 2177 (1975); C. S. Fadley, Chem. Phys. Lett. 25, 225 (1974); K. S. Kim and N. Winograd, Chem. Phys. Lett. 31, 312 (1975); T. A. Carlson, Faraday Disc. 60, 30 (1975); S. P. Kowalczyk et al., Phys. Rev. B11, 1721 (1975); Yu. G. Borodko et al., Chem. Phys. Lett. 42, 264 (1976); S. Larsen, J. Electron Spectrosc. 8, 171 (1976); G. A. Vernon, G. Stucky and T. A. Carlson, Inorg. Chem. 15 278 (1976); H. Berthou, C. K. Jørgensen and C. Bonnelle, Chem. Phys. Lett. 38, 199 (1976);

D. C. Frost, C. A. McDowell and B. Wallbank, Chem. Phys. Lett. 40, 189 (1976). For theoretical discussions see: R. Manne and T. Aberg, Chem. Phys. Lett. 7, 282 (1970); H. Basch, Chem. Phys. Lett 20, 233 (1973); J. Electron Spectrosc. 5, 463 (1974); S. Asada, C. Satoko and S. Sugano, J. Phys. Soc. Japan 38, 855 (1975); R. P. Gupta and S. K. Sen, Phys. Rev. B10, 71 (1974), ibid. B12, 15 (1975); S. Larson, Chem. Phys. Letts. 32, 401 (1975).

35. G. K. Wertheim and A. Rosencwaig, Phys. Rev. Lett. 26, 1179 (1971); J. M. Thomas, I. Adams, and M. Barber, Solid State Commun. 9, 1571 (1971; D. P. Spears, H. J. Fischbeck and T. A. Carlson, Phys. Rev. A9, 1603 (1974). See also S. Süzer, S.-T. Lee and D. A. Shirley, Phys. Rev. A13, 1842 (1976) for a related phenomenon.

36. J. Reader, Phys. Rev. A7, 1431 (1972); Y. Yafet and R. E. Watson, Intl. J. Quant. Chem. Symp. 7, 93 (1973).

37. J. Hedman, et al., Phys. Rev. Lett. A29, 178 (1969); C. S. Fadley, et al., Phys. Rev. Lett. 23, 1397 (1969); C. S. Fadley and D. A. Shirley, Phys. Rev. A2, 1109 (1970); R. L. Cohen, G. K. Wertheim, A. Rosencwaig and H. J. Guggenheim, Phys. Rev. B5, 1037 (1972); J. F. Herbst, D. N. Lowy and R. E. Watson, Phys. Rev. B6, 1913 (1972); J. C. Carver, G. K. Schweitzer and T. A. Carlson, J. Chem. Phys. 57, 973 (1972); G. K. Wertheim, S. Hüfner and H. J. Guggenheim, Phys. Rev. B7, 556 (1973); A. J. Freeman, P. S. Bagus and V. J. Mallow, Int. J. Magnetism 4, 35 (1973); P. S. Bagus, A. J. Freeman and F. Sasaki, Phys. Rev. Lett. 30, 850 (1973); Int. J. Quant. Chem. 57, 83 (1973); F. R. Mc Feely et al., Phys. Lett. 49A, 301 (1974); D. C. Frost, C. A. McDowell and I. J. Woolsey, Mol. Phys. 26, 1473 (1974).

38. A. J. Signorelli and R. G. Hayes, Phys. Rev. B8, 81 (1973); S. Suzuki, T. Ishii and T. Sagawa, J. Phys. Soc. Japan 37, 1334 (1974); S. P. Kowalczyk, et al., Chem. Phys. Lett. 29, 491 (1974); N. Spector, et al., Chem. Phys. Lett. 41, 199 (1976); W. C. Lang, et al., Faraday Disc. 60, 327 (1975).

39. S. P. Kowalczyk, et al., Phys. Rev. B7, 4009 (1973); E.-K. Viinikka and Y. Öhrn, Phys. Rev. B11, 4168 (1975).

40. S. Hüfner and G. K. Wertheim, Phys. Rev. B7, 2333 (1973).

41. G. D. Mahan, Phys. Rev. 163, 612 (1967); P. W. Anderson, Phys. Rev. Lett. 18, 1049 (1967); P. Nozières and C. T. DeDominicis, Phys. Rev. 178, 1097 (1969); K. D. Schotte and U. Schotte, Phys. Rev. 182, 479 (1969); D. C. Langreth, Phys. Rev. 182 973 (1969); M. Combescot and P. Nozières, J. Physique 32, 913 (1971).

42. S. Doniach and M. Sunjic, J. Phys. C: Solid State $\underline{3}$, 285 (1970).

43. P. Minnhagen, Phys. Lett. $\underline{56A}$, 327 (1976).

44. S. Hüfner, G. K. Wertheim and J. H. Wernick, Solid State
    Commun. $\underline{17}$, 417 (1975); G. K. Wertheim and S. Hüfner, Phys.
    Rev. Lett. $\underline{35}$, 53 (1975); P. H. Citrin, G. K. Wertheim and
    Y. Baer, Phys. Rev. Lett. $\underline{35}$, 885 (1975); Y. Baer, P. H.
    Citrin and G. K. Wertheim, Phys. Rev. Lett. $\underline{37}$, 49 (1976);
    G. K. Wertheim and P. H. Citrin in "Photoemission in Solids,"
    Cardona and Ley, Eds. Springer Tracts in Applied Physics (to
    be published). See also P. H. Citrin, Phys. Rev. $\underline{B8}$, 5545
    (1973); S. Hüfner, et al., Phys. Lett. $\underline{46A}$, 420 (1974); L. Ley,
    et al., Phys. Rev. $\underline{B11}$, 600 (1975); S. Hüfner and G. K.
    Wertheim, Phys. Rev. $\underline{B11}$, 678 (1975).

45. P. H. Citrin, G. K. Wertheim and Y. Baer (submitted to Phys.
    Rev. B.)

46. J. J. Hopfield, Comments in Solid State Physics $\underline{2}$, 40 (1969);
    G. D. Mahan, Phys. Rev. $\underline{B11}$, 4814 (1975); G. K. Wertheim and
    L. R. Walker, J. Phys. F: Metal Physics $\underline{6}$, 2297 (1976).

47. Y. Yafet and G. K. Wertheim, J. Phys. F: Metal Physics $\underline{7}$, 357
    (1977).

Current Problems in Auger Electron Spectroscopy

T E Gallon
Department of Physics
University of York
Heslington
York   UK

The technique of Auger Electron Spectroscopy has proved
extremely useful in the field of Surface Science where, after
ten years, it passes as an old, well established technique
almost rivalling LEED in its antiquity. However, with the
importation of high resolution spectrometers from, and the
exportation of good vacuum practice to, 'ESCA' the topic of
AES in solids has made considerable experimental and theoretical
advances in recent years. It has become apparent that there is
more information in AES than simple qualitative chemical analysis
and in these lectures I wish to discuss some of these recent
developments. My viewpoint will be that of the experimental
surface physicist interested in understanding the Auger process
from the point of view of using it to obtain information about
solids.

Firstly I wish to examine the effect of the solid environment
on Auger line energies and line profiles with a view to understanding solid state and 'chemical' shifts. Secondly I should
like to consider the possibility of obtaining density of state
information from the Auger process and to examine the quasi-
atomic processes which are found in some solids. Finally I will
discuss some of the more technical problems associated with
obtaining quantitative information from Auger spectra.

Free Atom Auger Energies
    Before considering the Auger process in solids it will be
informative if we first briefly consider the case of free atoms.
The Auger process in free atoms has been the subject of many
publications and the topic is well understood from a theoretical

viewpoint. I do not wish to discuss the calculation of relative
intensities or transition probabilities (see McGuire)[1] but I would
like to consider briefly the calculation of Auger energies as this
topic will play an important part in the understanding of the Auger
process in solids.

In an Auger process an atom is initially ionised in an inner
level A and Auger emission results in an atom with two final state
holes in levels B, C and an Auger electron of energy E(ABC). Three
approaches are possible in the calculation of E(ABC)-first principles,
semi-empirical and empirical methods.

(a) First principles calculations
      The Auger energy may be written

$$E(ABC;X) = E[A] - E[B,C;X] \tag{1}$$

where $E[A]$ is the energy of the atom ionised in level A and
$E[B,C;X]$ is the energy of the doubly ionised atom with appropriate
final multiplet state, both energies relative to the neutral atom
energy. Calculation of $E[A]$ and $E[B,C;X]$ requires SCF values for
one hole and two hole defect states with proper inclusion of
relativistic effects for inner levels[2]. In view of the difficulties
of such calculations this approach has not been widely used but it
is capable of predicting Auger energies correct to within a few
eV. Inclusion of correlation effects can improve this and very
good agreement with experiment is then obtained[2,5]

(b) Semi-Empirical Methods
      This approach uses the fact that accurate one electron binding
energies may be obtained from XPS or X-ray emission measurements
and has been the most widely used approach to date. The Auger
energy is written as

$$E(ABC;X) = E(A) - E(B) - E(C) - W[B,C;X] \tag{2}$$

where E_A E_B and E_C are experimental one electron atomic binding
energies and $W[B,C;X]$ is the recombination energy of the two final
state holes in configuration X, also termed the hole-hole interaction energy H. $W[B,C;X]$ is obtained from model calculations
using the approach of Hedin and Johansson[3] where the term is split

$$W[B,C;X] = F[B,C;X] - R_s^a (B,C) \tag{3}$$

$F[B,C;X]$ is the recombination energy calculated in the appropriate
coupling scheme based upon a frozen orbital method and $R_s^a(B,C)$ is
a calculated 'polarisation' term to allow for the effect of relaxation of the passive orbitals which will have the effect of reducing
the binding energy of the emitted electron. Thus this calculation

envisages Auger emission as a two step process. The term $R_s^a(B,C)$
was introduced by Shirley [4] and termed by him the 'static atomic
relaxation energy'. Shirley published a table of KLL Auger energies
based upon this method using the intermediate coupling scheme of
Asaad and Burhop[5] and calculating R_s^a by an equivalent cores method
based upon the results of Hedin and Johansson[3] Larkins[6] has published improved calculations based upon this approach but calculating
R_s^a from ∆RHF calculations for a range of Z values and interpolating
for the remaining Z. Both methods give good agreement with experiments with accuracies ranging from a few eV at low Z to tens of eV
at high Z where experimental values tend to be uncertain anyway.

(c) Empirical Methods
        If outer electrons are involved in the Auger transition the
Auger energy may be calculated entirely from experimental data
using tabulated binding energies for inner core levels[7,8], allowing
for solid state energy shifts if data are taken from solid specimens,
and optical data[9] for the final states.

$$E(A,B,C;X) = E(A) - E[B,C;X] \quad \text{as before}$$

The energy $E[B,C;X]$ is obtained from optical data by summing the
energies required to produce the final two hole state

$$E[B,C;X] = IP(I) + IP(II) + \varepsilon(X) \tag{4}$$

where $IP(I)$ and $IP(II)$ are the first and second ionisation potentials
and $\varepsilon(X)$ is the energy of the required two hole state above the
ground state of the doubly ionised species.

        This method will also give a value for the hole-hole interaction
energy, combining equations (2) and (4)

$$H = E[B,C;X] - (E(B) + E(C)) \tag{5}$$

in this case both $E(B)$ and $E(C)$ may be calculated from optical
data. This method gives very good agreement with experiment for
those cases where it is possible to obtain optical data - Table I[10]

TABLE I

Parameters for some transitions in the rare gases

ABC	Electron volts				
	E_A [7]	E_{BC} [9]	$E_A - E_{BC}$	E_{Auger}^{expt}	H
Ne $KL_{2,3}L_{2,3}{}^1D$	870.1	65.8	804.3	804.8	22.7
Ar $L_3M_{2,3}M_{2,3}{}^1D$	248.4	45.1	203.3	203.5	13.6
Kr $M_5N_{2,3}N_{2,3}{}^1D$	93.8	40.4	53.4	53.4	12.4

Finally mention should be made of the Z/Z+1 formula[11], for many years the standby of experimental surface physicists. This attempts to allow for the hole-hole interaction by using data from the element with the next highest atomic number.

$$E^Z(A,B,C) = E(A)^Z - E(B)^Z - E(C)^Z - \Delta Z(E(C)^Z - E(C)^{Z+1}) \quad (6)$$

with ΔZ taken as 1 this reduces to

$$E^Z(A,B,C) = E(A)^Z - E(B)^Z - E(C)^{Z+1} \qquad (7)$$

This gives relatively poor agreement with experiment for free atoms even in the symmetrised form[12], and cannot, of course, predict multiplet structure. Matthew[13] has shown that this approach can be improved if free ion data, i.e. $E_C(Z+1)^+$, is used and the method can be extended to the difficult problem of calculating Auger energies with multiple initial state ionisation.

Thus we conclude that semi-empirical methods and methods involving optical data, when available, provide a good basis for calculating free atom Auger energies.

When Auger emission occurs from an atom situated in a solid we can distinguish two cases. Auger emission from levels lying below the valence band and Auger emission involving one or two electrons originating in band states. I should like to deal with the former process first, although, as we shall see, the distinction is not so obvious as it might appear at first sight.

Auger Emission from Solids

Figure 1(a) shows the $M_{4,5}N_{4,5}N_{4,5}$ spectrum of Cd in the vapour phase[14] and Figure 1(b) and (c) show the same transitions in solid Cd and from Cd in CdS[15]. In Cd the 4d levels lie only 9 eV below the Fermi level so it might be doubtful to treat them as core levels but the results for Cd and CdS are very similar to those for Sn and SnO where the levels lie $\sim$ 31 eV below E_F. Comparison of the Cd metal and Cd vapour spectra shows that the multiplet structure survives in solid Cd but examination of the energy scale indicates that the peaks in the solid have been uniformly shifted by $\sim$ 11.8 eV. Measurement of the solid spectrum indicates that line broadening has also occurred and this is most obvious in the case of CdS where the multiplet structure has virtually disappeared. The energies of the peaks from CdS are again higher than the gas phase but only by $\sim$ 7.9 eV. Thus in the solid state there is shift in line energies and a broadening of the peaks and both effects seem to depend on the nature of the atomic environment.

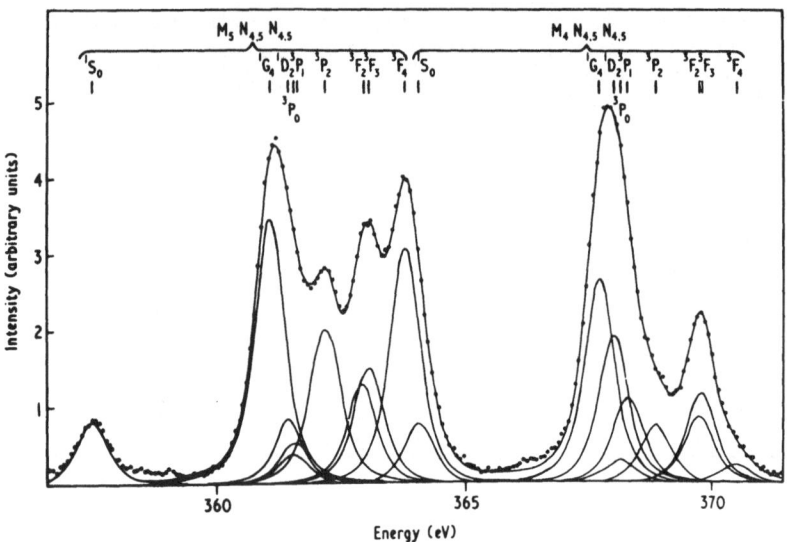

Fig.1(a)   The $M_{4,5}N_{4,5}N_{4,5}$ Auger spectrum of Cd vapour[14]

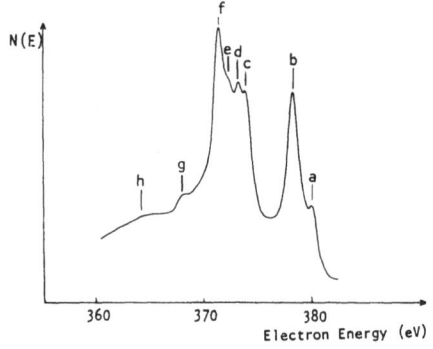

Fig.1(b) The Cd $M_{4,5}N_{4,5}N_{4,5}$ Auger spectrum from Cd metal [15]

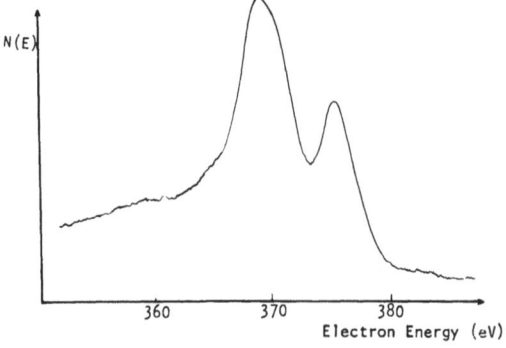

Fig. 1(c)   The Cd $M_{4,5}N_{4,5}N_{4,5}$ Auger spectrum from CdS[15]

Auger Energies in the solid state

In the free atom the final two holes couple to give multiplet structure and this effect is also observed in the solid. However in the solid the final doubly charged state will alter the local charge distribution and this polarisation will lead to a reduction in the final state energy and hence increase the energy of the emitted Auger energy[16]. We may understand this in terms of eqns (4) and (6) where, for a solid we may write

$$E_s(ABC;X) = E(A) - E(B) - E(C) - H + P \qquad (8)$$

where P (always positive) represents the reduction in the final state energy due to the response of the surrounding electrons to the two localised holes [17,18]. The sum of hole-hole and polarisation energies may be combined to give an effective potential U_{eff} so that equation (8) may be written

$$E_s(ABC;X) = E(A) - E(B) - E(C) - U_{eff} \qquad (9)$$

Of course if E(A), E(B) and E(C) are taken from XPS measurements on solids the measured values will also include a polarisation energy[19] but only that produced by a singly charged hole and for linear polarisation the polarisation energy will be proportional to the square of the charge, so that the polarisation energy of the doubly charged Auger final state will be approximately four times that of the singly charged XPS final state and thus P in equation (8) is approximately twice the polarisation correction in E(A), E(B) and E(C) included in the XPS data. If H, in equation (5), may be obtained from optical data the polarisation energy P may be determined empirically[18,20]. Table II shows the results of such calculations for the $KL_{23}L_{23}$ transition in free Na^+, free Na, solid Na and from Na in ionic solids. It may be seen that there is considerable variation in P with the largest value corresponding to Na metal where screening of the final state hole by conduction electrons is expected to be most efficient.

The two step model outlined above for the free atom has been further developed by Shirley and co workers [26,27,28] to include the effects of a solid state environment and has been extensively used in calculations on metals.

Table II XPS and $KL_{23}L_{23}$ D Auger data for Na in various forms, units eV

Na	E_{AUG}	E(1S)	E(2p)	P_{exp}	P_{theory}	P_{exp} - P_{exp} atom	Re
Ion	966.1[b]	1088.9[bc]	47.3[c]	-	-[f]	-	-
Atom	976.7[b]	1078.6[bc]	38.0[c]	2.3	2.8[f]	-	-
Metal	991.9[a]	1074.0[a]	32.7[a]	11.5	10.9[f]	9.2	9.2
NaI	991.9[d]	1070.3[d]	28.3[d]	6.4	5.4[g]	4.1	4.1
		(1069.6)[e]		(7.1)		(4.8)	
NaF	991.9[d]	1066.0[e]	24.8[d]	3.7	3.7[g]	1.4	(-0.2)
NaO	986.8[d]	1075.2[d]	32.8[d]	5.4	-	3.1	3.1

(a) Kowalczyk et al (1973)[27] (b) Hillig et al[21] (c) Moore[9]
(d) Kowalczyk et al[22,28] (e) assuming 1s-2p energy differences
as in Citrin and Thomas[23] (f) Hedin and Johansson[3] - twice XPS
polarisation energy (g) Citrin and Thomas[23] - twice XPS polarisation energy. Repulsive energy assumed the same in initial and
final state of the Auger process.

    In the case of a solid equations (2) and (3) are modified by
the inclusion of an additional term so that the Auger energy
becomes

$$E^s(A,B,C;X) = E^s(A)-E^s(B) - E^s(C) - F[B,C;X]$$
$$+ R_s^a(B,C) + R_s^{ea}(B,C) \qquad (10)$$

The term $R_s^{ea}(B,C)$ 'the static extra atomic relaxation energy' is
similar to the atomic term $R_s^a(B,C)$ in that it represents a reduction
in the binding energy of the emitted electron produced by the
attraction of electrons from outside the emitted atom towards the
resultant atomic hole. The terms $E^s(A)$, $E^s(B)$ and $E^s(C)$ will also
contain extra atomic relaxation terms produced by electrons attracted
towards the initial state hole so that

$$E^s(A) = E(A) + R_d^{ea}(A) \text{ etc.} \qquad (11)$$

assuming that the effect of 'chemical shifts', surface dipole
layer etc are small.

    Before considering the methods of estimating $R_s^{ea}(B,C)$ it will
be instructive to examine the results of this approach applied to
the $M_{45}N_{45}N_{45}$ spectrum of Cd by Weightman[29] for which free atom

and solid Auger data exist and where the final state is within the
range of optical data. Free atom binding energies were obtained
by using relative energies from X-ray data referenced by optical
data. Firstly Weightman calculated free atom Auger data using
optical data, equations (2) and (4) above and found calculated
multiplet energies in excellent agreement with the measured values[14].
The free atom energies were also calculated by a ΔH-F method which
gave good agreement apart from a systematic discrepancy of 6.4 eV,
attributed to neglect of relativistic effects in the initial state,
and a slightly large range of multiplet splittings. Finally the
free atom was treated by the two step semi empirical process,
eqns (2) and (3). The value of $F(N_{45}N_{45};X)$ was calculated by
standard coupling theory[30] using HF calculations for the required
direct and exchange integrals. The calculation for $R_S^a(B,C)$ used
the relationship derived by Shirley[4] for intra-shell static
relaxation

$$R_S^a(B,B) \ = \ 2\ R_D^a\ (B) \tag{12}$$

where $R_D^a(B)$ 'the atomic dynamic relaxation energy' is the
difference between the electron orbital energy and the binding
energy. The value of $R_D^a(N_{45})$ was calculated by ΔH-F and a value
of 6.6 eV obtained for $R_S^a(N_{45}N_{45})$. This enabled the free atom
$E(M_{45}N_{45}N_{45})$ value to be calculated and these gave excellent
agreement with experiment but with slight overestimate of multiplet
splitting.

For the solid case Weightman applied the two step semi-
empirical model, equation (10) and used the previous atomic
values to obtain a value for $R_S^{ea}(N_{45}N_{45})$ of 9.1 eV. To calculate
the energy shift relative to the free atom it is necessary to add
$R_d^{ea}(N_{45})$ to $R_S^{ea}(N_{45}N_{45})$ to allow for the fact that solid binding
energies each contain approximately equal extra electron contributions, compare equations (3) and (4) and equation (10) and (11).
This gave a calculated value for the energy difference between
vapour and solid of 12.4 eV compared with the measured value of
12.1 eV.

Shirley et al have made direct calculations of $R_S^{ea}(B,C)$
based on a localised excitonic model, a Friedel model[5], for the
screening processes involved. They assume that screening is by
electrons moving in to occupy localised states pulled below the
conduction band by the one hole and two hole atomic states involved
in the Auger process[27,28]. They estimate the interaction energy
by placing the screening electron in the lowest unoccupied level S
and calculating the interaction energy of this electron with
electron C using the appropriate Slater integrals in an equivalent
cores approximation.

Thus the expression they derive for $R_s^{ea}(B,C)$ is

$$R_s^{ea}(B,C) = F(C,S)_{Z+1}$$

where the screening orbital S is the lowest unoccupied atomic orbital and the choice of the element with atomic number Z+1 is to allow for the effect of the hole in B. Application of this method to Cu and Zn[27] gave satisfactory results predicting $L_3M_{4,5}M_{4,5}(^1D)$ values of 921.1 and 993.2 eV compared with experimental values of 918.0 and 991.8 eV. Comparison with a free atom value calculated using optical data gave rather poorer agreement with an estimated shift in Auger energy from free atom to solid of 11.1 eV from optical data and experimental value for $L_3M_{4,5}M_{4,5}$ (^1D) compared with an energy shift of 15.5 eV calculated from the two step model.

Calculation of the total relaxation energy for the $L_2M_{23}M_{23}$ transition in a number of first row transition elements have been performed by this group[28] and agreement with experimental values is poor for elements with partly filled d shells (e.g. Fe Re(Tot) calc = 41 eV, Re (Tot) expt = 28 eV). This is attributed to the 3 d screening states being more delocalised than atomic 3 d electrons thus reducing the extra atomic relaxation energies so that the above method of calculating R_s^{ea} will be an upper limit.

Kim, Gaarenstroom and Winograd[31] have objected to this latter explanation, they point out that 3 d screening orbitals should be more localised than the 4s and 4p orbitals used for Cu and Zn where the calculation appears to work quite well. They suggest that the disagreement arises from two sources, firstly in the implicit assumption that the dynamic atomic and extra atomic relaxation associated with binding energies in neutral and already ionised atoms are the same and secondly in the assumption that the screening electron always fills the lowest available free atom level. They show that the first of these effects will be relatively small if the two screening electrons have the same principal quantum number and calculate an additional term of -1.3 eV for solid Cu (screening electrons 4s, 4p) which leads to an improvement in calculated energies. However if the principal quantum numbers are different then this neglected term will be quite large. Kim et al also consider the screening orbitals in Ni. They point out that although the atomic configuration of Ni is $3d^84s^2$ this may not be appropriate in the solid state e.g. magnetic measurements suggest $3d^{9.46}4s^{0.69}$ to $3d^{8.85}4s^{1.15}$. They calculate the $L_3M_{23}M_{23}$ Auger energies for $3d^84s^2(S,S'\ 3d, 3d)$, $3d^94s(S,S'\ 3d, 4s)$ and $3d^{9.46}4s^{0.54}(S, S'3d^{0.54}4s^{0.46}, 4s)$ and find values for the ^3P peak of 732.9, 724.8 and 721.1 eV compared with 715.2 eV from experiment. Since the value calculated from the equation of Kowalczyk et al[28] is 731.4 eV for 3d screening the effects of placing the screening electrons in higher orbitals may clearly be seen.

The non equivalence of the dynamic relaxation energies
associated with E(C) and E(C*) has also been considered by
Hoogewijs, Fiermans and Vennik[32,33] for the case of the $L_3M_{45}M_{45}$

transition in zinc vapour and solid. They calculate by ΔH-F
that this leads to an additional relaxation energy of -1.38 eV in
the free atom leading to a lowering of the atomic dynamic relaxation energy involved in the emission of the second 3d electron.
This leads to a free atom value of 973.68 eV for $L_3M_{45}M_{45}$ ^{1}G compared
with a measured   value of 973.3 eV. In this calculation the value
for $R_s^a(M_{45}M_{45})$ was obtained by calculation and they find that the
value they obtain, 10.84 eV is not consistent with the value obtained[27] by assuming that equation (12) above is valid, 12.3 eV.
They show that equation (12) should be replaced by

$$R_s^a(B,B) = R_D^a(B)* + R_D^a(B)$$

where $R_D^a(B)*$ is the atomic dynamic relaxation energy in the once B
ionised atom. This formalism has been applied to solid zinc[33] where
an increase in the extra atomic relaxation energy results from an
explicit consideration of the once ionised state. This gives a
calculated value of the $L_3M_{45}M_{45}$ ^{1}G energy of 992.7 eV compared with
their experimental value of 992.3 eV.

Thus to summarise, free atom and metallic solid Auger energies
can be calculated by a two step process allowing for recoupling of
the final state holes, relaxation in the atom and, for solids,
screening of the final state. A localised excitonic, Friedel,
model, can be used successfully to calculate the extra atomic relaxation although both screening electrons should be considered
explicitly and the assumption that the excitonic states correspond
to the lowest available atomic orbitals may not always be correct.

An alternative approach to the calculation of relaxation
energies associated with localised core holes in metals has been
developed by Hedin, [50,3] Flynn and Lipari[51], Matthew[18,20] Gadzuk[52]
and Almbladh and von Barth[53]. In these investigations all electrons
outside closed cores are treated as an electron gas either within a
dielectric formalism or self consistently. Thus far relaxation
energies for single and double holes in nearly free electron metals,
such as Na Mg and Al, have been successfully calculated but further
work is needed to relate this approach to the Friedel exciton model
discussed here.

Chemical Shifts in Auger Spectra
    It is well known that the rearrangement of outer electrons
which occurs on formation of chemical compounds gives rise to
shifts in the binding energies of core electrons – chemical shifts
in XPS. On the simplest model, if all levels in an atom were shifted
downwards by an amount ΔE the energy of Auger transitions involving

these levels would also be reduced in energy by a similar amount.
Such a simple relationship is sometimes found[35] but from the
above discussion the relative shifts found in solids would not be
expected to behave in this manner. Measurements of Auger shifts on
oxidation of metals generally show larger values than shifts in the
XPS lines, Table II above, and Table III. This difference can be
understood in terms of the polarisation or extra atomic relaxation
energies discussed above.

TABLE III

Chemical shifts between element and oxide for selected
Auger and Photoelectron lines[17]

Ele-ment	Photoline shifts (eV)			Auger line shifts (eV)	
	$2p_{3/2}$	$3d_{5/2}$	4d	LMM	MNN
Zn	0.4	0.6		4.2	
Ga	1.7	2.2		6.2	
Ge	3.0	3.3		6.7	
As	2.3	3.6		6.4	
Cd		0.4	0.9		5.5
In		0.8	0.9		2.6
Sn		1.5	1.2		3.9

If we consider the case of Auger emission from a metal the
final state hole will be screened by conduction electrons and this
gives a large contribution to the Auger energy. On oxidation to
form a bulk oxide the final state hole will again polarise the
surrounding dielectric but the screening energy will be smaller
than in the metallic case leading to a displacement of the Auger
lines to lower energies, nearer the gas phase values. The XPS
lines will similarly be lowered in energy by polarisation but
the effect for the singly charged hole will be smaller. Again
in the simplest model if the shift in the energies of XPS lines
from core levels in oxidation is given by

$$\Delta E_X = \epsilon + \Delta P$$

where ϵ arises from charge exchange and ΔP as a result of changes
in polarisation energy, the change in the energy of an Auger line
involving these core levels will be

$$\Delta E_A = \epsilon + 3\Delta P$$

Of course this model is too simple and calculation of extra
atomic relaxation energies are required but the localised Friedel
model will not be appropriate.

TABLE IV   Experimental data from Fiermans, Hoogewijs & Vennik[36]

Sample	$E^F(L_3M_{45}M_{45}{}^1D)$ eV	$\Delta[E(3d) - E_B({}^1D)]$ eV	Refractive Index
Zn	992.3	ref level	-
ZnTe	991.3	-0.57	3.56
ZnSe	989.5	-1.9	2.89
ZnS	988.2	-3.3	2.37
ZnO	987.6	-2.59	2.02
ZnCl$_2$	986.6	-3.9	1.69
ZnBr$_2$	987.58	-3.2	1.55

Table IV shows a set of values of the Auger energies referred to
the Fermi level or top of the valence band. The quantity
$\Delta[E3(d) - E_B{}^1D]$ where $E_B{}^1D$ is the apparent binding energy of the
^1D Auger peak in the solid, is approximately equal to the change
in the extra atomic relaxation energy[28] relative to the metal.
Also shown are the refractive indices of these materials and while
correlation is not perfect those compounds showing the smallest
change in relaxation energy from the metal have the highest
optical dielectric constants while those showing the largest
shifts have the smallest dielectric constant, at least suggesting
a relationship between high frequency polarisability and final
state screening.

If the energy of an Auger line depends on local environment
as well as on chemical state it might be expected that the line
energy for thin overlayers should depend on layer thickness,
at least in the case of insulators when 'screening lengths' will
be much greater than atomic dimensions. Such effects have been
reported in the case of the oxidation of Zn[24] where the shift
between the Zn $L_3M_{45}M_{45}{}^1G$ peak in ZnO and in Zn metal varied from
2.6 eV for very thin oxide films to 4.2 eV for thick films, a value
close to the bulk difference.[28] A similar effect was observed in the
case of Ar condensed onto Ag[37], Fig. 2. The Ar $L_{23}MM$ Auger signal
is seen to vary in energy with film thickness. The energy from a
submonolayer film being 7.1 ± 0.5 eV above the gas phase energy
while for a 'bulk' film the shift was only 2.5 ± 0.5 eV. This
difference was explained as arising from efficient screening of
the final state holes by metal conduction electrons for the thinnest
film but a smaller shift produced by polarisation in the case of
solid Ar.

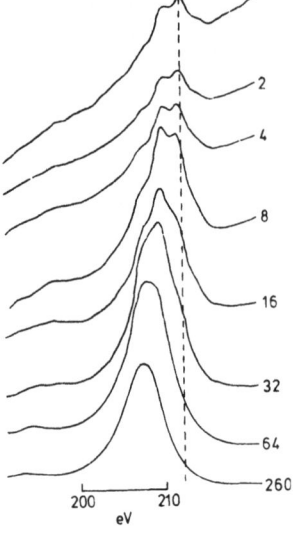

Fig. 2.   The L_{23}MM Auger spectrum of solid Ar for increasing film thickness.[37]

Line Broadening in the Solid State

It has been noted that Auger line breadths increase on going from the vapour phase to the solid. This effect varies from a relatively small amount for metals, where atomic fine structure is still well resolved, to quite large values in solid insulators, Table V. The origin of this broadening is not yet fully understood. Possible contribution might be increased lifetime broadening, phonon broadening, depth variation in polarisation energy or non uniform specimen charging.

Table V

Element	Transition	Breadth	
		Gas phase eV	Solid phase eV
Cd	$M_{45}N_{45}N_{45}$	$0.35 \pm .05$[14]	$1.03 \pm .05$[15]
Cd in CdS	$M_{45}N_{45}N_{45}$	−	2.3 ± 0.2[15]
Zn	$L_{23}M_{45}M_{45}$	0.50 ± 0.15[34]	1.0 ± 0.2[46]
Zn in ZnO	$L_{23}M_{45}M_{45}$	−	3.2 ± 0.3[46]
Ar	L_{23}MM	<0.4[47]	3.6 ± 0.2[15]
Thin film of Ar on Ag	L_{23}MM	−	2.0 ± 0.2[37]
Kr	M_{45}NN	<0.25[48]	4.6 ± 0.4[15]
Xe	$M_{45}N_{45}N_{45}$	$0.65 \pm .06$[49]	2.9 ± 0.2[15]

Lifetime broadening

The initial levels of the Auger processes shown in Table V all lie fairly deep so that the initial state lifetimes in the gas and solid should be very similar. However, there would be considerable differences in the final state lifetimes between the gas and solid phases where outer electrons are involved in the Auger process. Thus if the final state is long lived in the atom, formation of a solid will give rise to additional excitation channels via band states and this could lead to a considerable decrease in the final state lifetime[38]. Interatomic[39] or cross transitions[40] could also reduce lifetime as suggested by Citrin for the case of XPS lines from alkali halides[41], although detailed calculations[42] show that these processes are likely to be rather slow and unlikely to give significant lifetime broadening. These processes could well explain the broadening observed in metals but it is difficult to account for the large broadening observed in insulators and in the case of Xe where the final states lie $\sim$ 68 eV deep it is doubtful if the lifetime will be significantly different in the solid.

Phonon broadening

Matthew[43] has shown that coupling to the lattice via lattice vibrations generated by the Auger process can cause considerable broadening in the case of ionic solids. This model has also been used to explain broadening of XPS lines[44,45], it predicts that the line breadths should be temperature dependent and this effect has been found for the alkali halides[44]. Attempts have been made to find temperature dependent line breadth for the materials listed in Table V without success although in some cases the temperature range may have been too limited since a coth $(\theta_E/2T)$ behaviour is predicted, where θ_E is an effective phonon temperature. This suggests that either phonon broadening does not contribute to the observed breadths or that they are dominated by zero point breadths at the temperatures examined. It is doubtful if this model could explain line breadths in the solid rare gases where coupling to the lattice will be much smaller than in the case of ionic crystals.

Variation in polarisation energy

In insulators a localised excitonic screening model is no longer appropriate and the effects of the final two hole state will extend to large distances in the solid. The behaviour of Ar, Fig. 2 indicates that for very thin films the line breadths are much narrower than for thicker films, with quasi atomic fine structure being clearly resolved. The atomic environment, in terms of numbers of neighbouring atoms, will be different for atoms in the surface region compared with bulk atoms and a possible explanation of the larger line breadths found in insulators is that each layer in the surface region has a different polarisation shift from the layer above and below so that the effect of summing layer contributions to the Auger signal will be to give an overall broadening of the spectrum.

Non uniform charging

The largest line broadening is observed for insulators and a possible cause is that the electron beam used for excitation produces local charging which is either spatially or time dependent or both. Charging is observed in some insulators, for example in solid Kr, but others, ZnO,SnO and CdS, seem quite stable under the electron beam and give reproducible Auger spectra for a variety of excitation conditions. Comparison with published spectra obtained with X-ray irradiation where possible, e.g. ZnO, Fig. 3, indicate that the spectra are very similar and the X-ray excited spectra also show appreciable line broadening.

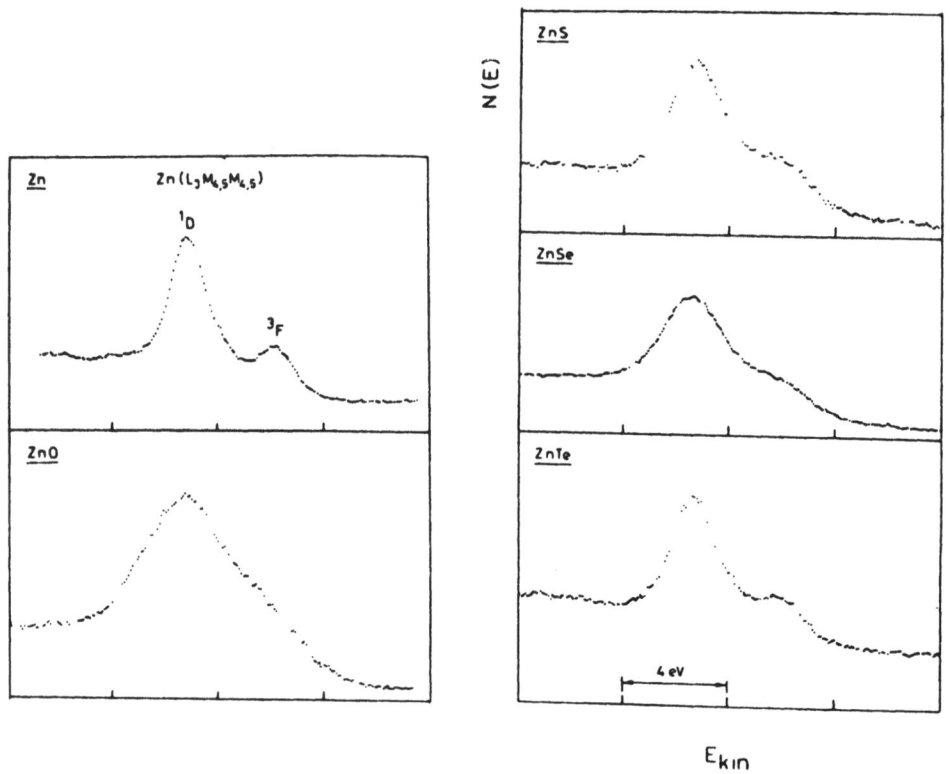

Fig. 3. The $Zn(L_3M_{4,5}M_{4,5})^1D$ and ^3F peaks of Zn metal and Zn chalcogenides [36].

One may conclude by observing that variation of line breadth has been little investigated to date. The effect is not instrumental for metals although in the case of insulators where the effect is much larger non uniform charging cannot be entirely eliminated.

I think that this area is worth investigation since it seems possible that 'chemical' information may be obtained from line profiles as may be seen by examining the Zn $L_3M_{4,5}M_{4,5}$ Auger lines for Zinc chalcogenides obtained by Fiermans, Hoogewijs and Vennik[36], Figure 3.

Of course there is a further possible cause of line broadening which has, so far, been ignored. On formation of solids the outer atomic energy levels broaden to form bands, could this be a source of the broadening referred to above? Can the Auger effect give useful information on local density of states? These are topics which I should now like to discuss.

Auger Spectra involving Band Electrons

So far the discussion on the Auger spectra of solids has ignored the possibility of one or both of the electrons participating in the Auger process originating in the conduction or valence band of the solid. Transitions of this type are conventionally labelled AVV or ABV. What effects might be expected on the simplest band approach? Firstly the energy of the Auger electrons will be affected. On a simple picture, where both electrons are delocalised, both the hole-hole interaction F and the relaxation energy R will be considerably reduced and in the limit will both vanish giving $U_{eff} = 0$.

The profile of the Auger line will be controlled by the band structure and if we take the case of an AVV transition the line profile will be given by (cf Amelio[54])

$$N(E) = A \times \begin{cases} \int_0^{\zeta} g(\zeta + \Delta) \, g(\zeta - \Delta) \, d\Delta & 0 \leq \zeta \leq \tfrac{1}{2}\zeta_1 \\ \\ \int_0^{\zeta_1 - \zeta} g(\zeta + \Delta) \, g(\zeta - \Delta) \, d\Delta & \tfrac{1}{2}\zeta_1 \leq \zeta \leq \zeta_1 \end{cases}$$

where $g(\zeta + \Delta)$ is the transition probability from a state lying at an energy $(\zeta + \Delta)$. ζ, Δ and ζ_1 are defined in Figure 4 and the Auger energy is given by $E = E_A - 2(\zeta + \phi)$
where E_A is the energy of the inner level participating in the transition and ϕ is the work function of the material

$g(\zeta + \Delta)$ and $g(\zeta - \Delta)$ depend on the density of states at $(\zeta + \Delta)$ and $(\zeta - \Delta)$, the density of the final states and also on the transition matrix element

$$< \psi_f \mid \frac{1}{r_{12}} \mid \psi_i > \qquad \text{where } \psi_i \text{ is the wavefunction}$$

appropriate to the initial state and ψ_f the wavefunction of the final two hole state plus the ejected electron.

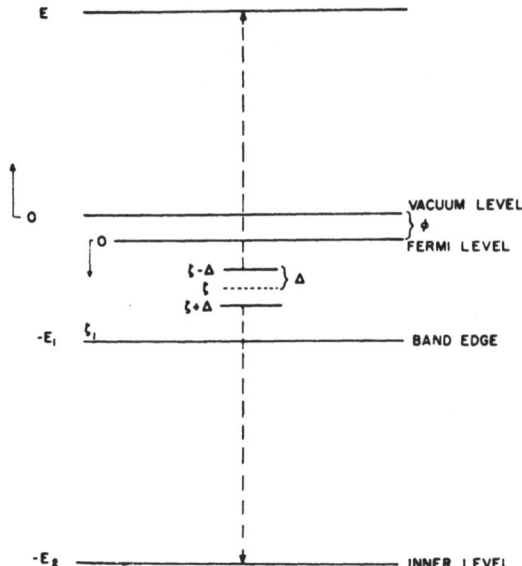

Fig. 4   Energy level diagram for a typical AVV Auger transition

    If we neglect final state effects, which is a good approximation
since Auger energies are generally quite high, and if we assume that
the transition matrix elements are constant across the band, the
above expression reduces to the self convolution of the density of
states. This suggests Auger processes of the type AVV might be
unfolded to give the local density of states – LDOS – which might
or might not, be the same as the bulk density of states. Similarly,
processes of the type ABV should give the density of states directly,
possibly broadened by the contribution of the core hole breadths.
The AVV Auger profiles should have a breadth $2\zeta_1$ and ABV processes a
breadth of ζ_1.

    Before considering the validity of assuming constant matrix
elements it will be instructive to examine experimental results on
different systems. In comparing the Auger profiles for an AVV
process with the DOS obtained by other methods, two approaches are
possible, firstly to unfold the experimental profile to obtain an
'Auger DOS' or secondly to convolute a known DOS and to compare it
with the measured profile. Both methods have been used but in view
of the difficulties and uncertainties of deconvolution processes the
latter approach is more common.

Experimental Results
    No clear overall picture emerges from experiment and it is
convenient to consider the results in a number of different categories

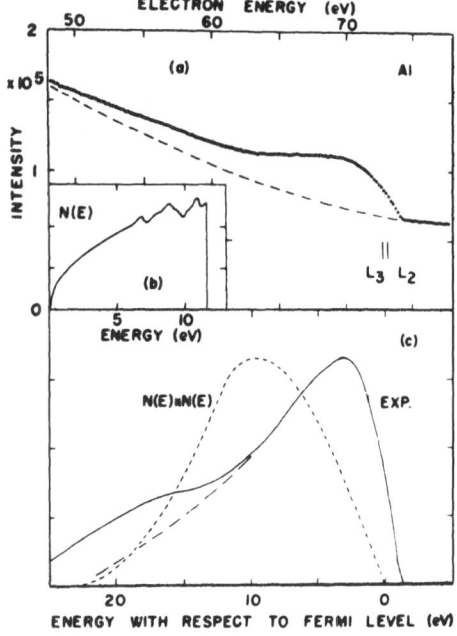

Fig. 5

(a) L_{23}VV Auger spectrum of Al

(b) shows a calculated density of states

(c) compares the self convolution of the density of states with the Auger profile minus background[55]

Simple Metals

In this category we may consider results on Al[55], Li[56], Na[57] and Mg[58].

Fig. 5 shows the results of Powell[55] for the L_{23}VV transition in Al metal. He compares his results with the self convolution of a calculated B DOS. It is apparent that the widths of the two distributions are very close, given the uncertainty in subtracting the experimental background, but there are important differences, in particular the maximum in the experimental distribution falls at higher energy than that of the self convoluted DOS. Very similar results were obtained by Jackson et al[56] for Li metal with, again, the 'pile-up' of electrons at higher energy in the experimental curves. Barrie and Street[57] have examined KL_1V and $KL_{23}V$ transition in Na metal and compared their measured profiles with band structures obtained by UPS, SXS and XPS. The measured Auger widths are in good agreement with those obtained by the other techniques but differences in detail were observed and the 'Auger band structure' given by KL_1V and $KL_{23}V$ transitions were different. Measurements on Mg by Fuggle et al[58], again involving KL_1V and $KL_{23}V$ processes, showed similar effects. The width of the Auger band transition agreed with SXS and XPS measurements but significant differences in detail were observed and again different structures were obtained for KL_1V and $KL_{23}V$ profiles.

Covalent Solids

Silicon, not surprisingly in view of the technical importance of its electronic structure, has been the subject of a number of attempts to examine the band structure by means of Auger spectroscopy. The $L_{23}VV$ peak in Si occurs at $\sim$ 89 eV and the $L_1L_{23}V$ at around 42 eV. Amelio[54] has published a full account of attempts to unfold the density of states from $Si(111)L_{23}VV$ transitions and his results are shown in curve (2) Figure 2. Arnott and Hanemann[59] and Ferrer, Baró and Salmerón[60] have examined the potentially simpler $L_1L_{23}V$ spectrum from Si(111). The results of Ferrer et al[60] and Amelio are shown in Figure 6, together with UPS, SXS and theoretical DOS.

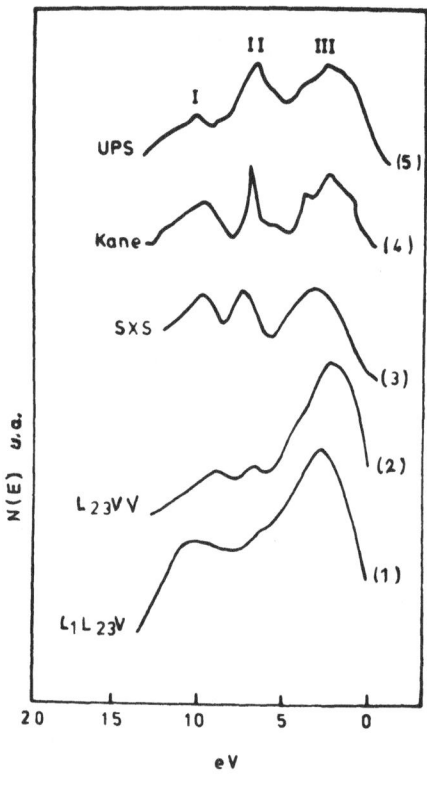

Fig. 6. Comparison of the $L_1L_{23}V$ Auger spectrum of Si(111) (1) with the results of $L_{23}VV$ deconvolution (2) S X S (3), theoretical DOS(4) and UPS (5)[60]

Comparison of the two Auger results with the DOS results shows that the widths of the $L_1L_{23}V$ and deconvoluted $L_{23}VV$ Auger profiles agree with the other methods but details differ. Both $L_1L_{23}V$ and $L_{23}VV$ spectra show the expected three peaks but their relative magnitudes differ from the DOS results and distinct differences may be seen between the $L_1L_{23}V$ and $L_{23}VV$ 'Auger density of states'. Amelio and Schiebner[61] have attempted a similar unfolding for the KVV Auger spectrum of graphite again with moderate qualitative agreement.

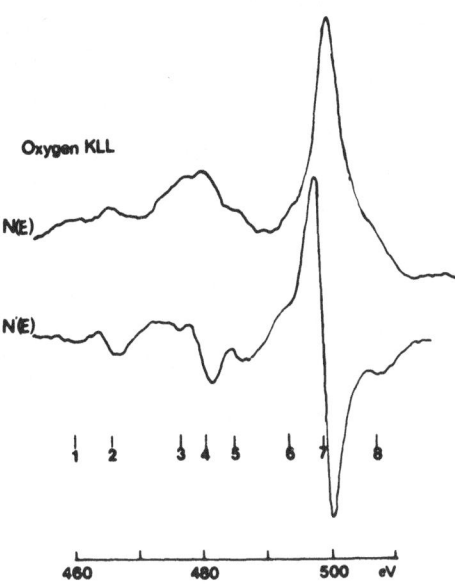

Figure 7. The KLL Auger spectrum from MgO(100)[62]. Peaks 4 and 5 are the KL_1L_{23} ^1P and ^3P.

Ionic Solids

The Auger spectra of a number of these shows behaviour that cannot be explained on a simple band picture. The oxygen KLL Auger spectrum from solid MgO[62] is shown in Figure 7. The spectrum has been analysed in terms of quasi-atomic fine structure. Peaks corresponding to the three possible final state configurations KL_1L_1, KL_1L_{23} and $KL_{23}L_{23}$ show fine structure which can be interpreted in terms of multiplet splitting in the final two hole state. This, together with loss processes in the solid, can account for all peaks except the small peak at highest energy which has been explained as arising from inter-atomic transitions[39,40] involving final state holes on neighbouring oxygen cores, a vestigial band-like transition. Similar quasi-atomic fine structure has been reported for oxygen in a number of transition metal oxides[36] for fluorine in fluorides[63] and for several adsorbates on transition metals[64]. Interestingly, analysis of the low energy ~ 34 eV Mg peak in MgO suggest that this peak does involve transition from the MgO valence band[65] and this has been explained as arising because the process involved in this peak is an inter atomic one involving ionised Mg^{++} and de-excitation via electrons mainly localised on neighbouring O^{--} cores.

d band metals

Measurements of the Auger spectra of Cu, Zn and Ni which involve one or two electrons from the d bands, M_{45} levels, show considerable fine structure. In Cu and Ni the 3d band lies about 2 - 3 eV below

the Fermi level while for Zn it lies about 10 eV below the Fermi
edge.   In all three cases XPS from the d band is rather featureless
with the deeper Zn band having the narrowest breadth[68].   Yin et al[66]
measured the $L_{23}M_{23}M_{45}$ and $L_{23}M_{45}M_{45}$ spectra of Cu and Zn.   They
found that the spectra from the different metals were very similar
in structure and that they could interpret the fine structure in
terms of multiplet splitting in the final state. Powell and Mandl[67]
measured $L_{23}M_{23}M_{45}$ transition in Cu and Ni and also found fine
structure which could be interpreted in quasi-atomic terms.   The
$L_{23}M_{45}M_{45}$ spectra of Cu and Zn have been studied in detail by a
number of authors and spectra taken from Antinodes et al[68] for
these elements together with $L_{23}M_{45}M_{45}$ spectra for Ga and Ge are
shown in Figure 8.   The vertical bars indicate the calculated
energies and intensities of the multiplet structure, these will be
discussed below.   It may be seen that the Auger spectra of these
elements is very similar even though in Ge the 3d level lies $\sim$ 29 eV
below the Fermi level.

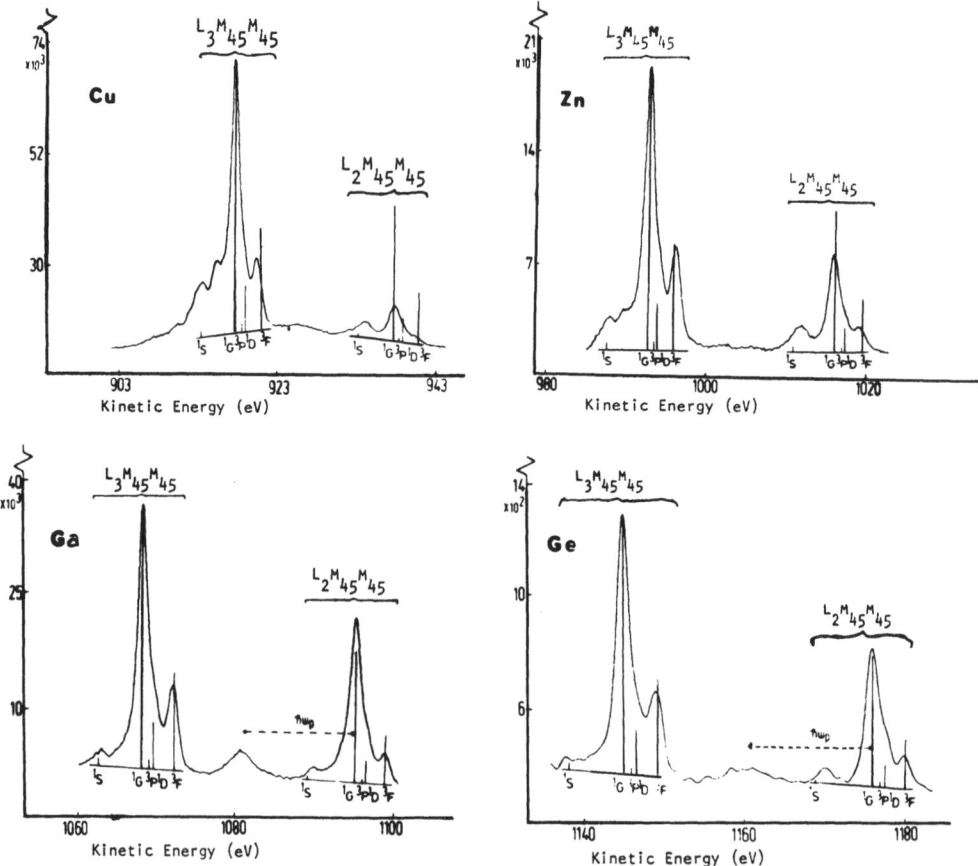

Figure 8.   The $L_{23}M_{45}M_{45}$ Auger spectra from solid Cu, Zn, Ga and Ge[68]

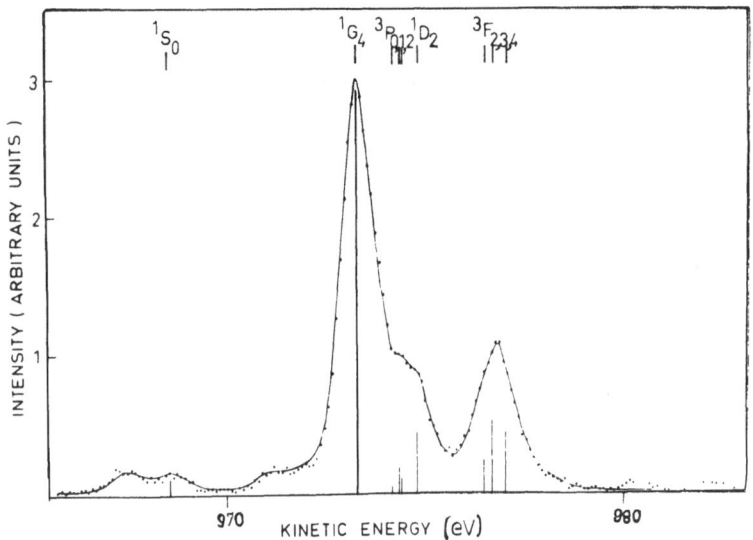

Figure 9. The $L_{23}M_{45}M_{45}$ Auger spectrum from Zn vapour[34]

Aksela, Vayrynen and Aksela[34] have published the $L_{23}M_{45}M_{45}$
Auger spectrum of Zn vapour, Fig. 9, and comparison of figures 8
and 9 shows that much of the structure present in the vapour phase
is present in the solid.

The 4d transition metals show similar effects, the Auger
spectrum of Ag can only be indexed in quasi-atomic terms[55,69] and
the spectrum is very similar to that of Cd[15] and In[69] where the
d levels lie $\sim$ 9 eV and $\sim$ 16 eV below the Fermi level.

Results of Experimental Measurements
        The experimental measurements of Auger line profiles when one
or both of the outer electrons originate in the band fail to give a
consistent picture of the contribution of the DOS to the Auger
process. It seems that in the extreme case of some transition metals
band structure plays no significant part in AVV or ABV transitions.
Even in the most favourable cases simple DOS ideas fail to predict
the measured Auger profiles accurately. Can these results be understood? In what circumstances will band structure control Auger
profiles and when can we expect quasi-atomic behaviour? The answers
to these questions have not yet been completely found but recent
theoretical work has suggested guidelines to the understanding of
band transition in solids and I should now like to consider some
recent theoretical contribution to the problem.

Theoretical Considerations

      The collected Auger electrons emenate from a thin surface
region and the bulk DOS might not be appropriate for such a layer.
Gadzuk[70] and Jackson et al[56] attempted to explain the differences
between experiment and the self convoluted DOS for the simple metals
Al and Li as arising from the effect of the pinning of the wavefunctions at the free surface of the solid. Both calculations used
free electron wavefunctions and both found that, in this approximation, including the effect of the transition matrix elements had
little effect on the predicted Auger profile and did nothing to
improve agreement with experiment. The difference does not
simply arise because of energy dependence of the matrix elements.
The calculations then assumed that the effect of the potential
step at the surface is to pin the wavefunctions with a node at the
surface giving standing waves in the Z direction, Figure 10.
Since the Auger electrons are only collected from the first few
atomic layers this has the effect of preferentially selecting
those wavefunctions of high k from the top of the assumed free
electron valence band. This will move the maximum in the convoluted DOS to higher energies as required by the experimental
results.

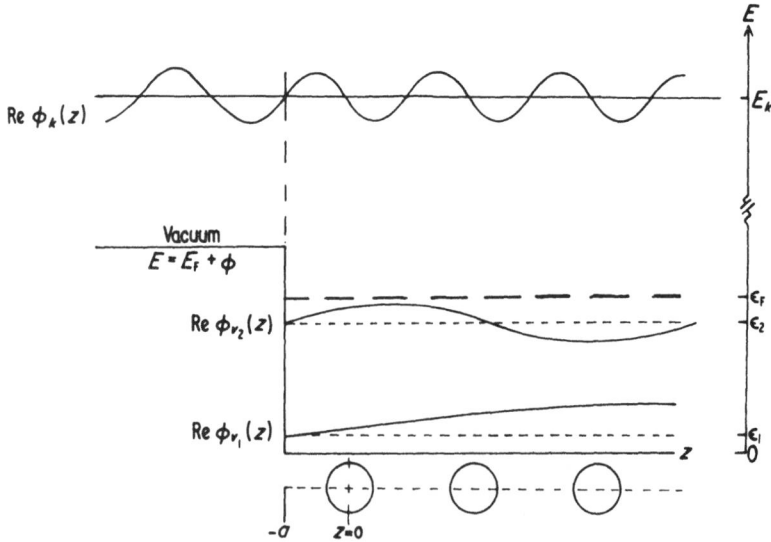

Figure 10. The quasi free electron states at a surface[56]

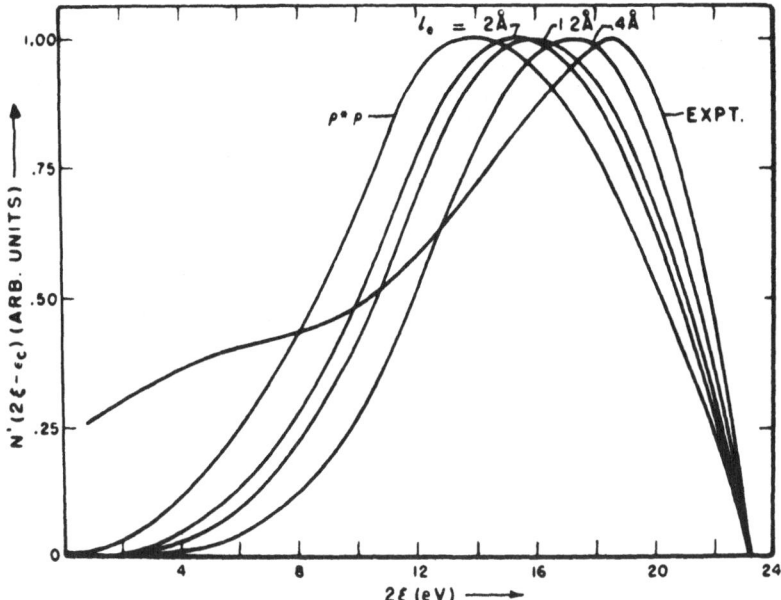

Figure 11. The $L_{23}VV$ Auger spectrum of Al using various mean free paths in a 'pinned free electron model'[70]

Jackson et al[56] found good agreement with experiment for a mean escape depth of 1.5 atomic layers but Gadzuk[70] was unable to obtain exact agreement even with an effective mean free path of 0.4Å, Figure 11. So the effect of including surface pinning is to improve agreement with experiment but other effects seem to contribute as well.

The most exact calculation to date, which includes both matrix element effects and uses a realistic band structure, is that performed by Feibelman, McGuire and Pandey[71] for Si Auger line shapes, They use a calculated tight binding DOS for a film of 20 layers and a mean free path of 7Å to obtain $L_{23}VV$ and $L_1L_{23}V$ Auger profiles for unreconstructed S(111),(100) and (110) surfaces with relaxation of the surface layers included. They found that the Auger matrix elements are strongly angular momentum dependent and the p like electrons near the top of the band are preferentially selected compared with the mainly s like electrons near the bottom of the band. Their calculation includes an interesting intermediate step which demonstrates the effect of including angular momentum conservation and parity selection. They calculate the effect of these selection rules on the multiplicity of the permitted decay channel and derive a 'kinematic' line shape by assuming a constant value for matrix elements of permitted transitions but weight the transition probability by the number of allowed channels.

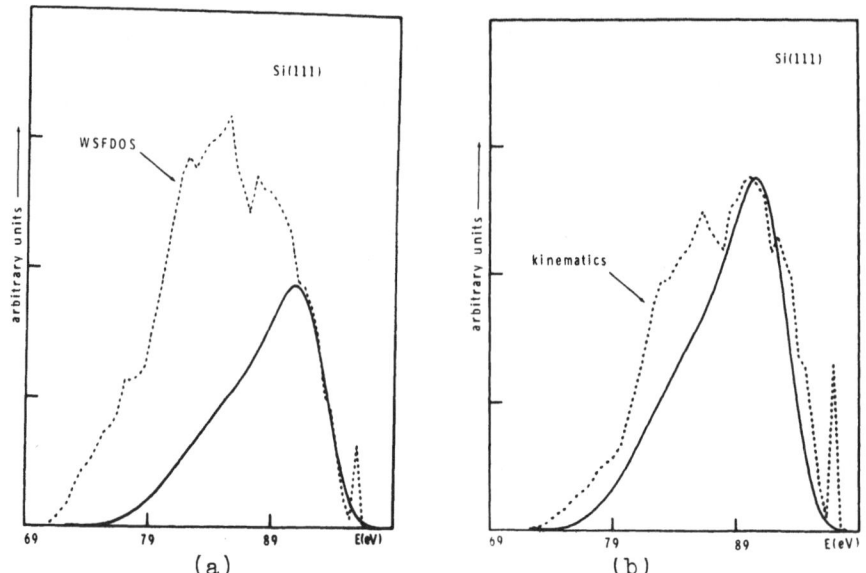

Figure 12 (a) Comparison of the weighted self-fold of the S(111) density of states with the $L_{23}VV$ Auger profile from Si(111) (solid line)

Figure 12 (b) The effect of including angular momentum and parity selection rules[71].

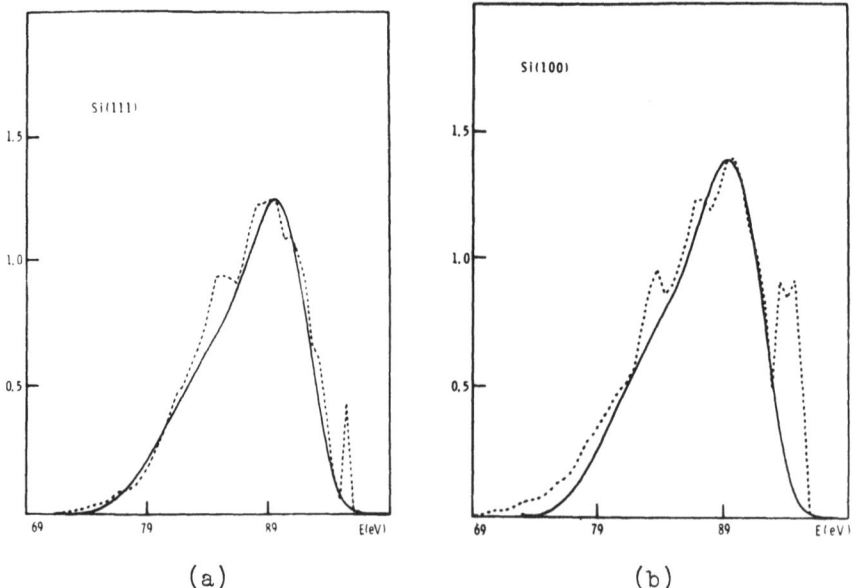

Figure 13 (a) Comparison of the calculated $L_{23}VV$ Auger line shape for Si(111) with the experimental profile (solid line)

Figure 13 (b) Comparison of the calculated $L_{23}VV$ line shape for Si(100) with the experimental profile (solid line) [1].

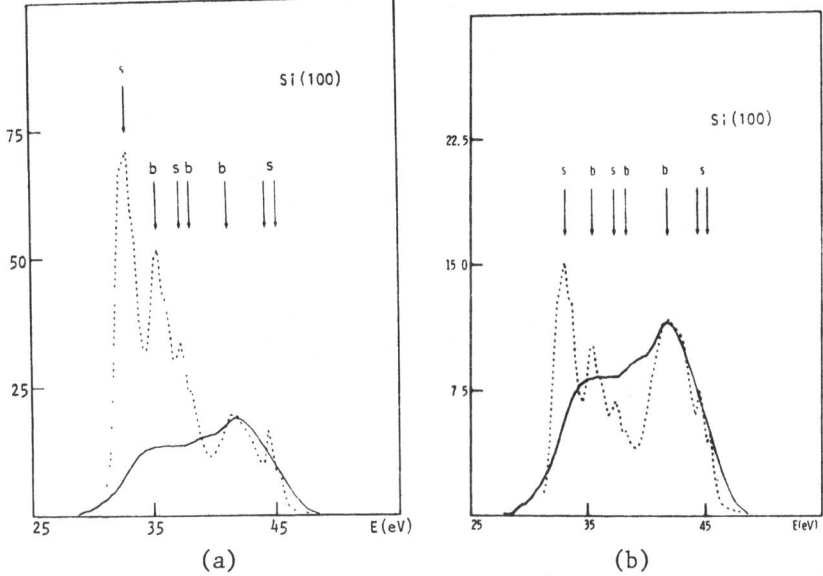

(a)   (b)

Figure 14(a) Comparison between the experimental (solid curve)
and theoretical $L_1L_{23}V$ Auger profiles for Si(100), s refers to
features arising from the first layer and b to features from the
bulk DOS.
Figure 14(b) The effect of reducing the matrix elements for s-li-
ke electrons by a factor of 2.5.

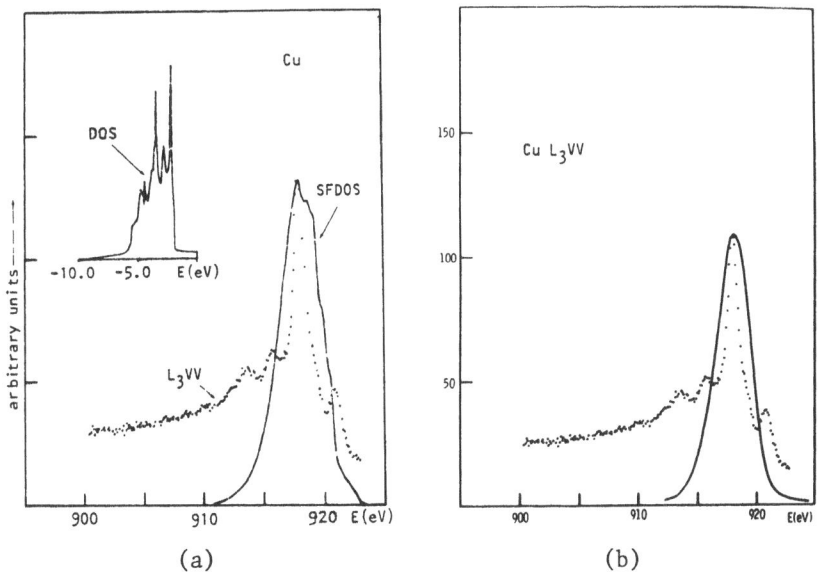

(a)   (b)

Fig. 15(a) Comparison of the measured Cu L_3VV spectrum with the self
convolution of the DOS   (b) The effect of including matrix elements
72.

Figure 12(a) shows the result of this calculation and agreement
with experiment is much improved. Inclusion of calculated matrix
elements further improves agreement with the unpublished results
of Houston, Lagally and Moore quoted by Feibelman et al. Figure 13
shows the now excellent agreement with experimental results for
Si(111) and (100) L_{23}VV transitions. The sharp spike at threshold,
predicted by the theory but not found experimentally, is explained
as arising from both valence electrons taking part in the Auger
process being located in the same dangling bond state and the
authors point out that Coulomb repulsion makes double occupation
of the dangling bond unlikely in practice. Calculations for the
L_1L_{23}V line shapes do not give such good agreement with experiment,
Figure 14. This is explained as possibly arising from surface
reconstruction and also because the low energy transition is much
more sensitive to inaccuracies in the tight binding matrix elements
used. It is suggested that many body effects might contribute
broadening to the predicted one electron surface features.
However, the excellent agreement between calculation and experiment
for the L_{23}VV transition indicates that matrix element effects are
extremely important in Auger band transitions and could probably
explain the discrepancies in the Li and Al results. It would be
expected that more realistic wavefunctions in the calculation would
also give emphasis to the p like states at the top of the band and
push the self convoluted maximum to higher energies as seen in the
experimental measurements. Since the angular momentum of both
electrons and core hole are included in the calculation, the differences between KL$_1$V and KL$_{23}$V Auger profiles referred to above [57,58]
might well be accounted for.

   This leaves the quasi-atomic case unexplained. Feibelman and
McGuire[72] have applied their matrix element calculation to Cu to
see if L_{23}VV and M_{23}VV Auger profiles could be explained purely
in band terms. Their results are shown in Figure 15; the quasi-
atomic theory stands! They speculate on why the transitions are
essentially atomic and suggest that this might be because the
effective lifetime of the two hole state is prolonged either
because of the effect of negative electron mass at the bottom of
the d band or because it is energetically impossible for the two
holes to separate simply. The latter could occur if the repulsive
energy of the two holes on the same atom is much greater than the
3d band width. If the holes are to separate spatially and the 3d
band width is too small electrons must be excited into unfilled
4s or 4p bands or phonons or photons must be created which would
suggest that the final two hole state might be relatively long
lived, giving sharp Auger peaks dominated by final state coupling.
This relationship between band width and quasi-atomic behaviour has
been independently proposed by Antonides, Janse and Sawatzky[68]
and considered in detail by Sawatzky[73].

TABLE VI. Values for the total relaxation energy, R, atomic relaxation energy R_{at}, calculated atomic Coulcomb energy F, and the effective Coulomb energy U_{eff} for the ^1G final-state term in the $L_3M_{45}M_{45}$ Auger process. All values are in eV.

	Cu	Zn	Ga	Ge
$R(M_{45}M_{45};^1G)$	19.0	20.7	22.2	23.3
$R_{at}(M_{45}M_{45};^1G)$	6.1	6.1	6.2	6.0
$F(M_{45}M_{45};^1G)$	27.0	30.2	33.3	36.2
$U_{eff}(M_{45}M_{45};^1G)$	8.0	9.5	11.1	12.9

Antonides et al[68] have published a full analysis of all the $L_{23}MM$ Auger spectra for Cu, Zn, Ga and Ge. They used an atomic model to calculate transition rates and term splittings with the relaxation energy determined to give best fit. They found total relaxation energies to be nearly independent of the final state, Table VII. They also performed a H-F calculation to estimate the atomic part of the relaxation energy and showed that this is small compared to the extra atomic relaxation, Table VI.

TABLE VII. Total relaxation energy values(eV) for Cu, Zn, Ga and Ge for the three main Auger processes.

	Cu	Zn	Ga	Ge
$R(M_{45}M_{45};^1G)$	19.0±1.0	20.7±1.0	22.2±1.0	23.3±1.0
$R(M_{23}M_{45})$	22.0±1.0	22.3±1.0	23.9±1.0	25.2±1.0
$R(M_{23}M_{23})$	21.3±1.0	20.7±1.0	22.2±1.0	23.3±1.0

The structure visible on the low energy side of the $L_3M_{45}M_{45}$ ^1G peaks in Cu and Zn has been explained as arising from a doubly ionised initial state produced by an initial ionisation of an L_2 level followed by an $L_2L_3M_{45}$ Koster-Kronig transition which leaves that atom ionised in L_3 and M_{45} levels[118,119,76]. A further Auger process involving two more M_{45} electrons follows, to give a final state with three M_{45} holes which couple in a quai-atomic manner.

Table VI also shows the value of the atomic Coulomb energy F and U_{eff} defined above. Their calculated transition rates give very good agreement with experiment, the calculation was based upon atomic wavefunctions and allowance had to be made for the effect of the final state hole on the outgoing electron to get good agreement. Figure 8 shows the results of their calculations for the $L_{23}M_{45}M_{45}$ transition and similar good agreement was also found for $L_{23}M_{23}M_{23}$ and $L_{23}M_{23}M_{45}$ transitions. The differences in intensities between theory and experiment for transitions originating on L_2 vacancies in Cu and Zn can be explained in terms of competing $L_2L_3M_{45}$ Koster-Kronig transition[76]. Sawatzky[73] has examined the relationship between atom like Auger spectra and valence band width in detail. His calculation shows that if $U_{eff} > 2\Gamma$, where Γ is the band width, the Auger process will be controlled by a bound excitonic-like final two hole state giving rise to quasi-atomic fine structure, while if $U_{eff} < 2\Gamma$ band like transitions will appear in the Auger profile. Antonides et al[68] publish a curve of U_{eff} and $2\Gamma_{45}$ for the first transition series, Figure 16, and it may be seen that the curves cross in the region of Z = 28 (Ni). This predicts that 3d transition metals with Z = 29 and above will show quasi-atomic fine structure while for Z ≤ 27 (Co) band like transitions should occur. Sawatzky[73] also predicts that vestigial band like transitions may also remain in quasi-atomic structure and explains a peak in the $L_{23}M_{45}M_{45}$ spectrum of Cu lying between the two main Auger groups as possibly arising from this process.

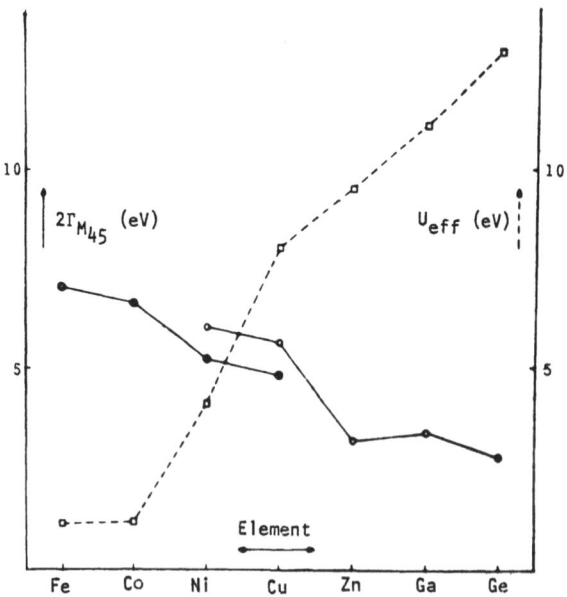

Figure 16. U_{eff} (dashed line) and $2\Gamma_{M_{45}}$ (full line) vs atomic number for Fe to Ge[68]

The values of U_{eff} have also been measured by Yin, Tsang and Adler[74] for V, Cr, Fe, Co, Ni, Cu and Zn and they report values of 0, 1.3, 2.6, 3.7, 5.3, 8.6 and 10.5, the latter two differing slightly from those of Antonides et al[68] shown in Table II. Yin et al comment that V and Cr have broad band like $L_3M_{4,5}M_{4,5}$ Auger peaks and that, in agreement with the small U_{eff} values, this represents band like behaviour. It is interesting to compare the Auger spectra of Co, Ni and Cu taken from another publication by Yin, Tsang and Adler[75], Figure 17. Clear differences may be seen between the $L_3M_{4,5}M_{4,5}$ spectrum of Co and those of Ni and Cu. The sharp peak found in Ni and Cu is absent in Co, suggesting that, in agreement with the prediction of Antonides et al, Ni and higher Z show quasi-atomic fine structure while Co and lower Z may not.

It is interesting to speculate on the effect of oxidation on the suggested stable excitonic like final state holes. As noted above the Auger profiles of transition metals having quasi-atomic fine structure broaden on oxidation although final state multiplet structures still remain. This may be due to the presence of the higher lying oxygen 2p states providing alternative decay channels and thus decreasing the lifetime of the final two hole localised state. Thus the multiplet structure will be retained but broadened.

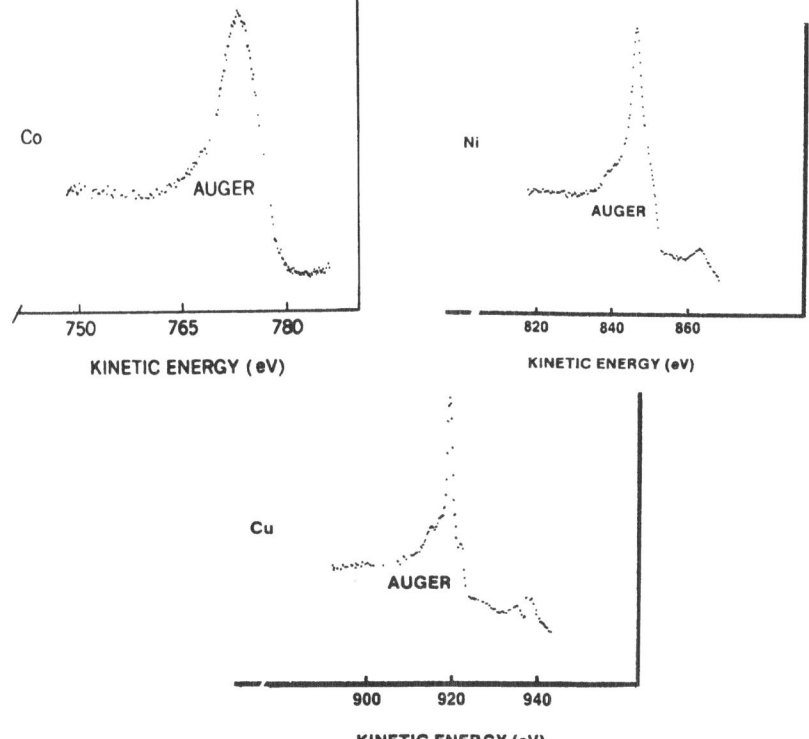

Figure 17. The $L_3M_{4,5}M_{4,5}$ Auger spectra for Co, Ni and Cu[75]

Quantitative Auger Spectroscopy

The early applications of AES were of a qualitative nature but recently quantitative methods have been developed. To understand the limitations of these it will be useful to start with an expression for the Auger current from a uniform solid of density N atoms/unit vol irradiated with a primary electron beam current I_p of energy E_p at an angle θ to the surface normal. If element i has concentration C_i we may write the total collected Auger current originating from ionisation of level energy E_A as[91]

$$I_i = \text{const.} \ I_p \ G_i \ Ci \ N \ \Phi_i(1-w_i) \ r_i \ \tau_i \ \sec \theta$$

when $G_i(E_{ABC})$ is a geometrical collection factor which will depend on the analyser and, since most analysers work at constant resolution, will depend on the Auger energy.

$\Phi_i(E_A,E_P)$ is the ionisation cross-section per atom

$w_i(E_A)$ is the fluorescence yield which we may take as zero for most Auger transitions.

r_i is the backscattering factor

τ_i is the 'equivalent escape depth'

This formula gives the total yield from ionisation of level E_A and if a single Auger peak is measured it must be multiplied by the branching ratio B_i. If the surface is not smooth a roughness factor R should also be included[77]. The L.H.S. gives the Auger current and if p-p heights in the $N'(E)$ spectrum are used they must be corrected for peak shape[78,79], or allowance must be made in the modulation method[80].

Ionisation cross-section

This subject has been discussed in detail by Powell[81]. No formula exists with universal validity but a number of theoretical or semi-empirical formulae exist and are generally of the form[82,83,84,85]

$$\Phi(E_A,E_P) = \frac{\sigma\text{max}}{E_A{}^2} \ f(U) \quad \text{where} \quad U = \frac{E_P}{E_A}$$

The form for f(U) differs slightly for each theory but rises sharply above threshold, peaks at U = 3 - 4 and then falls off slowly. Figure 18 shows f(U) for the Gryzinski[82], Worthington and Tomlin[83], Drawin[84] and Lotz[85] expressions.

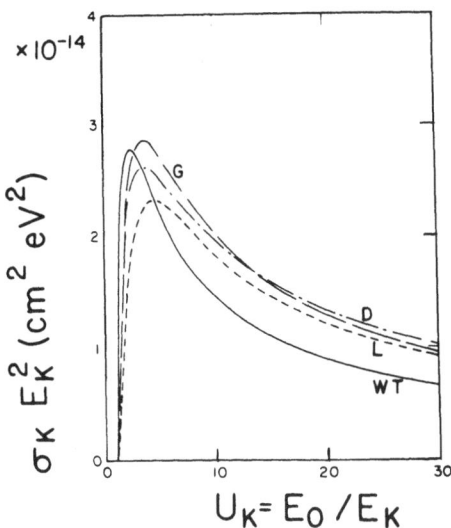

Figure 18. A plot of f(U) for some theoretical[82] and semi-empirical [83],[84],[85] electron-electron ionisation cross-sections[81].

The following questions arise, does the universal expression f(U) represent the energy variation reasonably and do cross-sections scale as E_A^{-2}? Data is rather scarce but it does seem to indicate that a universal curve is a reasonable approximation except for levels of high angular momentum[86]. Vrakking and Meyer have examined the question of scaling and found that a factor of E^{-d} is valid with $d = 2$ for K shells but with $d = 1.6$ for L_{23} shells.

Backscattering factor
        In addition to ionisation produced by the primary beam, ionisation will also be caused by the more energetic electrons which are scattered back through the escape region. This will lead to an enhancement of the Auger yield accounted for by the backscattering factor r. Vrakking and Meyer[88] and Smith and Gallon[89] have measured backscattering factors for a number of elements, the former by means of ellipsometric calibration, the latter by a self consistent measurement[90]. The results agree well with each other and are in fair agreement with theoretical values of Bishop[91]. Values obtained by Smith and Gallon[89] are shown in Figure 19. It may be seen that r increases with U and with atomic number. Measurements of relative backscattering factors may be obtained from measurements of the Auger currents from systems which grow in layer fashion[92].

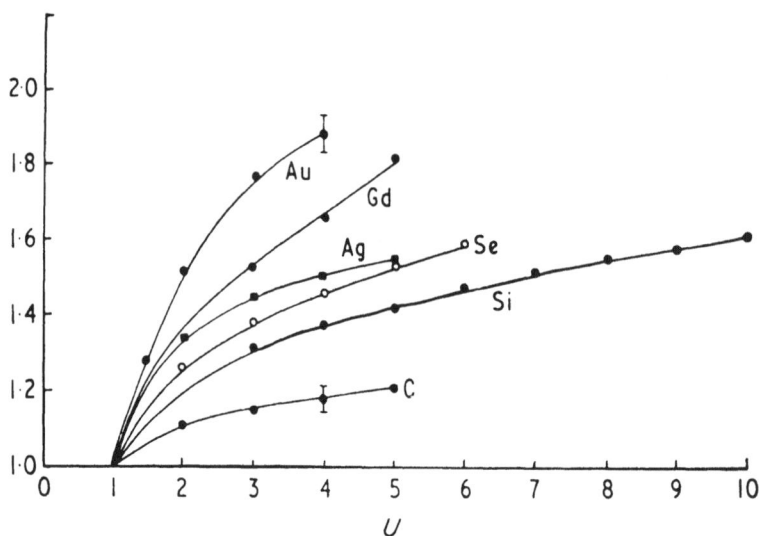

Figure 19.   Backscattering factors for elements with a range
of Z[89]

Escape Depth

This is probably the least well known factor contributing
to quantitative Auger analysis. Three different definitions of
'escape depth' seem to be used. The equivalent escape depth
used above may be understood by defining I_1 as the signal from
the top atomic layer of the solid and I_∞ as the signal from the
bulk solid

$$\text{then } \tau = \frac{I_\infty}{I_1}$$

The mean escape depth obtained from layer growth measurements
is usually defined as

$$d = \frac{1}{I_\infty} \int_0^\infty x \, dI$$

Finally the inelastic mean free path, λ, is often used. This is
defined by assuming exponential attenuation of the electron with
the depth so that

$$N = No \exp(-x/\lambda)$$

For large escape depths ($I_1 \ll I_\infty$) these definitions reduce
to equivalent expressions but for small escape depths they are
significantly different.

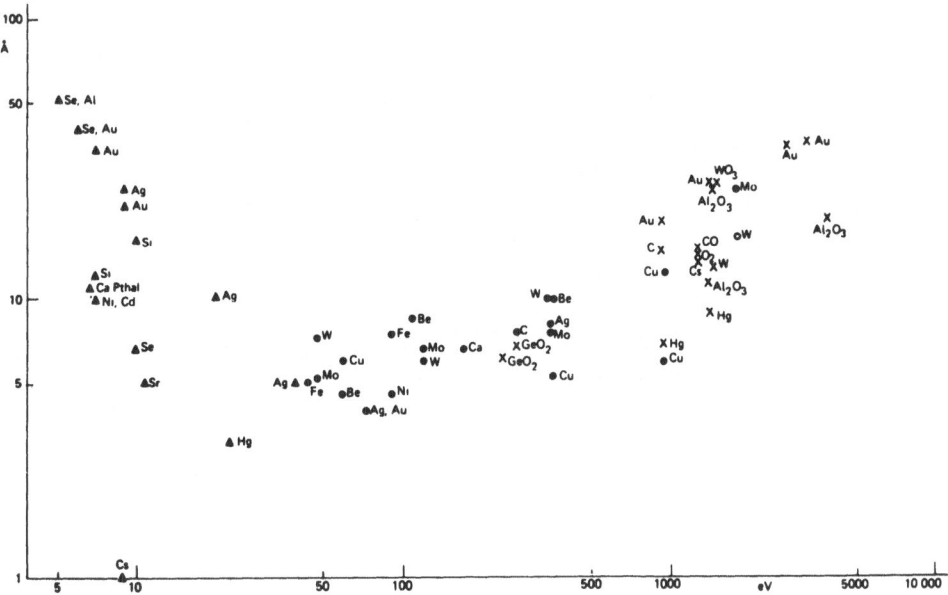

Figure 20.   Experimental values of electron inelastic mean free path[93].

Relatively few good experimental measurements of escape depths exist, some of these are shown in Figure 20 taken from a review by Brundle[94]. Powell[95] has examined measurements of mean free paths obtained from electron spectroscopy and he has related them to values obtained by other methods. His tabulated values are corrected for instrumental effects[95] and are probably internally consistent in definition. Attempts are sometimes made to relate the mean free paths to a universal curve, thus the high energy portion of Figure 20 is said to show an $E^{\frac{1}{2}}$ dependence[96]. However these attempts are unlikely to succeed as the mean free path is a function of both energy and the scattering properties of the matrix. Thus Carlson and McGuire[97] determined a mean escape depth of 13Å for 1455 eV electrons in W metal but this increased to 26Å for 1450 eV electrons in WO_3. Penn[120] has developed the free electron theory for mean free paths and extended it to cover most elemental solids[121]. He tabulates parameters which enable λ to be calculated as a function of energy and indicates how λ may be estimated for compounds. The agreement with experiment is only moderate for a number of the cases shown in fig. 20, but experimental uncertainties are large, and the minima in the theoretical $\lambda_v E$ curves seem to lie at too low an energy for a number of materials. However, the theory does contain two material parameters and should provide a basis for future experimental measurements.

To summarise, the factors which control Auger emission are known but accurate values are not yet available for a number of these at present. The largest uncertainty is probably in the value of τ. The true current is the best measure for use in quantitative AES and in the absence of this correction must be made to p-p heights obtained from $N'(E)$ curves especially if chemical changes affect the shape of the Auger peaks during an experiment.

Methods of quantitative analysis

A major problem in attempting quantitative AES lies in the homogeneity of the sample. In the case where the surface is significantly different from the bulk, either because of segregation or deposition of adlayers, it is very difficult to obtain exact information. Some information on the depth distribution may be obtained by variation of the angle of incidence of the primary beam[98] or by comparing Auger peaks at high and low energies[79] but if such a situation is suspected the best method of analysis is probably by simultaneous ion sputtering and Auger analysis – 'depth profiling'[100]

In the case of homogenous specimen application of the Auger yield equation gives

$$\frac{C_i}{C_j} = \frac{I_i^a}{G_i \Phi_i r_i \tau_i B_i} \qquad \frac{G_j \Phi_j r_j \tau_j B_j}{I_j}$$

which together with $\Sigma\, C_i = 1$ would enable the concentration to be determined from I_i and I_j if the other factors are known. This method has been attempted[101] but it must be regarded as the least accurate of the possible quantitative methods in view of the uncertainties discussed above.

Use of standards

The factors r τ and possibly B, if outer electrons which participate in bonding are involved, depend on both the atomic species and on the matrix in which it is situated. If we neglect this matrix variation we can define an 'elemental sensitivity factor'[102,99] by measuring the Auger signal from a pure element under standard conditions. This may be performed in situ or by using published values[103] if appropriate to the experimental set up. This gives the concentration simply as

$$C_i = \frac{I_i}{S_i} \left/ \Sigma \frac{I_j}{S_j} \right.$$

Chang[96] estimates the accuracy of this method as better than ±30% and any inclusion of matrix corrections should improve this accuracy.

The best method of using a standard is to prepare a specimen of known composition similar to the specimen of interest and to use this to obtain the relative sensitivity factors. This approach has proved quite successful although some doubt always remains about the surface composition of the standard. It is simple to prepare say an alloy of known composition but the surface of such a standard will have to be cleaned in situ. The usual method is to fracture in situ[104,105] and assume that the surface so formed is typical of the bulk composition but if the fracture is inter-granular the exposed surface may well present impurities which have segregated to the grain boundaries, a process well known in metallurgy which has been extensively studied with AES[102]. Alternative methods are mechanical scraping[106] or ion bombardment. The latter method is well known to alter surface composition[107] but Hall, Morabito and Conley[79] have discussed methods of correcting for this effect in studies on binary alloys.

Special Systems
In the case of simple deposit-substrate systems with simple growth modes AES can give quantitative results from internal measurements. Two such systems are the case of submonolayer deposits and systems which grow in a layered fashion.

Sub-monolayer systems
This growth mode is often found for gas adsorption on solids. For a fractional coverage θ the Auger signal from the deposit has the simple form

$$I_d = C\theta$$

If an Auger signal can be obtained from a known coverage, either at saturation with $\theta = 1$ or from a known chemical state, the constant C may be found. This method has been successfully applied to absorption studies[108,109] and to obtain information on electron beam interactions[110]. The ratio of substrate to overlayer peaks is sometimes used to obtain θ, this gives

$$\frac{I_d}{I_s} = \frac{I_d^1 \, \theta}{I_s^\infty - \theta I_s^1}$$

when I_d^1 is the signal from a monolayer deposit and I_s^∞ and I_s^1 are the signals from pure substrate and the substrate when covered by a monolayer of deposit.

Layer Growth systems

It is found that certain deposits grow in layer fashion
on particular substrates. If Auger spectroscopy is used to
monitor the growth process this morphology may be detected and
self consistent calibration is possible. The following
equations apply to the growth of deposit and the decay of
substrate signals[111].

$$I_d = I_d^1 \theta \qquad n \leqslant 1$$

$$I_d = I_d^\infty \left[1 - (1 - \frac{I_d^1}{I_d^\infty})^n \right] \qquad n \geqslant 1$$

$$I_s = I_s^\infty \left[1 - \theta(1 - \frac{I_s^1}{I_s^\infty}) \right] \qquad n \leqslant 1$$

$$I_s = I_s^\infty \left(\frac{I_s^1}{I_s^\infty}\right)^n \qquad n \geqslant 1$$

with a constant deposition rate R and sticking coefficients
S_1 for d on S and S_2 for d on d we have

$$\theta = R\, S_1 t$$

$$n = R\, S_2 t$$

These equations give the growth of Auger signals with time
for completed layers with linear growth in between. The key to
this method is that the slope of the Auger growth curves change
at each layer with the largest difference occurring at the
formation of the first monolayer. This enables I_1^s and I_1^d to be
determined[112].

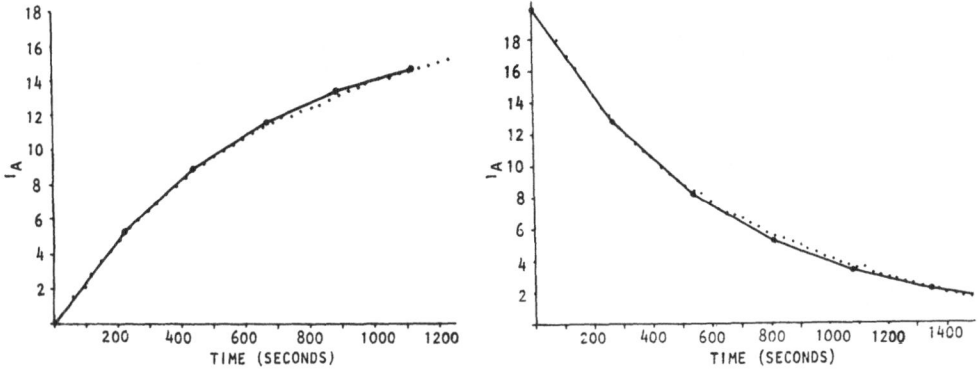

Figure 21.   The growth of the Ti (418 eV) peak and the decay of the W(169 eV) peak with time   during evaporation of Ti onto W[113].

Figure 21 shows the results of Armstrong[113] for the growth of Ti on W.   The mean escape depths may be obtained from such measurements.

$$\overline{x_d} = \frac{1}{I_d^\infty} \Sigma \, x \, d \, I = \frac{I_d^\infty}{I_1^\infty} - \tfrac{1}{2}$$

$$\overline{x_s} = \frac{I_s^1 + I_s^\infty}{2(I_\infty^s - I_1^s)}$$

Application of this approach will distinguish between growth in layer or nucleated modes and can also distinguish cases of growth in the Stranski-Krastonov[114] mode where island growth follows the formation of an initial uniform deposit.[115,116]   Finally this method has been used by Argile and Rhead[117] to obtain information about the surface coverage of one element by codepositing a second and observing the growth of the Auger signals.

Acknowledgements

     I should like to thank Dr J A D Matthew for many long
and illuminating discussions during the preparation of these
notes. I should also like to thank Dr J H Onsgaard for his
critical comments on the manuscript. I am indebted to
Dr G A Sawatzky and Dr F P Larkins for providing preprints of
their work. Finally I should like to thank Mrs J P Fearnehough
for deciphering my writing and preparing these notes and to
thank Mr A Gebbie for preparation of the figures.

References

1   McGuire E J in Atomic Inner-Shell Processes VI Ed B
     Craseman Academic Press, New York 1975
2   Larkins F P (as 1)
3   Hedin L & Johansson G, J Phys B 2 13 36, 1969
4   Shirley D A Phys Rev A 7 1520 1973
5   Asaad W N & Burhop E H S Proc. Phys Soc 72 369 1958
6   Larkins F P J Phys B 9 47 1976
7   Sevier K D Low Energy Electron Spectrometry Wiley
        Interscience New York 1972
8   Bearden J A & Burr A F Rev Mod Phys 39 125 1967
9   Moore C E Atomic Energy Levels Nat Bur Stn Circular
        No 467 vols 1 - 3 1949-58 (Washington DC)
10   Matthew J A D Surface Science 40 451 1973
11   Bergström I and Hill R D Ark Fys 8 21 1954
12   Chung M F and Jenkins L H Surface Science 22 479 1970
13   Matthew J A D   J Phys B 10 783 1977
14   Aksela H and Aksela S   J Phys B 7 1262 1974
15   Gallon T E & Nuttall J D Surface Science 53 698 1975
16   Matthew J A D (in 40)
17   Wagner C D and Biloen P Surface Science 35 82 1973
18   Matthew J A D Surface Science 40 451 1973
19   Citrin P H and Hamann D G   Chem Phys Lett 22 301 1973
20   Matthew J A D J Phys C 8 L136 1975
21   Hillig H, Cleff B, Melhorn W and Scmitz W   Z Phys 268
        225 1974
22   Kowalczyk S P, Ley L, McFeeley F R, Pollak R A & Shirley D A
        Phys Rev B 9 3573 1974
23   Citrin P H & Thomas J D J Chem Phys 57 4446 1972
24   Siegbahn K et al   ESCA Applied to Free Molecules (North
        Holland, Amsterdam 1971)
25   Kelly P H Phys Rev A 11 556 1975
26   Shirley D A   Chem Phys Lett 16 220 1972
27   Kowalczyk S P, Pollak R A, McFeeley F R, Ley L & Shirley D A
        Phys Rev B 8 2387 1973
28   Kowalczyk S P, Ley L, McFeeley F R, Pollak R A & Shirley D A
        Phys Rev B 9 381 1974
29   Weightman P J Phys C 9 1117 1976
30   Slater J C   Quantum Theory of Atomic Structure McGraw-Hill
        New York 1960
31   Kim K S, Gaarenstroom S W & Winograd N   Phys Rev B 14
        2281 1976
32   Hoogewijs R, Fiermans L & Vennik J Chem Phys Lett 38
        192 1976
33   Hoogewijs R, Fiermans L & Vennik J Chem Phys Lett 38
        471 1976

34   Aksela S, Väyrynen J & Aksela H   Phys Rev Lett $\underline{33}$
      999 1974
35   Carlson T A   Photoelectron and Auger Spectroscopy
      Plenum Press New York 1975
36   Fiermans L, Hoogewijs R and Vennik J   Surface Sci $\underline{47}$
      1 1975
37   Nuttall J D & Gallon T E   J Phys C $\underline{9}$ 4063 1976
38   Matthew   J A D   Phys Lett $\underline{32A}$ 361 1970
39   Citrin P H,   J Elec Spec $\underline{5}$ 273 1974
40   Gallon T E & Matthew J A D   Phys Stat Sol $\underline{41}$ 343 1970
41   Citrin P H Phys Rev Lett $\underline{31}$ 1164 1973
42   Matthew J A D & Komninos Y   Surface Science $\underline{53}$ 716 1975
43   Matthew J A D Surface Science $\underline{20}$ 183 1970
44   Citrin P H, Eisenberger P, Hamann D R Phys Rev Lett
      33 965 1974
45   Matthew J A D & Devey M G   J Phys C $\underline{7}$ L335 1974
46   Fox J H, Nuttall J D & Gallon T E Surface Science $\underline{63}$
      390 1977
47   Werme L O, Bergmark T & Siegbahn K   Phys Scripta $\underline{8}$
      149 1973
48   Werme L O, Bergmark T & Siegbahn K Phys Scripta $\underline{6}$
      141 1972
49   Hagmann S, Hermann G & Mehlhorn W   Z Physik $\underline{226}$
      189 1974
50   Hedin L   Arkiv för Fysik $\underline{30}$ 231 1965
51   Flynn C P & Lipari N O   Phys Rev B $\underline{7}$ 2215 1973
52   Gadzuk J W   Phys Rev B $\underline{14}$ 2267 1976
53   Almbladh C O & von Barth U   Phys Rev B $\underline{13}$ 3307 1976
54   Amelio G F   Surf. Sci $\underline{22}$ 301 1970
55   Powell C J   Phys Rev Lett $\underline{30}$ 1179 1973
56   Jackson A J, Tate C, Gallon T E, Bassett P J & Matthew J A D
      J Phys F $\underline{5}$ 363 1975
57   Barrie A & Street F J   J Elec Spec $\underline{7}$ 1 1975
58   Fuggle J C, Watson L M, Norris P R & Fabian D J
      J Phys F $\underline{5}$ 590 1975
59   Arnott D R & Hanemann D   Surf Sci $\underline{45}$ 128 1974
60   Ferrer S, Baró A M & Salmerón M   Sol Stat Comm 16 651
      1975
61   Amelio G F & Schiebner E J   Sur Sci $\underline{11}$ 242 1968
62   Bassett P J, Gallon T E, Prutton M & Matthew J A D
      Surf Sci $\underline{33}$ 213 1972
63   Siegbahn K et al   ESCA Atomic, molecular and solid state
      structure studied by means of electron spectroscopy
      (Almqvist and Wiksells, Uppsala) 1967
64   Salmerón M, Baró A M & Rojo J M Phys Rev B $\underline{13}$ 4348 1976
65   Salmerón M, Baró A M & Rojo J M Surf Sci $\underline{53}$ 689 1975
66   Yin L, Tsang T, Adler I & Yellin E   J App Phys $\underline{43}$ 3464 1972

67   Powell C J & Mandl A Phys Rev Lett 29   1153 1972
68   Antonides E, Janse E C & Sawatzky G A Phys Rev B 15
        1669 1977
69   Bassett P J, Gallon T E, Matthew J A D & Prutton M
        Surf Sci 35 63 1973
70   Gadzuk J W   Phys Rev B 9 1978 1974
71   Feibelman P J, McGuire E J & Pandey K C   Phys Rev B
        15 2202 1977
72   Feibelman P J & McGuire E J   Phys Rev B 15 3575 1977
73   Sawatzky G A Phys Rev Lett 39 504 1977
74   Yin L, Tsang T & Adler I   Phys Lett 57A 193 1976
75   Yin L, Tsang T & Adler I   J Elec Spec 9 67 1976
76   Antonides E, Janse E C & Sawatzky G A Phys Rev B 15, 4596,
        1977
77   Holloway P H J Elec Spec 7 215 1975
78   Taylor N J   Rev Sci Instr. 40 792 1969
79   Hall P M, Morabito J M & Conley D K Surf Sci 62 1 1977
80   Grant J T, Hooker M P, Springer R W & Haas T W
        Surf Sci 60 1 1976
81   Powell C J Rev Mod Phys 48 33 1976
82   Gryzinski M   Phys Rev 138 336 1965
83   Worthington C R & Tomlin S G Proc Phys Soc A69 401 1956
84   Drawin H W   Z Phys 172 429 1963
85   Lotz W   Z Phys 232 101 1970
86   Smith D M, Gallon T E & Matthew J A D J Phys B 7 1255 1974
87   Vrakking J J & Meyer F Phys Rev A 9 1932 1974
88   Vrakking J J & Meyer F Surf Sci 47 50 1975
89   Smith D M & Gallon T E J Phys D 7 151 1974
90   Gallon T E J Phys D 5 822 1972
91   Bishop H E & Rivière J C. J Appl Phys 40 1740 1969
92   Tarng M L & Wehner G K   J App Phys 42 1539 1973
93   Brundle C R Surf Sci 48 99 1975
94   Powell C J Surf Sci 44 29 1974
95   Seah M P Surf Sci 32 703 1972
96   Chang C C Surf Sci 48 9 1975
97   Carlson T A & McGuire G E J Elec Spec 1 161 1972
98   Palmberg P W Appl Phys Lett 13 183 1968
99   Chang C C in Characterisation of Solid Surfaces Ed Kane P F
        & Larrabee G B   Plenum Press, New York 1974
100 Morabito J M Thin Solid Films 19 21 1973
101 Gallon T E, Prutton M & Wray L J Vac Sci Technol. 9
        911 1972
102 Joshi A, Davis L E & Palmberg P W in Methods of Surface
        Analysis ed Czanderna A W Elsevier Scientific Pub Co
        Amsterdam 1975
103 Palmberg P W, Riach G E, Weber R E & MacDonald N C
        Handbook of Auger Electron Spectroscopy, Physical
        Electronics Industries, Edina 1972

104   Von Santer R A, Toneman L H & Bouwman R   Surf Sci
       $\underline{47}$ 64 1975
105   Rynd J P & Rastogi A K   Surf Sci $\underline{48}$ 22 1975
106   Braun P & Farber W   Surf Sci $\underline{47}$ 57 1975
107   Wehner G K   in Methods of Surface Analysis   ed Czanderna A W
       Elsevier Scientific Pub Co Amsterdam 1975
108   Florio J V & Robertson W D   Surf Sci $\underline{18}$ 398 1968
109   Housley M & King D A Surf Sci $\underline{62}$ 81 1977
110   Musket R G   Surf Sci $\underline{21}$ 440 1970
111   Gallon T E   Surf Sci $\underline{17}$ 486 1969
112   Jackson D C, Gallon T E & Chambers A   Surf Sci $\underline{36}$ 381
       1973
113   Armstrong R A   Surf Sci $\underline{50}$ 615 1975
114   Bauer E & Poppa H   Thin Solid Films 12 167 1972
115   Matthews J W, Jackson D C & Chambers $\overline{\text{A}}$ Thin Solid Films
       $\underline{26}$ 129 1975
116   Chambers A & Jackson D C   Phil Mag $\underline{31}$ 1357 1975
117   Argile C & Rhead G E Surf Sci $\underline{53}$ 659 1975
118   Robert, E D, Weightman, P & Johnson C E   J Phys C $\underline{8}$, L301
       1975
119   Weightman, P, McGilp, J F & Johnson, C E   J Phys C $\underline{9}$,
       L585, 1976
120   Penn, D R   Phys Rev B $\underline{13}$, 5248, 1976
121   Penn, D R   J Elec Spec $\underline{9}$, 29, 1976

STUDIES OF ADSORBATE ELECTRONIC STRUCTURE USING ION NEUTRALIZATION

AND PHOTOEMISSION SPECTROSCOPIES

Homer D. Hagstrum

Bell Laboratories

Murray Hill, N. J. 07974 USA

I.   INTRODUCTION

Electron spectroscopies have become the principal means for investigating the electronic structure of the surface monolayer of a solid. Chemisorption has played an important role in this work since by it one can in a controlled way vary this electronic structure. Thus chemisorption plays an analogous role in the study and modification of surface electronic structure to bulk doping in the study and modification of the electronic structure of bulk solids. The study of adsorbate electronic structure is important in its own right since it illuminates the chemistry of surface bonding and touches the important field of chemical reactions at surfaces.

Electron spectroscopy is peculiarly suited to the study of surfaces by virtue of its surface sensitivity and several electronic spectroscopies have been devised. In this paper we shall discuss two such spectroscopies: ion neutralization spectroscopy (INS) and ultraviolet photoemission spectroscopy (UPS). Our particular emphasis will be on INS, discussing not only its specific development and characteristics but also the electronic transition process upon which it is based. UPS will be discussed as a contrasting electron spectroscopy that illuminates the specific characteristics of each, thus enabling us to demonstrate how more can be learned by the use of two spectroscopies that focus on the same problem than by either spectroscopy alone.

In this paper we shall spend a considerable amount of time discussing the electronic transition processes that can occur between a solid surface and an excited and/or ionized atom slowly

incident on the surface. This body of surface investigation is of
interest in its own right since it comprises the low energy end of
a spectrum of particle-solid interactions. The Auger neutralization
process upon which INS is based is one of a family of interrelated
electronic transition processes. The potential energy stored
within the ionized or excited atom is the driving force for these
electronic interactions. The electronic transitions are studied by
detecting and measuring the yield and kinetic energy distributions
of electrons in those processes, Auger in character, that are
capable of ejecting electrons from the solid surface. We shall be
interested in presenting both our theoretical understanding of low
energy, particle-solid electronic interactions as well as the
experimental evidence supporting our conclusions.

The material to be presented here falls into three main
categories. First, the nature of the electronic interactions
between a solid surface and an excited and/or ionized atom. Second,
the development of INS as a viable, two-electron spectroscopy based
on the Auger neutralization process. Third, the use of INS and UPS
in the study of adsorbate electronic structure. This organization
of the material follows essentially the path of the historical
development of the subject. There remain, however, many intriguing
problems to be solved in each of these aspects of the field.

More specifically, this paper is organized as follows: The
electronic transition processes are presented in Sec. II, followed
by discussions in Sec. III of what we may expect to learn from the
studies of such processes and in Sec. IV of experimental apparatus
and techniques. Transition rate and probability functions needed
in our understanding of the electronic processes are derived in
Sec. V. This is followed by detailed discussions of the Auger
neutralization process in Sec. VI, the evidence for atomic energy
level variation near a solid surface in Sec. VII, and the specific
transition rate parameters that we can derive for the Auger
neutralization process in Sec. VIII. The process of Auger de-
excitation of an incident excited atom is discussed in Sec. IX, the
tunneling processes for unexcited ions in Sec. X, and the two-stage
process of electron ejection for unexcited ions in Sec. XI. Some
work on two-stage ejection processes for excited ions is presented
in Sec. XII and on oscillatory ion scattering from surfaces in
Sec. XIII. The phenomenon of "kinetic ejection" of electrons by
ions is discussed in Sec. XIV and the various reasons for energy
broadening in the Auger neutralization process in Sec. XV. The
method of INS is presented in Sec. XVI and the question of what
INS and UPS specifically measure is considered in Sec. XVII. Experimental studies of INS and UPS of some adsorbate systems are
summarized in Sec. XVIII. The use of INS to detect wave function
variation outside a solid surface is discussed in Sec. XIX. The
paper is concluded in Sec. XX with a consideration of the
uniquenesses and limitations of INS.

II. ELECTRONIC TRANSITION PROCESSES

The two types of electronic transition in which we are most interested are the adiabatic, resonant tunneling processes and the Auger ejection processes.[1] These are illustrated, respectively, by electron energy diagrams in Figs. 1 and 2. In these figures are shown the filled levels in a metal below the Fermi level, FL, which, in turn, lies below the vacuum level, VL, by an energy interval equal to the work function, ϕ. Electronic transitions are indicated by arrows, initially filled electron states in atom or solid by filled circles, initially empty states by open circles. The initial state lifetime, τ, for each process occurring for a particle moving slowly toward the surface is also indicated on the figures.

The resonance neutralization of an ion to an excited level of the neutralized atom that lies below the Fermi level is indicated at the left in Fig. 1. In this process an electron tunnels from a filled level in the solid into the excited level at the same energy. Resonance neutralization to an atomic ground state is, of course, also possible if the ground state level lies opposite the filled band of the solid. Resonance ionization of an atom at a surface is possible if the filled electronic level in the atom lies above the Fermi level near the surface and thus lies opposite unfilled levels in the metal. Again this may be the filled ground level if the effective ionization energy, E_i', of the atom near the surface is less than the work function ϕ. For electrons in excited levels the energy condition required for resonance ionization is $(E_i' - E_x') < \phi$, where E_x' is the effective excitation energy near the surface. If each of the processes of Fig. 1 is considered to be truly resonant the kinetic energy of the incoming particle may be altered by the

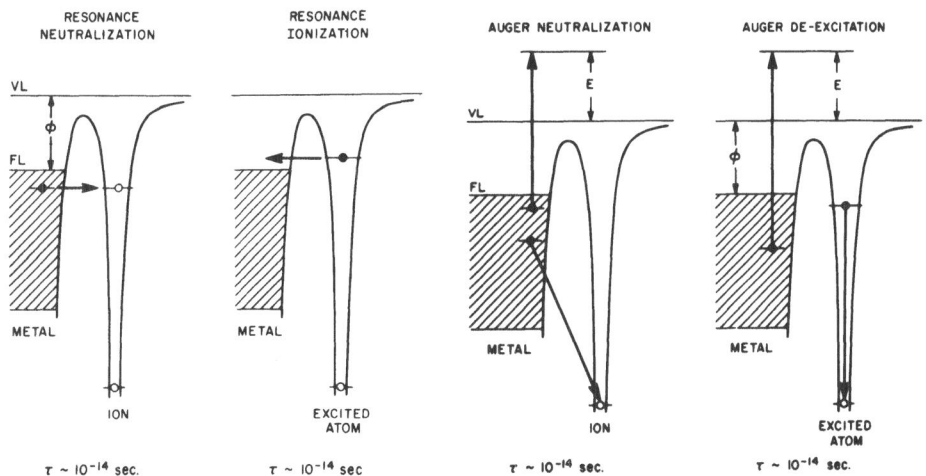

| RESONANCE NEUTRALIZATION | RESONANCE IONIZATION | AUGER NEUTRALIZATION | AUGER DE-EXCITATION |

$\tau \sim 10^{-14}$ sec. $\tau \sim 10^{-14}$ sec $\tau \sim 10^{-14}$ sec. $\tau \sim 10^{-14}$ sec.

Fig. 1   Fig. 2

particle-solid interaction but kinetic energy as such is not
involved in the electronic transition process directly. In the
low energy range any non-adiabaticity of the so-called resonance
tunneling processes may be thought to broaden the range of
continuum levels in the solid that are involved in the process
(Sec. XV).

The second category of fast, and thus important, electronic
transition process is Auger in character and involves the simultaneous transition of two electrons. Two such processes are
illustrated in Fig. 2. In Auger neutralization (left of Fig. 2) an
electron from the filled band of the solid tunnels into the atomic
well and drops in energy into the atomic ground state. A second
electron picks up the energy thus released and becomes a fast
excited electron that may, if properly directed leave the solid.
The so-called "down" and "up" electrons lose and gain, respectively,
the same amount of energy in this radiationless process. The
initial states of each of these electrons may lie anywhere in the
filled band of the solid.

In Auger de-excitation (right of Fig. 2) an excited atom is
de-excited with the simultaneous ejection of an electron from the
system. There are two energetically equivalent processes, one,
that shown in Fig. 2, and the other, that in which the up electron
originates in the excited level in the atom and the down electron
in the filled band of the solid. In either case one of the two
electrons must originate in a specific level, namely the atomic
excited level, and cannot originate in a band of levels as does the
other electron in Auger de-excitation or both electrons in Auger
neutralization. Thus Auger de-excitation, although a two-electron
Auger process, is, in fact, quasi one-electron in character. This
property causes Auger de-excitation to have fundamentally different
energetic character resulting in a quite different kinetic energy
distribution of the excited electrons from that of Auger neutralization. The excited or up electrons in each Auger process may
appear outside the solid if they possess momentum normal to the
surface sufficient to surmount the surface potential barrier. Each
of the Auger processes, as the name implies, is a radiationless,
two-electron reorganization of the particle-solid system in which,
in a sense, the system de-excites itself with the production of a
fast excited electron in the continuum.

Two possible competing processes to those of Figs. 1 and 2 are
shown in Fig. 3. Radiative neutralization (left of Fig. 3) in
which the neutralization energy of the ion is emitted as a photon
is a possibility of low probability because the initial state lifetime of 10^{-8} sec is about 10^6 times longer than that of either the
resonance tunneling or Auger processes. Near resonant non-adiabatic
electron tunneling between the ground level in the atom and a core

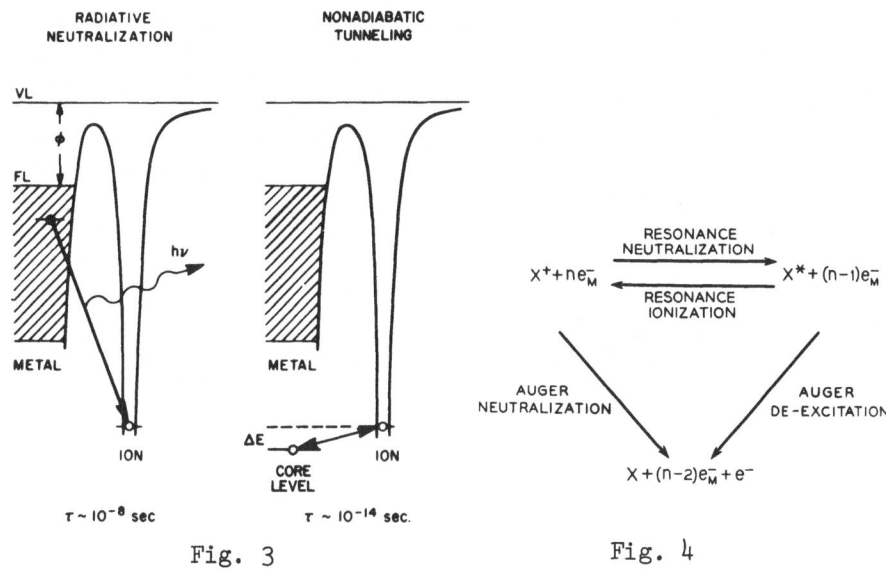

Fig. 3   Fig. 4

level in the solid is possible and determines the oscillatory character of elastic ion scattering (Sec. XIII). This process, shown schematically at the right of Fig. 3, is possible for certain atom-solid systems in which the atomic ground level and the solid core level lie within approximately 10 eV of each other. However, for the systems of particular interest to us here the resonance and Auger processes of Figs. 1 and 2 completely dominate the picture. Although they are roughly equally fast for equivalent wave function overlap, the resonance tunneling and Auger processes occur sequentially for a slow ($\sim$ 10 eV) incoming particle. This results from the fact that because the tunneling processes occur principally high in the band near the Fermi level, appreciable wave function overlap occurs first for these processes. The Auger processes become probable nearer the surface as discussed in Sec. XI.

The specific processes discussed here and illustrated in Figs. 1 and 2 by no means exhausts the possibilities of particle-solid electronic interactions. As we shall see in Sec. XII the use of metastably excited ions as incident particles provides another set of processes. The study of these illuminates our theoretical understanding of electronic interactions in general.

III. WHAT WE CAN LEARN FROM STUDYING PARTICLE-SOLID ELECTRONIC INTERACTIONS

At this point we pause to ask: What can we learn from the study of the electronic interactions involved in slow ion-solid encounters? Of what use is this knowledge in the study of solid surfaces?

First, we shall find that the tunneling and Auger processes
are interrelated in interesting ways. In fact, they come in
families as is illustrated by the "triangle" diagram of Fig. 4.
Here the atomic particle and the electrons, in the solid or free,
are indicated in a chemical-like notation. If the system of the X^+
ion and the n electrons resident in the metal, $X^+ + ne_M^-$, is
resonance neutralized to an excited state, the system $X^* + (n-1)e_M^-$
results. Similarly this latter system can be returned, if energies
are right, to the former via resonance ionization. The two Auger
processes transform their respective starting configurations to the
final state $X + (n-2)e_M^- + e^-$. It should be noted, however, that
the kinetic energy of the free electron e^- and the energy levels of
the two holes left in the system differ between these Auger
processes.

The triangle of Fig. 4 indicates that there are two distinguishable processes by which an unexcited ion incident on a solid
can be neutralized with the ejection of an electron. These are
the direct process of Auger neutralization first discussed by
Shekhter[2] and the two-stage process of resonance neutralization
followed by Auger de-excitation discussed by Massey,[3] Shekhter,[2]
and Cobas and Lamb.[4] Oliphant and Moon[5] had earlier suggested the
possibility of resonance neutralization of an ion to an excited
state.

A second type of understanding we achieve in the study of
these electronic transition processes concerns the relative
kinetics of the various processes. In some instances, as we shall
see in Sec. XI, two types of process in a sense come into direct
competition enabling us to draw conclusions about their relative
kinetics. We can also determine absolute values for transition
rate parameters (Sec. V).

A third area of understanding encompasses the energetics of
the interaction of the atomic particle in various excitation and/or
ionization states with the solid surface. We obtain direct evidence
for the variation of energy levels caused by such particle-solid
interactions (Sec. VII).

Finally our understanding of these processes have enabled us
to develop a viable two-electron, Auger-type electron spectroscopy
(Sec. XVI). This spectroscopy (INS) has its own peculiar transition
probability and surface sensitivity characteristics. Use of this
spectroscopy, particularly in conjunction with a one-electron
spectroscopy (UPS) has led to a greater understanding of certain
aspects of surface electronic structure (Secs. XVII, XVIII, XIX).

IV. EXPERIMENTAL APPARATUS AND TECHNIQUES
The study of any of the transition processes we have been
discussing is accomplished via the detection of the electrons

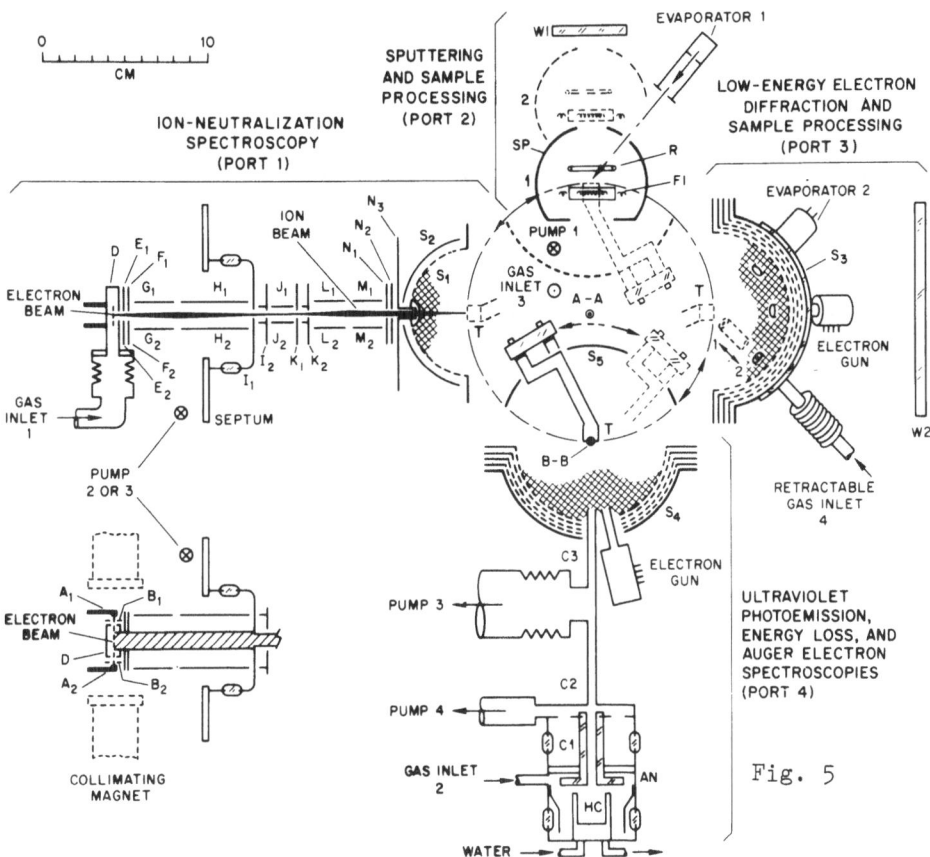

Fig. 5

ejected in one or the other of the possible Auger processes. Thus
the experimental apparatus must include means for generating a well
focussed slowly moving ($\sim$ 5 to 10 eV) ion beam and for measuring
the kinetic energy distribution of the ejected electrons.[6] Means
for processing and characterizing the solid surface and for studying it using other electron spectroscopies must also be provided.[7]

A succession of increasingly complex instruments has been
developed over the years to study electron ejection processes at
surfaces induced by slowly moving incident ions.[8] The latest
version is shown schematically in Fig. 5. This is a top view of
an apparatus whose central element is contained in a three-dimensional X-Y-Z cross with four ports in the horizontal plane to which
are attached the equipment indicated in Fig. 5 at ports 1, 2, 3,
and 4. The top port holds the sample positioning mechanism that
permits motion of the single crystal sample T from port to port by
rotation of the entire mechanism about the axis A-A and enables
the variation of incidence angle at T for ions (port 1) or photons
(port 4) by rotation of T about axis B-B.[9] The bottom port is
connected to the vacuum pumps.

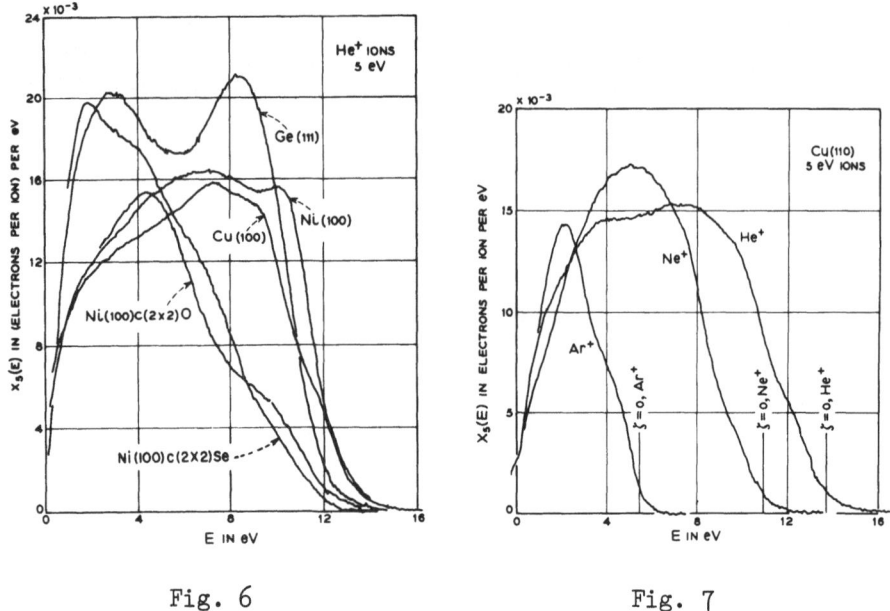

Fig. 6   Fig. 7

The target is cleaned by sputtering at port 2 and its surface
crystallography studied by low energy electron diffraction at
port 3. The clean surface can be modified by the adsorption of
atoms provided by evaporation (ports 2 and 3) or by gas inlet
through a retractable tube (port 3). The apparatus provides the
possibility of comparing electron ejection from solids by ions
produced by electron impact in a gas (port 1) and by photons
generated in a capillary resonance lamp (port 4). Ejected electrons
are analyzed with respect to kinetic energy in the gridded retarding
potential analyzers of ports 1 and 4. Chemical characterization of
the surface is accomplished by Auger electron spectroscopy using the
electron gun and grid system at port 4.

As we have said, our means of study is via the measurements
made on the electrons ejected in the Auger processes. A number of
such distributions [here called $X_5(E)$] are shown in Fig. 6 for
5 eV He^+ ions incident on several different well characterized
surfaces.[10] Those labelled Ge(111), Ni(100), and Cu(100) are
atomically clean. The other two Ni(100) surfaces have ordered
layers of chalcogen atoms adsorbed upon them. In Fig. 7 are shown
distributions from the same surface for the singly-charged noble
gas ions.[11] In Fig. 8 distributions are plotted for a given ion
incident at various kinetic energies $E_k(He^+)$ on clean, polycrystalline Mo.[12] It is also possible to plot the total yield γ_i
in electrons per incident positive ion as a function of the ion's
kinetic energy [$E_k(X^+)$ or K]. This is shown in Fig. 9 for the
singly-charged noble gas ions incident on polycrystalline Mo.[12]

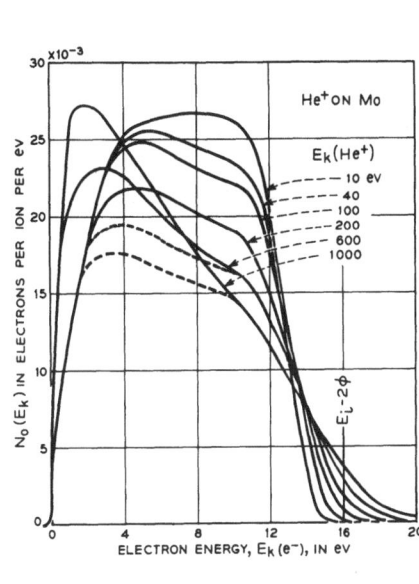

Fig. 8

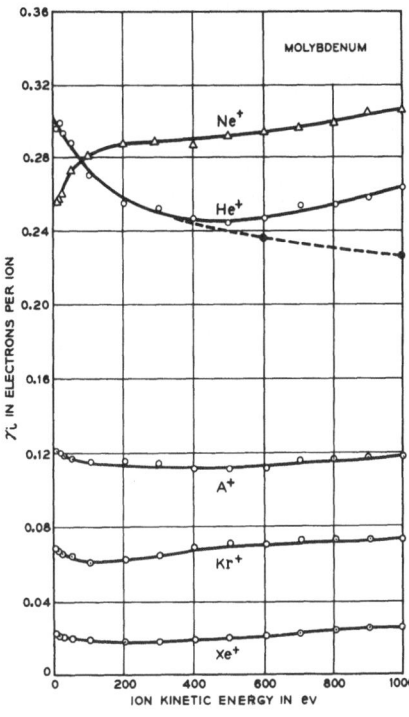

Fig. 9

γ_i is, of course, the integrated area under the kinetic energy distribution curves [$N_0(E_k)$ or $X(E)$] shown in Figs. 6, 7, and 8.

It is clear that the resonance tunneling processes at the top
of the triangle of Fig. 4 cannot be detected directly since no free
electron is produced by either of them. When they occur they occur
as precursors to Auger ejection processes thereby reducing the
number of incident particles participating in a direct process.
Resonance neutralization is the possible precursor of Auger de-
excitation for incident ions and resonance ionization is the
possible precursor of Auger neutralization for incident excited
atoms. Whether one or the other of these two-stage processes is
possible relative to its companion direct process· is determined by
the position of the excited level relative to the Fermi level in
the metal. Each two-stage process has both a different γ_i and a
different $N_0(E_k)$ from its companion direct process, which facts
enable us to detect its occurrence.

How the ion neutralization data presented above are processed
by the methods of INS is discussed in Sec. XVI. Examples of data
obtained with the photoemission apparatus are presented and discussed in Sec. XVII and XVIII.

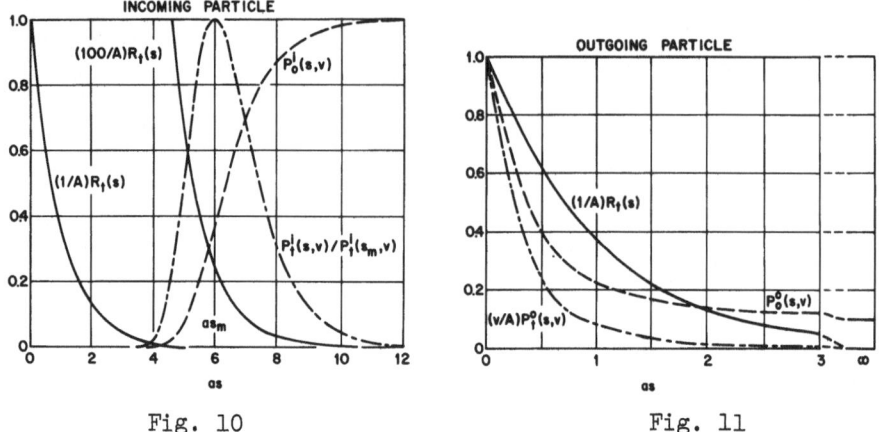

Fig. 10   Fig. 11

V.   TRANSITION RATE AND PROBABILITY FUNCTIONS

The basic function specifying the rate at which a process occurs is the transition probability per unit time or the so-called transition rate $R_t(s)$. It is a function of the process and of the distance of the particle from the surface. For particles held at a specified separation s it gives the rate at which they become involved in a specific process. There are two probability functions that are convenient to define.[1] These have been discussed by many authors as far back as 1930.[3] They are the probability $P_o(s,v)$ that a particle of normal velocity v will reach s in its original charge or excitation state and the probability $P_t(s,v)ds$ that a particle of normal velocity v will undergo an electronic transition in the distance element ds at the distance s. These probabilities can be defined for incoming particles incident on the surface, P_o^i and P_t^i, and for outgoing particles leaving the surface, P_o^o and P_t^o. These functions may be derived from the basic rate function $R_t(s)$ as indicated in Table 1.

For the specific choice of an exponential rate function $R_t(s) = A\exp(-as)$ the resulting P_o and P_t functions are given for incoming particles in Table 2 and for outgoing particles in Table 3. The functions of Tables 2 and 3 are plotted, respectively, in Figs. 10 and 11. Note the quite different forms of these functions. For the incoming particle the P_t^i function is peaked at the separation $s = s_m$ with s_m given, for $R_t(s) = A\exp(-as)$, in Table 2. Thus as v is changed both P_o^i and P_t^i in Fig. 10 move relative to $R_t(s)$. The same is true for the P_o^o and P_t^o functions of Fig. 11. The specific velocity chosen in each case is that which for Fig. 10 makes $as_m = 6$, and for Fig. 11 makes the survival probability at $s = \infty$, $P_o^o(\infty,v) = \exp(-A/av)$, equal 0.1. In the case of an exponential rate function for incoming particles

TABLE 1

Derivations of Probability Functions from the Rate Function

INCOMING	OUTGOING
$dP_o^i/ds = (R_t/v)P_o^i$	$dP_o^o/ds = -(R_t/v)P_o^o$
$P_o^i = \exp[-\int_s^\infty (R_t/v)ds]$	$P_o^o = \exp[-\int_0^s (R_t/v)ds]$
$P_t^i = (R_t/v)P_o^i$	$P_t^o = (R_t/v)P_o^o$

TABLE 2

Specific Functions for an Incoming Particle

	$s=0$	$s=s_m$	$s=\infty$
$R_t(s) = A\exp(-as)$	A	$A\exp(-as_m)$	0
$P_o^i(s,v) = \exp[-(A/av)\exp(-as)]$ $\quad = \exp\{-\exp[-a(s-s_m)]\}$	$\exp(-A/av)$	$1/e$	1
$P_t^i(s,v) = (A/v)P_o^i(s,v)\exp(-as)$ $\quad = aP_o^i(s,v)\exp[-a(s-s_m)]$	$(A/v)P_o^i(0,v)$	a/e	0

$$s_m = (1/a)\ln(A/av)$$

TABLE 3

Specific Functions for an Outgoing Particle

	$s=0$	$s=\infty$
$R_t(s) = A\exp(-as)$	A	0
$P_o^o(s,v) = \exp\{(A/av)[\exp(-as)-1]\}$	1	$\exp(-A/av)$
$P_t^o(s,v) = (A/v)P_o^o(s,v)\exp(-as)$	A/v	0

the functions for P_o^i and P_t^i written in terms of $(s-s_m)$ indicate that these two functions do not change form as v is varied but move "bodily" closer to or farther away from the surface as v increases and decreases, respectively.

Figures 10 and 11 illustrate graphically a basic difference between the cases for incoming and outgoing particles. We note that the incoming processes are limited by the $P_t^i(s,v)$ function to distances where $R_t(s)$ is a small fraction (< 0.01) of its value at the surface. On the other hand for outgoing particles transition processes can occur at all distances from the surface. This circumstance makes the exponential rate function a better approximation for incoming particles than for outgoing particles since $R_t(s)$ is closest to an exponential far from the surface and deviates most from an exponential close to the surface. We shall have occasion to calculate the parameters A/a for the Auger neutralization process in Sec. VIII.

VI.   THE AUGER NEUTRALIZATION PROCESS

We discuss now the process of Auger neutralization depicted as the left leg of the triangle of Fig. 4. It is the only process in the family that can occur for incident unexcited ions if the work function, ϕ, of the solid is greater than the effective ionization energy of the atom's lowest lying excited state near the surface.

Consider an elemental Auger neutralization process in which the two electrons involved lie initially α and β below the vacuum level (VL). Inspection of the electron energy diagram[1] for this process in Fig. 12 leads us to conclude that the kinetic energy of the excited electron outside the solid is $E_k(e^-) = E_i' - \alpha - \beta$. The maximum of E_k, $(E_k)_{max} = E_i' - 2\phi$, occurs when $\alpha = \beta = \phi$ and both electrons lie initially at the Fermi level of the metal. The minimum kinetic energy is $(E_k)_{min} = E_i' - 2\varepsilon_o$ when both electrons lie initially at the bottom of the conduction band ε_o below VL. The unknown quantity among these energies is the effective ionization energy of the parent atom, or the neutralization energy of the ion, at the distance from the surface where the electronic transition occurs.

The energy level variations occurring in an atom near a surface is best depicted by the potential energy diagram. Figure 13 is such a diagram for Auger neutralization.[1] Here the energies of initial and final states are plotted as functions of separation of the atomic particle from the surface. The initial state, $He^+ + ne_W^-$, specified as a He^+ ion plus n electrons in the metal here taken as tungsten W, varies in energy at larger distances by virtue of the image force interaction, $-3.6(eV)/s(Å)$. Closer to

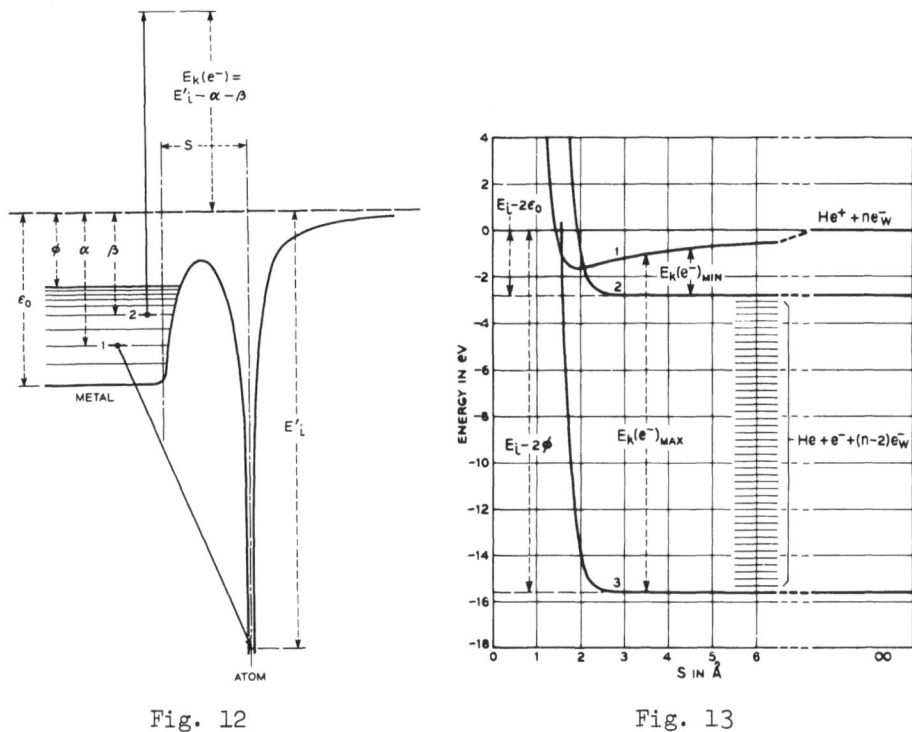

Fig. 12   Fig. 13

the surface a repulsive interaction sets in. We do not know exactly
where such interactions become significant. Those shown in Fig. 13
are estimates based on atom-atom interactions and particle
diameters.[1] There is a band of final states in Fig. 13, He + e$^-$ +
(n-2)e$_{\overline{W}}$ corresponding to the removal of the 2 electrons from
positions in the conduction band varying from ϕ to ε_o below the
vacuum level. This results in a band of final states $2(\varepsilon_o-\phi)$ in
width. The final state atom He interacts with the solid via the
van der Waals interaction whose magnitude is so small as to be
invisible on the scale of the figure.

If the Auger process is essentially adiabatic we expect
electronic transitions between potential curves for initial and
final states to obey the Franck-Condon principle. This states that
during the electronic transition neither position nor momentum of
the nucleus of the atomic particle can change appreciably. This
means that transitions, as for diatomic molecules, occur on the
potential diagram vertically and that the kinetic energy of He
after the electronic transition equals that of He$^+$ before the
transition. From this latter fact we conclude that the energy
distance between curve 1 and a curve in the manifold between curves
2 and 3 must equal the kinetic energy of the free electron, $E_k(e^-)$.
$E_k(e^-)_{max}$, the distance between curves 1 and 3, when equated to

$E_i - 2\phi$ yields a means of estimating E_i'. This was done in Ref. 1 with the result:

$$E_i - E_i' = B_n \exp(-b_n s) - B_i \exp(-b_i s) + 3.6/s, \tag{1}$$

where the first and second exponentials are the repulsive terms for the neutral and ionized atom, respectively, and the third term is the image force interaction of He^+ with the solid in eV with s in Å. The van der Waals interaction of He with the solid is negligible with respect to the other terms.

We may translate some of the conclusions we have drawn above from the potential energy diagram of Fig. 13 to an electron energy diagram. This is done in Fig. 14 where the effective ionization energy $E_i'(s)$ is shown as a function of particle-solid separation, s. On this electron energy diagram[1] are shown energy distribution curves determined under various assumptions for the specific simple case of a constant density of initial states $N_c(\varepsilon)$ and constant Auger transition probability through the energy band. The Auger transitions take place at distances from the surface specified by the $P_t(s,v_0)$ function, shown here with its maximum at s_t (s_m of Table 2), the distance at which the transition most probably occurs.

If we neglect energy level change and energy broadening we obtain curve 1 for $N_i(\varepsilon_k)$, the kinetic energy distribution of electrons inside the solid before any have crossed the surface barrier. $N_i(\varepsilon_k)$ is proportional to the probability for producing electrons in the energy range $d\varepsilon_k$ at ε_k. For the elemental process shown this must be proportional to the product of the initial state electron densities at $\varepsilon = \varepsilon'$ and ε'', that is, to $N_c(\varepsilon')N_c(\varepsilon'') = N_c(\varepsilon+\Delta)N_c(\varepsilon-\Delta)$. We note that there is an infinite set of paired initial states yielding electrons at ε_k, namely those symmetrically disposed $\pm\Delta$ from ε, the halfway point between the levels at ε_k and $-E_i$. Thus:

$$N_i(\varepsilon) \propto \int_{-(\varepsilon-\varepsilon_F)}^{(\varepsilon-\varepsilon_F)} N_c(\varepsilon+\Delta)N_c(\varepsilon-\Delta)d\Delta, \tag{2}$$

with $N_i(\varepsilon_k)$ obtained from $N_i(\varepsilon)$ using the transformation:

$$\varepsilon_k = E_i' - \varepsilon_o + 2\varepsilon. \tag{3}$$

Equation (3) results from equating the magnitudes of the transitions of the down and up electrons. Since $E_k = \varepsilon_k - \varepsilon_o$ we may also write:

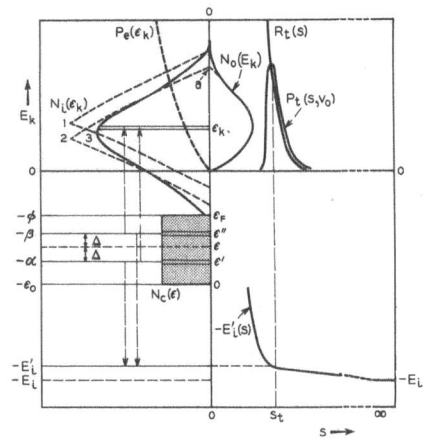

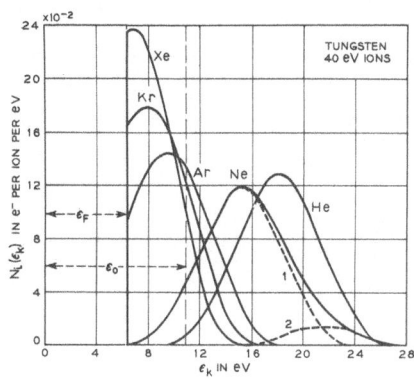

Fig. 14   Fig. 15

$$E_k = E_i' - 2(\varepsilon_o - \varepsilon). \tag{4}$$

The function $N_c(\varepsilon)$, which is constant between $\varepsilon = 0$ and ε_F and is
zero outside these limits, folds to a triangle of base width $2\varepsilon_F$.
Curve 1 for $N_i(\varepsilon_k)$ in Fig. 14 is slightly distorted from a simple
equilateral triangle by the variation of the free-electron-like
density of final states which enters into the transition
probability.[1]

If we now use E_i' at $s = s_t$ instead of E_i we obtain curve 2 of
Fig. 14 for $N_i(\varepsilon_k)$. It lies lower in energy than curve 1 because
of the reduced ion neutralization energy. Finally, if we broaden
curve 2 by convolution with a Lorentzian or a Gaussian to account
approximately for the energy broadening inherent in the Auger
process[13] we obtain curve 3 as the internal energy distribution of
Auger electrons. $P_e(\varepsilon_k)$ is the probability of electron escape
over the surface barrier and $N_o(E_k)$ is the external distribution
in kinetic energy measurable outside the solid (Fig. 6). The total
yield γ_i is, of course,

$$\gamma_i = \int_0^\infty N_o(E_k)dE_k. \tag{5}$$

$N_i(\varepsilon_k)$ functions obtained as illustrated for He^+ ions in
Fig. 14 are plotted for all the singly charged noble gas ions in
Fig. 15.[1] Curve 1 for Ne^+ ions is that obtained for pure Auger
neutralization. We shall discuss curve 2 of Fig. 15 in Sec. X.
It represents a small admixture of electrons from Auger de-
excitation in the partition between these two processes.

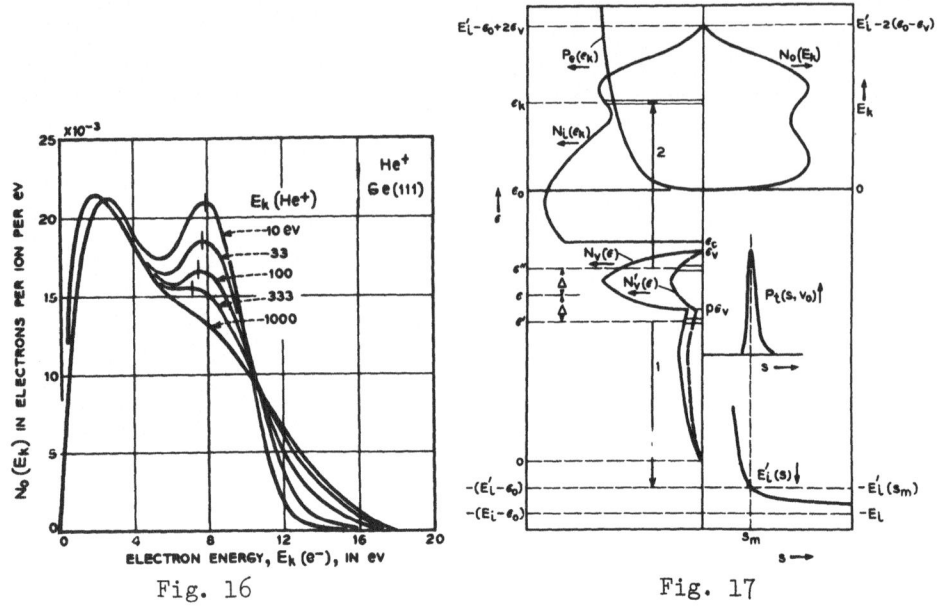

Fig. 16   Fig. 17

VII.   EVIDENCE FOR VARIATION OF ATOMIC ENERGY LEVELS NEAR A SURFACE

The simple theory underlying Fig. 14 predicts that, as E_i' is reduced, characteristic features of $N_i(\varepsilon_k)$ that are reflected in the measured $N_o(E_k)$ will move to lower E_k, that is, will appear closer to the vacuum level, and that the total yield γ_i will decrease. Data in Fig. 16 for He$^+$ ions of varying kinetic energy incident on a germanium surface clearly show that the higher energy peak, a feature resulting from the form of $N_c(\varepsilon)$, moves to lower energy as $E_k(\text{He}^+)$ is increased.[14] We note also in Fig. 16 the greater energy broadening as $E_k(\text{He}^+)$ increases. That total yield decreases with increasing ion energy is seen in the He$^+$ data of Fig. 9 in the energy range $0 < E_k(\text{He}^+) < 400$ eV.[12] The rise in γ_i for ion energies above this range is explained in Sec. XIV. The anomalous behavior of γ_i for Ne$^+$ ions is discussed in Sec. XI.

A second and more quantitative determination of energy level shift is obtained by fitting the measured kinetic energy distribution of electrons ejected by He$^+$ incident on Ge(111) starting from a reasonable valence band density $N_v(\varepsilon)$ and including the known transition probability factors.[14] As seen in Fig. 17 we start from an $N_v(\varepsilon)$ made up of four parabolas to simulate the s-p band between $\varepsilon = 0$ and ε_v as well as the degenerate p band between the energies $\varepsilon = p\varepsilon_v$ and ε_v. As parameters we have p in the expression $\varepsilon_v - p\varepsilon_v$, the width of the degenerate p band, and r in the expression $(1-r)(\varepsilon/\varepsilon_v)N_i(\varepsilon)$ which decreases the transition probability as one moves up in the band. This effect results from the fact that the p orbitals, which predominate near the top of the band, do not project as strongly from the surface as do s orbitals

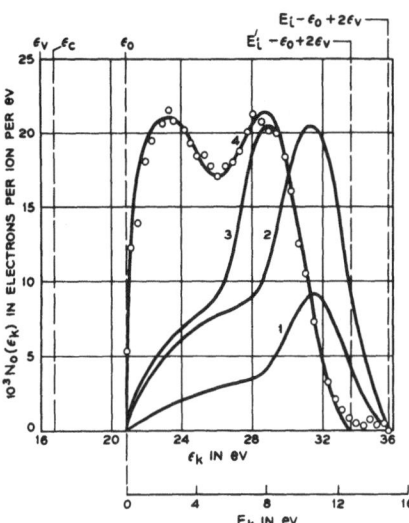

Fig. 18

of the same energy. Thus p wave function magnitude is relatively smaller at the ion than s wave function magnitude resulting in a relatively smaller transition probability for electrons initially in a p state as compared to an s state. Two other parameters are $P_e(E_k)$, the probability of escape, and E_i', the effective ionization energy near the surface.

The procedure was then to determine $N_i(\varepsilon_k)$ from Eqs. (2) and (3) and to multiply by an assumed $P_e(\varepsilon_k)$ to get $N_o(E_k)$. Several such theoretical $N_o(E_k)$ are shown in Fig. 18 where they are compared with experimental data points. Curve 1 corresponds to using $E_i' = E_i$, the free-space ionization potential, and a $P_e(\varepsilon_k)$ calculated from an isotropic distribution of excited electrons. We see that it is too small and predicts faster electrons than are actually observed. Curve 2 is obtained by increasing $P_e(\varepsilon_k)$ appropriately and curve 3 by, in addition, using $E_i' = E_i - 2$ eV. Finally, curve 4, in good agreement with the data points, is achieved by adjusting the parameter r discussed above. The important thing to note about this procedure is that the parameters are basically orthogonal to one another and that there is no way to bring coincidence between the high energy sides of the theoretical and experimental distributions without making $E_i' = E_i - 2$ eV. Although $E_i - E_i'$ could depend on the specific particle-solid combination, 2 eV has turned out to be a remarkably universal figure for, say, 10 eV He$^+$ ions on a variety of solid surfaces.

VIII. TRANSITION RATE PARAMETERS FOR AUGER NEUTRALIZATION

The determination of the change in neutralization energy, ΔE_i, between infinite separation and that separation at which Auger neutralization takes place enables us to calculate approximate values for the parameters A and a in the expression $R_t(s) = A\exp(-as)$ for the transition rate of the process. If we equate $\Delta E_i = -2$ eV to the image force interaction at the separation s_m corresponding to the maximum of the $P_t^i(s,v)$ function we obtain

$$\Delta E_i = -2 \text{ eV} = -3.6/s_m (\overset{o}{A}); \; s_m \overset{\sim}{=} 2 \overset{o}{A}. \tag{6}$$

The parameter a is determined by the rate of fall off of the wave functions outside the surface. If, at larger distances from the surface the wave function has the form $\psi \propto \exp(-\lambda s)$ then $a = 2\lambda$ because $R_t(s)$ involves the square of the wave function. λ is expressible in terms of the ionization energy, E_i, of the wave function state, yielding:

$$a = 2\lambda = 2[2mE_i/\hbar^2]^{1/2} = 4[2E_i(\text{Hartrees})]^{1/2}(\overset{o}{A}{}^{-1}). \tag{7}$$

For a state lying 7 eV below the vacuum level, that is for which $E_i = 7$ eV $\cong 0.25$ Hartrees, we calculate $a \sim 3 \overset{o}{A}{}^{-1}$. Using the expression $s_m = (1/a)\ln(A/av)$ and taking $v = 2.2 \times 10^6$ cm/sec for a 10 eV He^+ ion we obtain:

$$A/a = v \exp(as_m) = 2.2 \times 10^6 \exp(6) \cong 9 \times 10^8 \text{ cm/sec.} \tag{8}$$

With $a \cong 3 \overset{o}{A}{}^{-1}$ as calculated above $A \cong 2.7 \times 10^7$ sec^{-1}. This leads to an estimate of the transition rate at $s = s_m$:

$$R_t(s_m) = A \exp(-as_m) = 2.7 \times 10^{17} \exp(-6)$$

$$\cong 6.7 \times 10^{14} \text{ sec}^{-1}. \tag{9}$$

This is a rate close to that expected for a radiationless Auger process and to the value quoted in Ref. 1. In Table X of Ref. 1 theoretical values for A/a by Shekhter[2] and Cobas and Lamb[4] are given for the Auger and resonance processes. In general the experimental estimates are considerably larger than the theoretical. The theoretical estimates for s_m appear to be unrealistic, however.

Estimates of A/a have been obtained by others using the expression $P_t^o(\infty,v) = \exp(-A/av)$ for the survival probability at infinity of an excited or ionized particle leaving the solid (Table 3). Van der Weg and Bierman[15] have obtained A/a $\cong 1.2 \times$

10^6 cm/sec from the survival probability of 30-90 keV reflected Ar^+
ions with respect to Auger neutralization on the outward trip.
$A/a \cong 2 \times 10^6$ cm/sec is estimated by the same authors[16] from the
line profile of CuI 3247 Å observed in the collision of 80 keV Ar^+
on Cu. The same number is estimated by White and Tolk[17] from
radiationless de-excitation processes of receding particles of
energies 10-3000 eV. However, higher values of A/a in the range
1.2 to 1.5×10^8 cm/sec have been obtained by two groups[18,19]
studying Doppler broadened line shapes. No work involving the de-
excitation of outgoing particles has yielded as high a value for
A/a as that estimated above for the Auger neutralization of incoming
slow ions.

IX.   TUNNELING PROCESSES FOR UNEXCITED IONS

There are ion-solid combinations for which the Auger
neutralization process AN (left vertical leg of triangle in Fig. 4)
is not the only possible process. For some ion-solid combinations
and the appropriate incident ion velocity the ejection processes
are partitioned between AN and the two stage process of resonance
neutralization (RN) followed by Auger de-excitation (AD) (the top
and right legs of the triangle of Fig. 4). The study of two-stage
electron ejection has been particularly informative concerning the
details of particle-solid electronic interactions and the kinetics
of the processes involved. We shall discuss resonance tunneling
first, then Auger de-excitation (Sec. X), and follow this with a
discussion of the two-stage process of electron ejection, how it is
detected and how it explains the experimental results for Ne^+ ions
incident on W or Mo (Sec. XI).

An electron energy diagram illustrating the resonance tunneling
processes of neutralization (transition 1) and ionization (transition
2) is shown in Fig. 19. Transition 1 can occur if $(E'_i - E'_x) > \phi$ near
the surface and transition 2 if $(E'_i - E'_x) < \phi$. $(E'_i - E'_x)$ is the
effective ionization energy of the excited state near the surface.
α, which specifies the level in the solid involved in the transition,
is the quantity used also in Fig. 20.

The potential energy diagram appropriate to the resonance
tunneling processes involving the noble gas singly charged ions and
metastable atoms is given in Fig. 20. The ionic state ION + $e_M^-(\phi)$
is indicated by curve 1 with its asymptote at $s = \infty$ placed at zero
on the energy scale. Here the metal electron, considered to be a
part of the system, is at the Fermi level, ϕ eV below the vacuum
level. For $s > 5$ Å the ionic interaction is primarily the image
interaction and so is the same for all the ions. The potential
curves for the ions will differ closer to the surface as is
illustrated schematically in the figure. Potential curves for the
metastable atoms are those labelled 2 through 6. The asymptote of

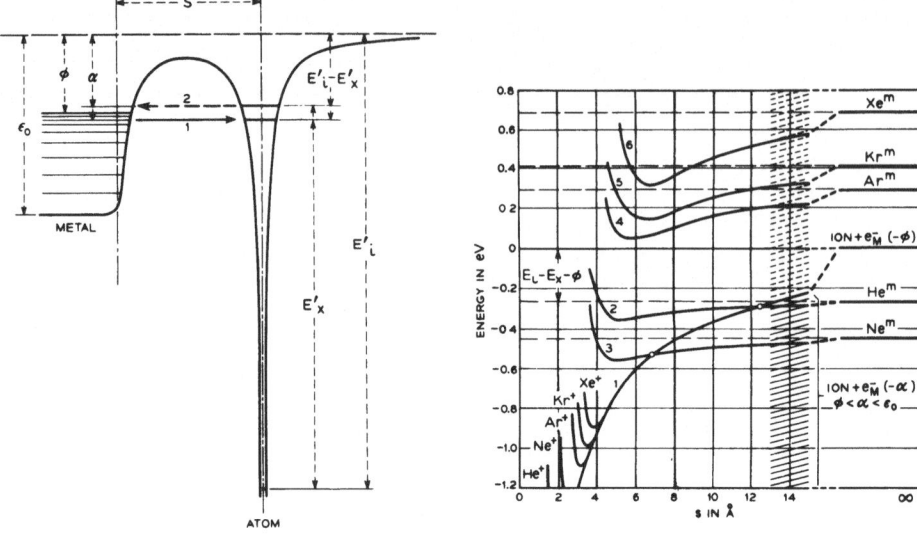

Fig. 19   Fig. 20

such a curve lies below that of the state ION + $e_M^-(-\phi)$ by the
energy $E_i - E_x - \phi$, the free space ionization energy of the ground
state atom, E_i, minus the excitation energy of the metastable
state, E_x, minus the work function of the surface, ϕ. (E_i-E_x) is
the free space ionization energy of the metastably excited atom.
The energy interval $E_i - E_x - \phi$ is indicated in the drawing only
for the case of He^m. At separations s near the surface the interval
between the ionic and the metastable potential curves is $E_i' - E_x' -$
ϕ, where E_i' and E_x' are the effective energies at the separation s.

    Potential curves for the state ION + $e_M^-(-\alpha)$, $\phi < \alpha < \varepsilon_0$, lie
below curve 1 as indicated in Fig. 20 and correspond to the
electron inside the metal, e_M^-, having an energy in the range of
the filled band. States ION + $e_M^-(-\alpha)$ with $\alpha < \phi$, for which the
metal electron is in the normally unoccupied levels, lie above
curve 1. The resonance processes $X^+ + e_M^-(-\alpha) \overset{\leftarrow}{\to} X^m$ must occur at the
crossing point of the X^m potential curve and one of the infinite
set for $X^+ + e_M^-(-\alpha)$. This is so because the Franck–Condon
principle must be obeyed for an adiabatic resonance process and
there is no free electron to carry off excess energy. Curves 4,
5, and 6 lie entirely above curve 1 in the region. Thus α is
less than ϕ for any ionic potential curve crossing these curves
and only resonance ionization of Ar^m, Kr^m, and Xe^m is possible.
Curves 2 and 3 each cross curve 1 in the range of s shown in
Fig. 20. At such a crossing $E_i' - E_x' = \phi$ and the separation s = s_c
at the crossing is a critical distance separating spatial regions
in each of which one of the resonance processes but not the other
is possible. For s > s_c resonance neutralization of the ion to

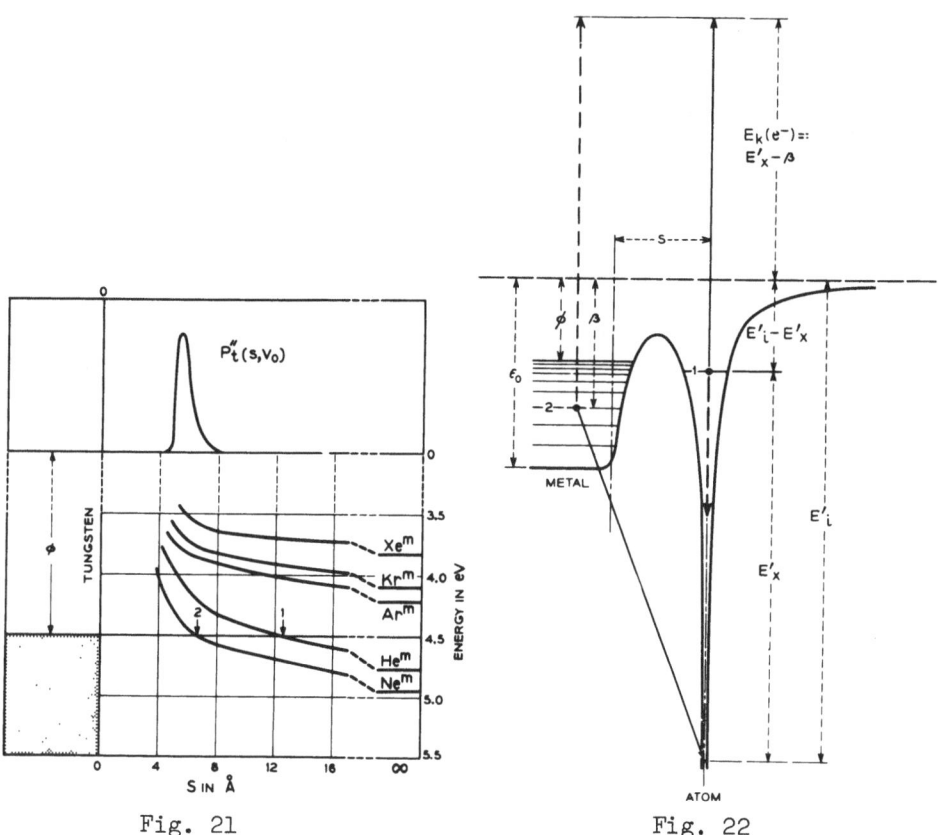

Fig. 21   Fig. 22

the metastable is possible but not resonance ionization of the
metastable. For s < s_c the reverse situation prevails.

    Translating energies from the potential curve diagram of
Fig. 20 to an electron energy diagram showing metastable levels we
obtain Fig. 21. Here the metastable level is plotted below the
vacuum level (E=0 in Fig. 21) by the energy $E_i' - E_x' = \alpha$ obtained
from Fig. 20. Note that the metastable levels rise relative to
the levels in the solid as the particle approaches the solid. We
see that the He^m and Ne^m levels cross the Fermi level (ϕ=4.5 eV)
at the points labelled 1 and 2, respectively. These correspond to
the crossing points indicated by the open circles in Fig. 20. The
Ar^m, Kr^m, and Xe^m levels are above the Fermi level at all
separations as expected. The He^m level crosses the Fermi level at
such a large distance that no electronic transition can occur
there. The Ne^m crossing at point 2, however, occurs where wave
function overlap between atom and solid is such that resonance
transitions are possible when the particle is at separations
either greater than or less than s_c. This, as we shall see,
provides us with an interesting possibility for studying the
kinetics of the tunneling processes.

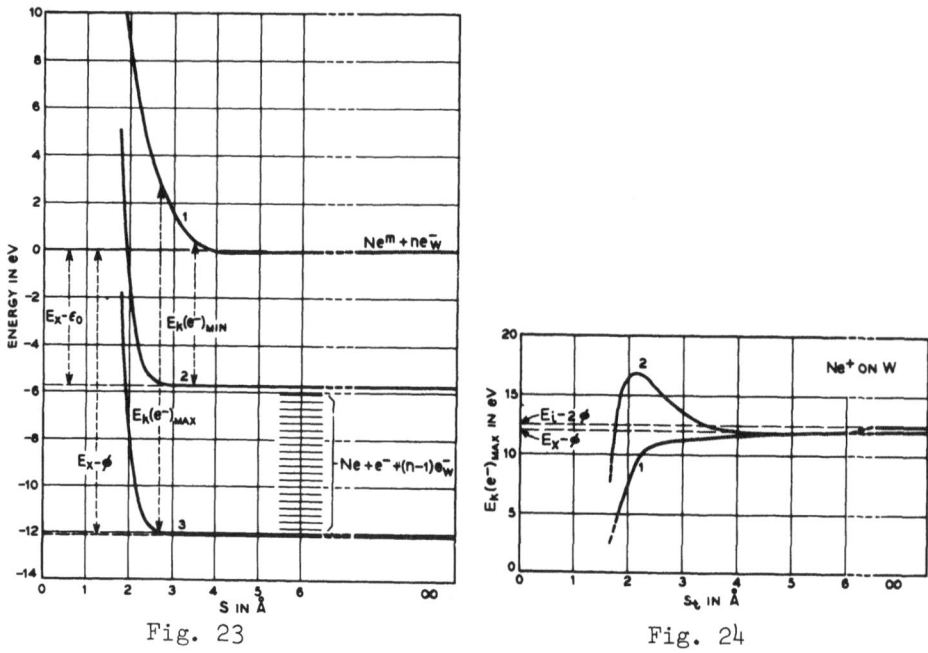

Fig. 23   Fig. 24

X. THE AUGER DE-EXCITATION PROCESS

An electron energy diagram for the Auger de-excitation process
is presented in Fig. 22. The corresponding potential energy
diagram for the de-excitation of a neon metastable is given in
Fig. 23.[1] The variation of the final state energy with atom-solid
separation varies as it did for the Auger neutralization process
(Fig. 13). The initial state behavior is considerably different,
however. The potential curve for the state $Ne^m + ne_W^-$ is compounded
of two terms: a van der Waals attraction, which, although larger
for Ne^m than for Ne, is still small ($\sim$ 0.1 eV), and a repulsive
interaction which, because of the larger size of the metastable,
must cause the potential energy curve to rise above the asymptote
at considerably larger distances than does that for Ne. The magnitude of this effect was estimated in Ref. 1 and is shown in Fig. 23.

The form of the potential curves in Fig. 23 indicate that, in
contradistinction to Auger neutralization, the separation of the
curves for the initial state (curve 1) and a final state (a curve
between curves 2 and 3) initially increases as s decreases. Since
the kinetic energy of the Auger electron outside the solid is the
separation of such curves we expect the maximum kinetic energy
$E_k(e^-)_{max}$ to increase above the free space value before decreasing
close to the surface, as does curve 2 of Fig. 24 for the Auger de-
excitation process. Curve 1 in Fig. 24 shows that $E_k(e^-)_{max}$ for
Auger neutralization decreases monotonically as the atom-surface
separation at which the process occurs decreases.

The differences in electron kinetic energy maxima just discussed will form the basis for an experimental demonstration of the occurrence of Auger de-excitation (Sec. XI). In anticipation of this we can investigate for the simplified solid of Fig. 14 what an admixture of a small percentage of Auger de-excitation with the predominant process of Auger neutralization will do to the observed kinetic energy distribution of ejected electrons. The kinetic energy distributions in Fig. 15 for He, Ar, Kr, Xe ions and curve 1 for Ne ions correspond to $N_i(\varepsilon_k)$ distributions like curve 3 of Fig. 14. The energy positions of the broadened triangles are dictated by the neutralization energy of the ion. Smaller neutralization energy drops the position of the $N_i(\varepsilon_k)$ distribution in Fig. 14. All of the $N_i(\varepsilon_k)$ distributions inside the solid are cut off at the Fermi level (ε_k = 6.3 eV in Fig. 15). The external distributions are cut off at the vacuum level (ε_k = 10.8 eV in Fig. 15).

If one considers the case of no energy level shifts and no broadening that gave curve 1 for $N_i(\varepsilon_k)$ in Fig. 14, it is readily seen that the Auger de-excitation process will result in an $N_i(\varepsilon_k)$ distribution that reproduces $N_c(\varepsilon)$ without folding it [Fig. 25(a)][20] Although it is a two-electron Auger process, Auger de-excitation is quasi one-electron since one of the two electrons transits

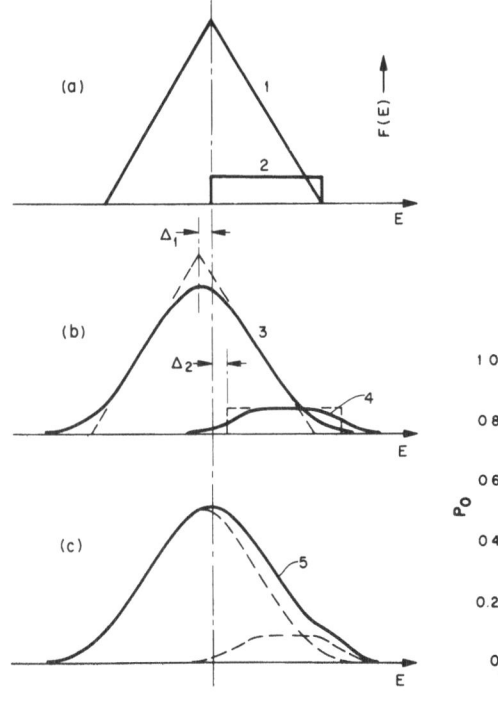

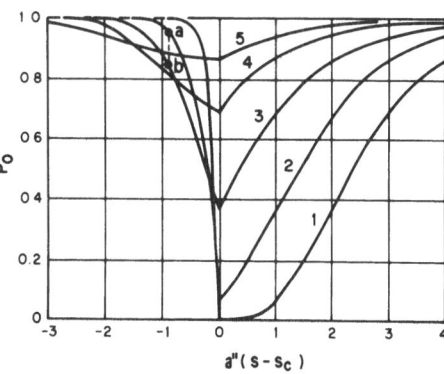

Fig. 25   Fig. 26

between specific atomic levels in the incident particle. Furthermore, for no level shift or broadening the triangular $N_i(\varepsilon_k)$ of
base width $2\varepsilon_F$ from neutralization and the rectangular $\bar{N}_i(\varepsilon_k)$ of
base width ε_F from de-excitation have the same maximum point
[Fig. 25(a)].

When one admits energy level variation in each process we see
by Fig. 25(b) that the de-excitation distribution moves up in
energy whereas the neutralization distribution moves down. When
broadening is also included we see how the component $N_i(\varepsilon_k)$
functions 1 and 2 in Fig. 15 are obtained for an 8% admixture of
the de-excitation process [Fig. 25(c)]. We would thus predict
that such an admixture would distort the high energy end of the
distribution producing a measurable shift of its maximum to higher
energy. It also increases the yield because the faster electrons
in curve 2 leave the solid with greater probability than they
would if distributed in energy as specified by curve 1.

XI.   TWO-STAGE EJECTION PROCESSES FOR UNEXCITED IONS

We are now prepared to discuss the two-stage electron ejection
process in which an incident ion is neutralized to an excited state
followed by Auger de-excitation of the excited atom thus formed with
the ejection of an electron. This process takes us across the top
of the triangle in Fig. 4 and down the right-hand side. In our
discussion of Sec. X we saw that Auger de-excitation has a
significantly different energy distribution from Auger neutralization
and in Fig. 15 we saw the effect of an 8% admixture of the de-
excitation process. In Sec. IX we concluded that for Ne^+ the
critical point at s_c most likely lies close enough to the surface
to separate regions in which the two resonances process can occur.

Our ability to detect the partial occurrence of the Auger de-
excitation process depends on the above facts and, importantly, on
the further fact that the partition of processes for Ne^+ incident
on W, for which s_c is properly placed, is a function of incident
ion velocity. How this comes about is as follows. If we plot the
probability $P_0(s,v)$ that an incident ion remains an ion for various
incident velocities we get a graph like Fig. 26.[1] Here s_c is the
critical distance discussed above, and a'' is the parameter in the
rate $R_t''(s) = A'' \exp(-a''s)$ of the resonance neutralization and
ionization processes. For a sufficiently slow ion curve 1 of
Fig. 26 would be appropriate. In this case all Ne^+ ions would
become Ne^m metastable atoms before s_c is reached from larger
distances. After the critical separation s_c is crossed these
metastables can revert to ions via the resonance ionization process.
As ion velocity toward the surface is increased we progress from
curve 1 to curve 5 on Fig. 26. We note that at $s = s_c$ a larger
fraction remains ions as v is increased. However, the space rate

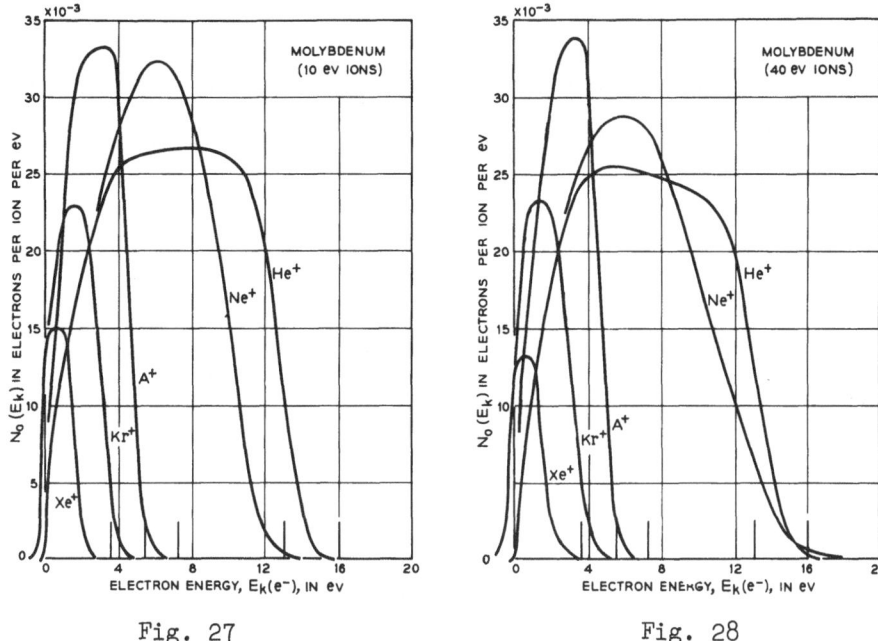

Fig. 27   Fig. 28

of reversion to metastables is similarly slowed for the faster
ions so the curves cross each other in the region $s < s_c$. If we
assume that the Auger process, whichever is possible, occurs at
$a''(s-s_c) \cong 0.9$, the partition between the Auger processes for curve
2 is given by point a, and for curve 3 by point b. Thus for faster
particles a smaller number of ions have converted to metastables at
$s = s_c$ but a greater number remain as metastables at the separation
where the Auger processes occur. Thus we expect the partition
between the Auger processes to shift from essentially 100% Auger
neutralization at very low velocities to an increasing percentage
of Auger de-excitation, near 10% say, as velocity increases.
According to our discussion in Sec. X we expect one result of this
to be an increase in the yield γ_i since the yield per ion of de-
excitation is greater than that for neutralization. The former
process produces, on the average, faster electrons more of which
can leave the solid. This is seen to occur for Ne^+ in Fig. 9 and
explains the Ne^+ anomaly. A second consequence of this shifting
partition is to increase the maximum kinetic energy of ejected
electrons as predicted by Fig. 15. Figures 27 and 28 demonstrate
that this occurs.[12] The vertical lines along the $E_k(e^-)$ ordinate
axis indicate the values $E_i(X^+) - 2\phi$ for each ion. The Ne^+ distribution in Fig. 27 is normal for 10 eV ions for which the yield γ_i
in Fig. 9 is also normal. At 40 eV ion energy the Ne^+ distribution
in Fig. 28 clearly has an anomalously high maximum kinetic energy
and the yield γ_i in Fig. 9 is also anomalously high. Thus we are
able to understand in essentially complete detail the nature of the

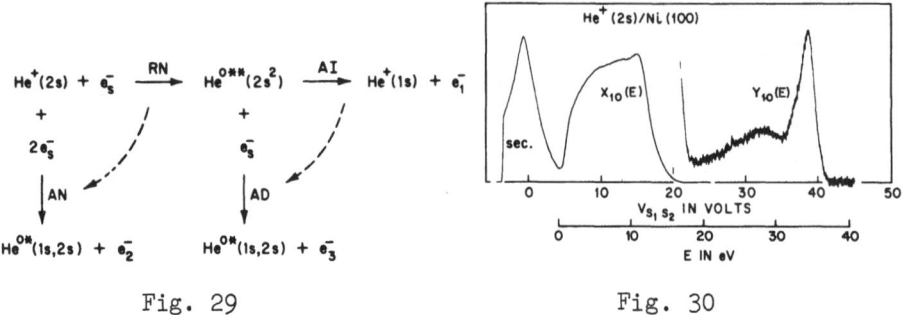

Fig. 29   Fig. 30

Auger neutralization and de-excitation processes and how they are
partitioned by the kinetics of the resonance processes that are
the precursors of de-excitation in the two-stage ejection
mechanism.

XII. TWO-STAGE EJECTION PROCESSES FOR EXCITED IONS

An example of another family of interrelated and competing
electronic transition processes clearly distinguishable from those
of Fig. 4 is shown in Fig. 29.[21] Here the ion initially incident
on the solid is in an excited rather than the ground state. The
ion shown at the upper left hand corner of the diagram is
metastably excited $He^+(2s)$, in which the single extranuclear
electron is in the metastable 2s orbital. Across the top of the
diagram two sequential processes are shown. The first is
resonance neutralization (RN) of $He^+(2s)$ by an electron from the
solid, e_s^-, to the doubly excited neutral species $He^{o**}(2s^2)$. The
second is autoionization (AI) of this doubly excited species to
the ground state ion $He^+(1s)$ with the ejection of a free electron
e_1^-. Competing with the components of this two-stage process are
the Auger neutralization (AN) of $He^+(2s)$ to a singly excited ion,
$He^{o**}(1s,2s)$, and a free electron, e_2^-, and the Auger de-excitation
(AD) of $He^{o**}(2s^2)$ to $He^{o*}(1s,2s) + e_3^-$. The curved, dashed-line
arrows indicate the direction in which the partition shifts as the
velocity toward the surface of the incident $He^+(2s)$ ion is
increased.

As we started our studies with the metastably excited ion
$He^+(2s)$ the transition we had hoped to see was the ionic de-
excitation process:

$$He^+(2s) + ne_s^- \rightarrow He^+(1s) + (n-1)e_s^- + e^-, \qquad (10)$$

in which a metastably excited helium ion would be de-excited to

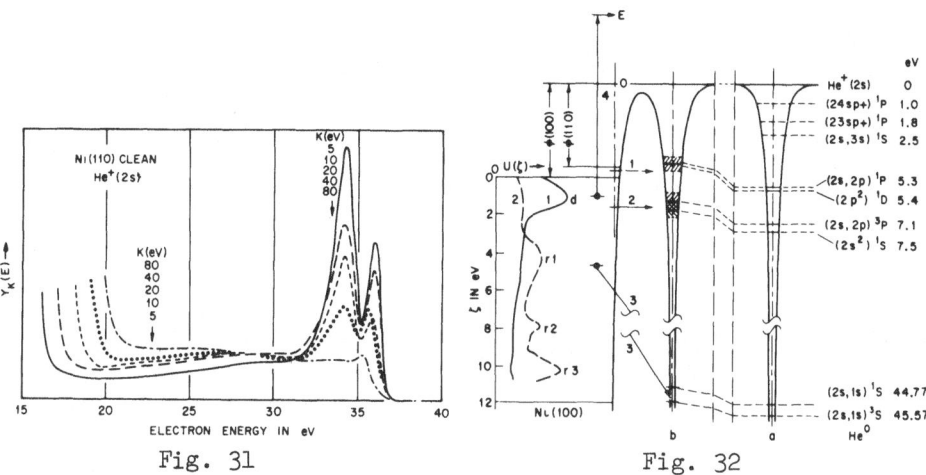

Fig. 31   Fig. 32

its ground state with the release of a metal electron. The process
would be energetic — the excitation energy is 40.8 eV — and like
all de-excitation processes it would be quasi two-electron with the
kinetic energy distribution not a fold but reflecting directly the
local density of states.

The data were taken with a beam of He^+ ions, 99.9% being
$He^+(1s)$ and 0.1% $He^+(2s)$. The possible processes for these two
ions produce electrons in quite different energy ranges so they
could be studied separately. When this mixed beam was incident on
the clean Ni(100) surface the observed electron energy spectrum
was that of Fig. 30. Near $V_{S_1S_2} = 0$ a peak of secondary electrons
from the grids of the analyzer is observed. Electrons ejected
from the Ni surface are accelerated by 4 eV to separate them from
these secondaries. This results in the electron energy scale
labelled E in Fig. 30. The $X_{10}(E)$ distribution is that ejected by
10 eV $He^+(1s)$, the $Y_{10}(E)$ distribution that ejected by 10 eV
$He^+(2s)$. The sensitivity of the apparatus was increased 300 times
for the $Y_{10}(E)$ measurement.

The $Y_{10}(E)$ distribution of Fig. 30 looks superficially like
what one would expect if the process were that given in Eq. (10).
However, the high energy peak is too broad and reduces in
magnitude as incident ion energy is increased, the reverse of what
should occur. When a Ni(110) surface was used the results of
Fig. 31 were obtained for Y(E). Here we see the interesting
result that two high-energy peaks are observed and that these
peaks decrease in intensity as ion energy increases. The latter
phenomenon clearly points to competition with processes that occur
with greater probability nearer to the surface and are thus
greatly favored as ion energy is increased.

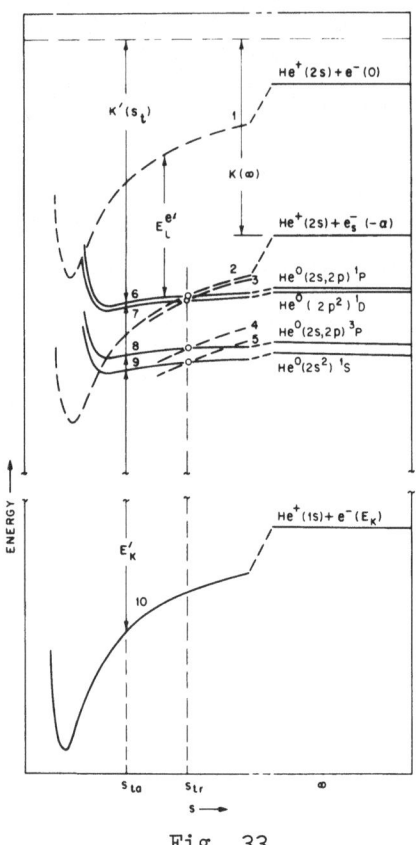

Fig. 33

The processes that are in fact being observed are those given across the top of Fig. 29. First, the He$^+$(2s) is neutralized to a doubly excited state by a resonance tunneling process such as:

$$He^+(2s) + ne_S^- \rightarrow He^{O**}(2s^2) + (n-1)e_S^-. \tag{11}$$

As can be seen from Fig. 32 there are four possible levels, $(2s^2)^1S$, $(2s,2p)^3P$, $(2p^2)^1D$, and $(2s,2p)^1P$, into which electrons near the Fermi level can tunnel. These amalgamate into two broader groups of states near the surface. Tunneling into these is indicated by arrows 1 and 2 of Fig. 32. After the formation of the He^{O**} particle in any one of these states autoionization occurs, an example of which is:

$$He^{O**}(2s,2p) \rightarrow He^+(1s) + e^-. \tag{12}$$

This is an Auger de-excitation process in which all the electrons

are in the atom. It occurs near the surface where the atomic
levels are shifted as is evidenced by the fact that the electrons
ejected in each of the high energy peaks of Fig. 31 are 0.6 to
0.7 eV faster than those observed when these states autoionize in
free space.[21]

Figure 32 indicates why we observe one autoionization peak
for Ni(100) and two for Ni(110). In the former case only one of
the two groups of excited He^{o**} states lies below the Fermi level.
In the latter case at least part of the upper of the two groups
lies below the Fermi level where tunneling from filled states in
the solid is possible. Curves 1 and 2 at the left of Fig. 32 are
experimental local densities of states used in the detailed
discussions of Ref. 21. Transitions 3 and 4 are those of the
process of Eq. (10).

The reason why the electrons from autoionization are faster
when the process occurs near the surface than in free space can be
seen from the potential energy diagram of Fig. 33. Here the
resonance neutralization process occurs at the crossing of an
initial potential curve 2, 3, 4, or 5, with one of the potential
curves representing the interaction of He^{o**} with the surface 6,
7, 8, or 9. The subsequent autoionization process involves a
transition from one of these curves to the final state curve, 10,
representing the de-excited ion and an ionized electron. An
electron produced in this process near the surface is faster than
one produced in a process in free space because of the nature and
configuration of the potential curves of Fig. 33. Here, unlike
Auger neutralization (Fig. 13), the ionic state is the final
state. Its potential curve (curve 10) lies below the flatter
potential curve (curve 9, say) for the neutral initial state.
Nearer the surface these curves lie farther apart, not closer
together as in Fig. 13, resulting in a faster electron, that is,
greater E_k'.

The amount by which the ejected electron is faster than that
from a free-space process is the result, principally, of the image
force interaction of the ion with the surface. The same statement
can be made about the magnitude of the upward shift of the doubly
excited levels (Fig. 32) into which the tunneling electron transits
in the first stage of the process. This shift must be just enough
at the particle solid separation where the tunneling process occurs
to put the ^1P and ^1D states above the Fermi level for the Ni(100)
surface and at or just below the Fermi level for the Ni(110)
surface. This consistency requirement between the energetics of
the two components of the two-stage process is a particularly good
test of the general picture of the role of particle-solid interaction we have been using. Equating the level shift to the image
force and van der Waals terms yields a transition separation of
about 5 Å. Thus the autoionization process occurs farther from
the surface than does Auger neutralization.

XIII.   OSCILLATORY ION SCATTERING FROM SURFACES

We turn now to a brief discussion of the interesting phenomenon of oscillatory quasi-elastic scattering of ions from solid surfaces that has recently been discovered and studied experimentally[22,23] and treated theoretically.[24] The subject matter, without undue simplification, divides itself quite naturally into nonresonant and near-resonant scattering. We shall discuss these in turn.

Nonresonant scattering occurs when there is no energy level in surface atoms of the crystal which lie within ±10 eV of the un-filled atomic ground level of the incident ion (Fig. 3, right). In He^+ near a surface this level lies about 24.5-2 = 22.5 eV below the vacuum level as previously discussed in Sec. VII. Despite this condition of nonresonance the ion can be neutralized via the Auger neutralization process discussed in Sec. VI. If the overall probability of ion neutralization over both inward and outward trips is $P_n(E,\theta)$ and the differential cross section for elastic scattering at the surface is $\sigma(E,\theta)$, then the intensity $I(E,\theta)$ of backscattered ions at energy E and angle θ is proportional to the product

$$I(E,\theta) \propto \sigma(E,\theta)[1-P_n(E,\theta)]. \tag{13}$$

Classical calculations of σ for gas phase collisions[25,26] and for atom-surface collisions[24] show it to decrease monatonically with increasing energy. Tables 2 and 3 show that the ion survival probabilities P_0 for both inward and outward trips vary as $\exp(-A/av)$. Thus with increasing E and v, P_n should decrease and $(1-P_n)$ increase monatonically. These facts, by Eq. (13), lead to an intensity function with a single broad maximum in the range 100 < E < 1000 eV. The top curve labelled Class I in Fig. 34 is an experimental function of this type. In Cu the highest lying core levels, 3p and 3d, lie, respectively, 78 and 6 eV below the vacuum level of Cu.[23] Each lies considerably more than 10 eV from the He^+ empty ground level.

When there is a core level in the solid less than 10 eV from the He level an interestingly different phenomenon is found. Then the yield of backscattered ions may be oscillatory as energy is varied as is the case for the curve labelled Class II in Fig. 34. Erickson and Smith[22] conclude that the phenomenon has to do with near resonant oscillatory electron exchange between the core level in the solid and the ground level in the incident or receding atomic particle outside the surface as well as the magnitude of the electron transfer time relative to the time the atom interacts with the solid (Fig. 3, right). The variation of the latter with incident ion velocity cyclically varies the phase of the electron

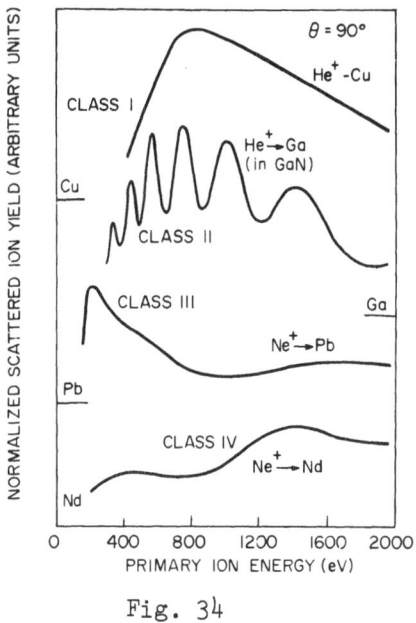

Fig. 34

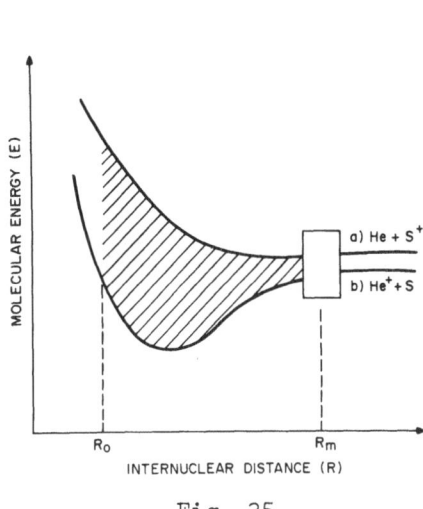

Fig. 35

exchange at the point when the atomic particle leaves the vicinity
of the surface.[27]

Tully has produced a detailed theory of the phenomenon and
has compared it directly with experiment.[24] His qualitative
description of the mechanism in terms of his schematic figure,
reproduced here as Fig. 35, follows. A wave packet representing
the He$^+$ ion approaches the surface on the potential energy curve
labelled He$^+$ + S. This designation prescribes that the incident
He atom is ionized and the surface atom S with which it interacts
in a binary collision is neutral. In the region indicated by the
rectangular box at $R = R_m$ in Fig. 35 the exchange interaction
between the states He$^+$ + S and He + S$^+$ becomes comparable with
the asymptotic splitting ΔE. Here the wave packet splits between
the two states, each of which evolves independently along its
respective potential curve with phase $\exp(iEt/\hbar)$. After scattering
at the surface the two paths recombine in the vicinity of R_m on
the outward trip. The oscillatory ion scattering results from
interference between the two terms whose relative phase is velocity
dependent.

Tully's detailed theory involves solution of the coupled
equations along the trajectory determined for nonresonant scattering using a Thomas-Fermi-Moliére interaction potential. These
equations are:

$$i\hbar\dot{a}_1 = a_1(H_{11}-i\Gamma_{11}) + a_2 H_{12} \tag{14}$$

$$i\hbar\dot{a}_2 = a_2(H_{22}-i\Gamma_{22}) + a_1 H_{12} \tag{15}$$

in which a_1 and a_2 are, respectively, the amplitudes of the states $He^+ + S$ and $He + S^+$, H_{11} the effective ionization energy of He near the surface, H_{22} the ionization energy of the surface atom state S, $\Gamma_{11}/\hbar$ the exponential rate function $A\exp(-as)$ of Tables 2 and 3 for the Auger neutralization process, Γ_{22} the rate at which the hole in the surface ion S^+ is filled by solid-state electronic processes (taken as 0.4×10^{15} sec^{-1}), and H_{12} the exchange interaction resulting from the overlap of the wave functions for the states $He^+ + S$ and $He + S^+$. Γ_{11} and Γ_{22} are the only adjustable terms but have been fixed just once for a series of ion-solid systems. On recombination at R_m the ionic state has amplitude

$$|a_1|^2 \sim A + B\cos\delta \tag{16}$$

in which the phase shift δ has the form:

$$\delta \underset{\sim}{\sim} (2/\hbar\bar{v})\int_0^{R_m} \Delta E dR. \tag{17}$$

Here the integral is the area between the potential curves of Fig. 35 and $\bar{v}$ is the average velocity of the atomic particle near the surface. The result of the application of this theory to the system $He^+ + Pb$ is shown in Fig. 36 where it is compared with the experimental observation of Rusch and Erickson.[23] The agreement in curve shape is remarkable. The small shift in energy position between the two curves results from the approximate nature of the calculated interactions H_{11}, H_{22}, and H_{12} employed in Eqs. (14) and (15). The theory gives oscillations in backscattering intensity when experiment does and none when experiment does not. Tully concludes from his work that the phenomenon is correctly described as charge transfer in a binary ion-atom interaction.

XIV. KINETIC EJECTION OF ELECTRONS BY IONS

In all the electron ejection processes discussed to this point the electron is ejected from the solid by the potential energy carried by the incident ion or excited atom. For this reason this class of electron ejection has been referred to as potential ejection. We expect that, despite the disparity of ion and electron masses, it should become possible at sufficiently high

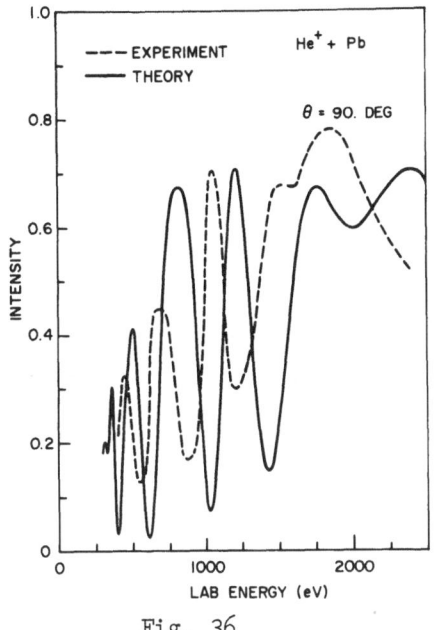

Fig. 36

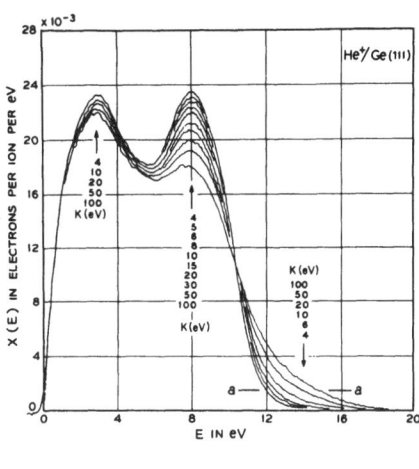

Fig. 37

incident kinetic energy for an ion to eject an electron by transfer
of kinetic energy. Such ejection, called kinetic ejection, is, in
fact, possible and we shall point out here the evidence of its
occurrence in data already presented.

We recall that in our discussion of Fig. 9 in Sec. VII we
could account for the decrease in γ_i for $0 < E_k(He^+) < 400$ eV in
terms of our theory of the Auger neutralization process. The
subsequent rise in γ_i for $E_k(He^+) > 400$ eV is unaccounted for. We
note, however, that in this energy range a new low energy component
appears in the kinetic energy distributions $N_0(E_k)$ in Fig. 8.[12]
This does, in fact, signal the onset of kinetic ejection. If the
Auger component of the distributions is extrapolated to the distributions for $E_k(He^+) = 600$ and 1000 eV as suggested by the dashed
lines in Fig. 8 and the Auger yields determined, we obtain the
dashed-line extrapolation of γ_i shown on Fig. 9. The expected
downward trend of γ_i for the Auger component is thus confirmed.

XV. ENERGY BROADENING IN AUGER NEUTRALIZATION

We have noted in Figs. 8 and 16 that as incident ion energy,
K, is increased the kinetic energy distributions of Auger ejected
electrons broaden. This broadening smooths the characteristic
features of the distribution and produces an extension of the high
energy tail that increases with K. These modifications of the

distribution are strikingly similar to what one expects if distributions at higher K are obtained from a distribution characteristic
of very slow ions by convolution with a broadening function.
Figure 37 shows a series of experimental distributions for He^+ on
Ge(111) with K varying from 4 to 100 eV. In this section we shall
recall briefly the principal findings and conclusions of a study
of energy broadening of electron distributions from both Ge(111)
and Ni(111).[13]

For quantitative discussion we define a single parameter that
characterizes the broadening of a particular kinetic energy distribution. We define z, termed the extension, as the distance at the
level a-a indicated in Fig. 37 from z = 0 at the second maximum at
E = 8 eV to the distribution curve. We shall then proceed on the
initial assumption that the internal distribution $N_i(E) = N_i(\varepsilon)$ of
Auger excited electrons [Eq. (2); Figs. 14, 15, 17] for ions of
kinetic energy K_2 is related to the distribution for ions of energy
K_1 by the integral convolution:

$$N_{iK_2}(E) = \int_{-\infty}^{\infty} B(x, L_{K_1K_2}) N_{iK_1}(E-x)dx. \tag{18}$$

Here $B(x, L_{K_1K_2})$ is the broadening function taken to be the
Lorentzian:

$$B(x, L_{K_1K_2}) = (L_{K_1K_2}/\pi)/(L_{K_1K_2}^2 + x^2), \tag{19}$$

where the half width at half height of the Lorentzian is $L_{K_1K_2}$.

If we take $N_{iK_1}(E)$ to be the step function:

$$N_{iK_1}(E) = 1, \ E < E_1; \ = 0 \ \ E > E_1 \tag{20}$$

we find that the extension $z = E - E_1$ at the level f relative to
the magnitude $N_{iK}(E) = 1$ is

$$z = E - E_1 = L_{K_1K_2} \tan(\pi/2 - f\pi). \tag{21}$$

It thus appears reasonable to expect that for experimental distributions as well the extension z may be closely proportional to
$L_{K_1K_2}$. It is shown in Ref. 13 that the conclusions drawn are

independent of the level at which z is measured.

The next step in elucidating the nature of the distribution
broadening with increasing K is to compare the extensions z for
the experimental distributions of Fig. 37 with those of a set of
theoretical distributions obtained by broadening the experimental
4 eV curve of Fig. 37 using Eqs. (18) and (19). The broadening
parameter used in producing these theoretically broadened distributions is L_{4K}. Plotting L_{4K} versus z_t for the theoretical curves
and z_e versus ion velocity v for the experimental curves we can
eliminate z obtaining

$$L_{4K} = f_t(z_t) = f_t[f_e(v)] = a + bv = a + cK^{\frac{1}{2}}. \tag{22}$$

Using the equation expressing the additivity of broadening
parameters,

$$L_{K_1 K_3} = L_{K_1 K_2} + L_{K_2 K_3}, \tag{23}$$

we determine

$$a = -L_{OK},$$

$$L_{OK} = bv = cK^{\frac{1}{2}}, \tag{24}$$

demonstrating that energy broadening varies linearly with ion
velocity.

Defining an incremental extension z′ such that z′ = 0 when
v = 0 we may plot extensions from a series of distributions for
various surfaces as in Fig. 38. Here we see that broadening can
vary greatly from one surface to another. Later data on a
covered surface show that the slope dz′/dv for Ni(111) is reduced
by a factor 10 when a c(2×2)Se overlayer is adsorbed on the surface.
This result and the results in Fig. 38 appear to indicate that
broadening varies with the local density of electron states above
the Fermi level in that spatial region in which the up electron in
the Auger neutralization process is excited. We note that Cu
having a lower density of states at the Fermi level than Ni has
about 1/3 the broadening. The semiconductors, whose states
immediately near the Fermi level, E_F, are surface states in the
band gap also show smaller broadening than Ni. Adsorption of an
overlayer in which the surface resonances lie considerably below
E_F moves the region in which Auger electrons are excited about one
atomic diameter from the substrate surface. This effect also
reduces the local density of states above the Fermi level in the

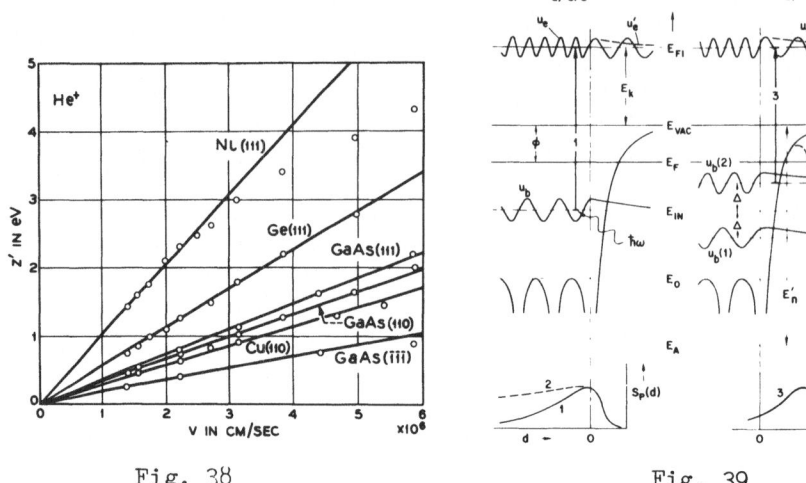

Fig. 38   Fig. 39

region where Auger electrons are excited. We shall see below that
the magnitude of one of the two principal types of energy broadening
depends directly on this factor.

We can enumerate five types of energy broadening operative in
the Auger neutralization process. These result from (1) initial
state lifetime, (2) final state lifetime, (3) atomic level shift
near the surface, (4) variation of impact parameter, and (5) nonadiabatic excitation of electrons in the solid. It is shown in
Ref. 13 that broadening types (2), (3) and (4) are small,
independent of ion velocity, and cannot contribute to the high
energy tail that appears on the distribution as ion velocity
increases. It is further shown in Ref. 13 that the broadening
factors for initial state lifetime, (1), and nonadiabatic broadening,
(5), are each expected to vary with ion velocity and to be approximately equal in magnitude for K < 100 eV.

Initial state lifetime broadening is that which occurs as the
result of the Heisenberg uncertainty principle. Its broadening
parameter $b_i = 2(L_{0K})_i = \Delta E$ can be expressed as

$$b_i = \hbar/\tau_i = \hbar R_t(s_m) = \hbar av, \tag{25}$$

using the expressions for the rate function $R_t = A\exp(-as)$ and for
s_m given in Table 2. Nonadiabatic broadening results from the
fact that the electrons in the solid cannot remain in their ground
states in the presence of the moving positive charge of the incoming
positive ion. Consideration of the Fourier analysis of the potential
at a point lying a distance d from the line of motion of the ion

leads to the conclusion that the broadening parameter, b_n, for the nonadiabatic broadening should have the form:

$$b_n = \Delta E \propto \hbar v/d, \tag{26}$$

thus depending on ion velocity as does b_i. In addition we expect that excitation of ground state electrons to states above the Fermi level should also depend on the local density of states above the Fermi level in the region in space from which Auger electrons are excited. This could well account for the evidences of such a dependence in the experimental data.

XVI. THE METHOD OF ION NEUTRALIZATION SPECTROSCOPY (INS)

We turn now to discuss the development of an electron spectroscopy of solid surfaces based on the Auger neutralization process.[11,28,29] Of the processes depicted in Fig. 4 Auger neutralization is the only one that does not involve the excited states of the incoming atomic particle. In order that Auger neutralization be the only electron ejection process possible for an incident ion it is necessary that resonance tunneling processes cannot occur. Thus the work function of the metal or electron affinity of the semiconductor to be studied must be greater than the ionization energy of the lowest lying excited state in the ion used. This is a basic limitation on the solid-ion systems to which INS can be applied. In INS the sole role of the incoming ion is to provide the low lying vacant electronic state, the ground atomic state, that permits the Auger neutralization process to proceed. The only atomic parameter that enters into the energetics of the process is the effective neutralization energy of the ion near the solid surface. In this section we shall present the steps of the method leaving for Sec. XVII the discussion as to what the spectroscopy measures and the interpretation and justification of these procedures.

In Figs. 39 and 40 are presented two electron energy diagrams in terms of which we shall discuss the method and characteristics of INS. We shall find it convenient to compare the two-electron spectroscopy, INS, with the one-electron spectroscopy, UPS. In Fig. 39 are shown wave functions for initial and final states of each electron ejection process. Figure 40 includes energy distribution functions needed in the discussion.

It is clear that any electron spectroscopy is based on an electronic transition process and that each spectroscopy yields an electron energy distribution function that is in fact an energy density function of electronic transitions. The definition of the transition density function for UPS is easily made since it is a

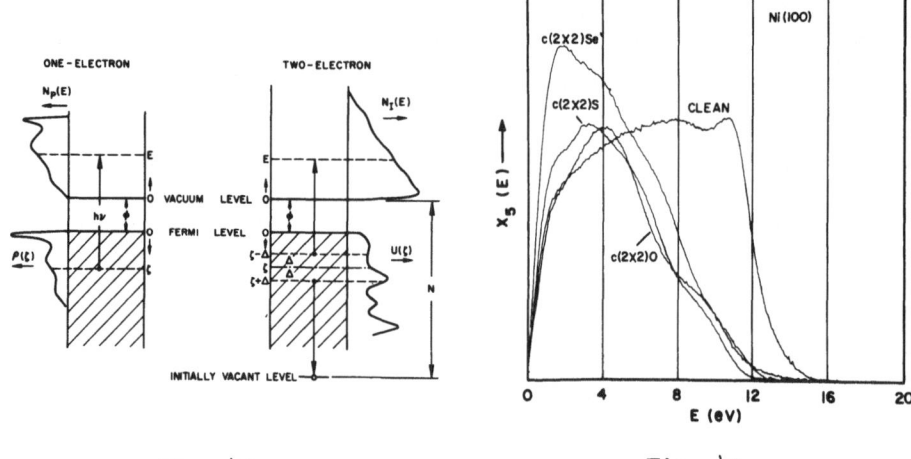

Fig. 40   Fig. 41

one-electron process in which an electron in an initial state at E_{IN} (Fig. 39) absorbs a photon and moves upward in energy by $\hbar\omega$ to the final energy state E_{FI}. The measured electron energy distribution $N_P(E)$ is not identical to the density of initial states $\rho(\zeta)$ since transition probability depends on energy. Note that ζ is an energy variable increasing positively downward into the filled band from $\zeta = 0$ at E_F. E is the external kinetic energy measured from zero at E_{VAC}.

In the case of INS it is necessary to define two transition density functions. The initial state transition density is the probability that an electron in an initial state will be involved in the process. It is not the same as the final state transition density which gives the probability that an electron will be excited to a particular final state. In UPS the initial and final state transition densities are identical. They are not equal in INS because there is an infinite set of paired initial levels whose involvement in the Auger process yields an excited electron at the same energy. Since the transitions 2 and 3 of Fig. 39 must be equal in energy it is clear that all pairs of levels $\zeta \pm \Delta$, symmetrically disposed relative to the initial level ζ that lies halfway between the final level E_{FI} and the ground atomic level E_A, will produce electrons at E_{FI}.

The simplest expression for the final state transition density is obtained by assuming that transition probability is independent of energy and that the initial state transition densities for the down and up electrons are identical. Then the elemental Auger process with initial states at $\zeta \pm \Delta$ and the final state at E_{FI} should have a relative probability of occurrence that is proportional to the product of the densities of initial states, namely $\rho(\zeta + \Delta)\rho(\zeta - \Delta)$

and the final state transition density $F(\zeta)$ may be written

$$F(\zeta) \propto \int_{-\zeta}^{\zeta} \rho(\zeta+\Delta)\rho(\zeta-\Delta)d\Delta. \qquad (27)$$

Equating the lengths of the down and up transition vectors (Fig. 40) yields the expression

$$\zeta = (N-E-2\phi)/2 \qquad (28)$$

relating the mean initial level ζ to the final level E (= E_{FI} of Fig. 39), the ion neutralization energy N and the work function ϕ.

It is clear, however, that the matrix element of the Auger process is not energy independent and that the transition densities for the down and up electrons are not identical. A more general expression than Eq. (27) for $F(\zeta)$ is

$$F(\zeta) \propto \int_{-\zeta}^{\zeta} |M_I|^2 \rho_d(\zeta+\Delta)\rho_u(\zeta-\Delta)d\Delta \qquad (29)$$

in which M_I is the matrix element and ρ_d and ρ_u are the initial state transition densities for the down and up electrons, respectively. ρ_d and ρ_u should be approximately the local state densities in the spatial regions in which the down and up electrons make their transitions. In view of the localized character of the final state wave function $u_g(1)$ of the down electron (Fig. 39) ρ_d should be the local density of states at the ion position. Since the up electron is excited between the ion and the surface ρ_u is an average local state density in this region.

The method of INS is based upon two assumptions. The first of these is that the term $|M_I|^2$ can be factored into relative transition probability terms for the up and down electrons that can be combined with ρ_d and ρ_u to form the functions $V(\zeta+\Delta)$ and $W(\zeta-\Delta)$, respectively. Then $F(\zeta)$ becomes:

$$F(\zeta) = \int_{-\zeta}^{\zeta} V(\zeta+\Delta)W(\zeta-\Delta)d\Delta = V*W \qquad (30)$$

where V and W may now be regarded as the initial state transition densities for the down and up electrons, respectively. The second assumption underlying the method of INS is that if we write $F(\zeta)$ as the self convolution

$$F(\zeta) = \int_{-\zeta}^{\zeta} U(\zeta+\Delta)U(\zeta-\Delta)d\Delta = U*U \qquad (31)$$

of a single function $U(\zeta)$, $U(\zeta)$ will be interpretable as a mean initial state transition density, namely the convolution mean of V and W, and will reveal the essential features of V and W.

The method of INS involves two basic deconvolutions. First $F(\zeta)$ is derived from an effective electron energy distribution for "zero velocity" ions obtained by "debroadening" measured distributions $N_K(E)$. Secondly the integral expression, Eq. (31) for $F(\zeta)$ is inverted to obtain $U(\zeta)$ the mean initial state transition density function. It is the ability to obtain $U(\zeta)$ that makes INS a viable two-electron spectroscopy since $U(\zeta)$ is the function that is to be compared with the transition density function of a one-electron spectroscopy such as UPS. We shall discuss each of these steps illustrating them with data for He^+ ions incident on the Ni(100)c(2×2)Se surface.

The basic data with which we start are electron energy distributions such as those of Fig. 41. Each of these distributions, which we shall call $X(E)$ in the following, includes some energy broadening. It is first necessary to remove this in order to obtain an $X_0(E)$ function characteristic of an unbroadened function for a ion moving very slowly. This debroadening is done by one of two methods based on the extrapolation scheme of Van Cittert.[30]

Van Cittert started with a function X_1 that included broadening via the convolution [Eq. (18)] of the unbroadened function X_0 and the known broadening function B. Thus

$$X_1 = \int_{-\infty}^{\infty} B(x)X_0(E-x)dx = B*X_0. \qquad (32)$$

He then broadened X_1 with B to obtain

$$X_2 = B*X_1 \qquad (33)$$

and assumed the incremental functions X_0-X_1 and X_1-X_2 to be equal to a first approximation. This yielded the equation for X_0:

$$X_0 = X_1 + (X_1 - X_2). \tag{34}$$

The procedure could be iterated to yield better approximations for X_0.

Since in INS data reduction we do not know the broadening function B two methods have been developed from Van Cittert's extrapolation method specifically for INS. In the first of these, the so-called two-curve method we assume we have two experimental distributions X_1 and X_2 measured for ions of energies K_1 and K_2, respectively, and that $B \propto K^{1/2}$. Then we may write

$$X_0 = X_1 + R(X_1 - X_2) \tag{35}$$

where the parameter R is given by

$$R = (X_0 - X_1)/[(X_0 - X_2) - (X_0 - X_1)] = K_1^{\frac{1}{2}}/(K_2^{\frac{1}{2}} - K_1^{\frac{1}{2}}). \tag{36}$$

Thus an initial value of R can be calculated from known ion energies. Or R may be determined so that the zero of X_0 ($\zeta = 0$) appears where a UPS distribution measured on the same surface indicates it should be. R may also be tuned to vary slightly the origin of X_0 as needed in later stages of the data reduction.

The second INS extrapolation method, the so-called one-curve method, uses only one measured distribution, X_1, and takes the energy level $\zeta = 0$ as given by UPS. X_0 is then written

$$X_0 = X_1 + R(X_1 - B*X_1) \tag{37}$$

in which:

$$R = R_1 - R_2 \times \zeta. \tag{38}$$

B is an assumed narrow broadening function. R_1 and B are chosen so that X_0 has the UPS origin. R_2 which adjusts the amplitude of X_0 deeper in the band is determined so that the one- and two-curve methods give the same result for the clean surface.

The two-curve extrapolation method really mixes energy broadening with wave function variation outside the solid surface since variation of K to obtain X_1 and X_2 changes not only broadening but also the distance from the surface at which the ion is neutralized. The one-curve method enables us to study the change in wave function composition outside a solid surface as is discussed in Sec. XIX.

Having obtained X_0 by one of the extrapolation methods just discussed we divide it by the probability, $P_e(E)$, of electron escape over the surface barrier to obtain the internal distribution, $N_i(E)$. Changing variables from E to ζ via Eq. (28) yields $F(\zeta)$, the final state transition density. Several parametric $P_e(E)$ functions have been derived whose parameters may be determined so that the final $F(\zeta)$ obtained by He[+] and Ne[+] ions agree[11] Use of $P_e(E)$ = constant modifies the level of $F(\zeta)$ deep in the band but does not alter the positions of resonances in the final $U(\zeta)$ function.

The $F(\zeta)$ function we have now obtained is assumed to represent the integral self convolution of an unfold function $U(\zeta)$ defined by Eq. (31). Equation (31) is inverted sequentially using digital data. The formalism is obtained by inverting a digital integration procedure derived by approximating $U(\zeta)$ with a stepped function. The expressions for the digital U_{2n-2} from digital F_n are

$$U_0 = (F_1/2\Delta\zeta)^{\frac{1}{2}}, \tag{39}$$

$$U_2 = (1/2U_0)(F_2/2\Delta\zeta), \tag{40}$$

$$U_{2n-2} = (1/2U_0)[(F_n/2\Delta\zeta) - \sum_{p=1, n-2} U_{2n-2p-2}U_{2p}], \quad n \geq 2. \tag{41}$$

This sequential unfold procedure will not work well for all $F(\zeta)$. It works exceedingly well and rapidly for the class of $F(\zeta)$ functions that lie near a straight line corresponding to $U(\zeta)$ functions that lie near the step function having its step at E_F, $\zeta = 0$. Deconvolution, being an ill-posed mathematical problem, must be checked in some way. For the class of functions specified above $U(\zeta)$, a nonlocal function, will have the same features as the local derivative of $F(\zeta)$, $dF/d\zeta = F'(\zeta)$.

Examples of the functions we have been discussing are given in Figs. 42, 43, and 44. The $X_5(E)$ of Fig. 42 is a digital plot of several averaged runs of the $X_5(E)$ function for $c(2\times2)$Se of Fig. 41. $F(\zeta)$ of Fig. 43 is derived from $X_5(E)$ functions like that of Fig. 42. The $U(\zeta)$ function at the top of Fig. 44 is the digital unfold of the $F(\zeta)$ of Fig. 43. At the bottom of Fig. 44 is plotted $F'(\zeta)$, the derivative of $F(\zeta)$. Note that $F'(\zeta)$ and $U(\zeta)$ have identical structural features.

XVII.   WHAT INS AND UPS MEASURE

The spectroscopic function that any electron spectroscopy produces is the final state transition density function of the process

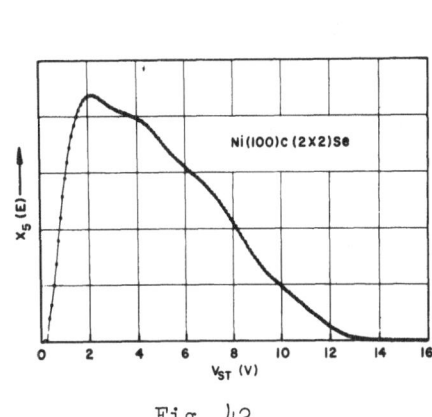

Fig. 42

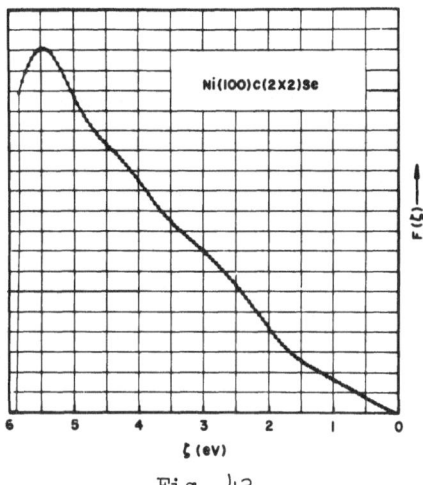

Fig. 43

upon which the spectroscopy is based. Each spectroscopy has
specific transition probability or matrix element characteristics
and each has its peculiar volume of sensitivity near the surface of
the solid. We shall now discuss briefly how these factors are
combined for UPS and for INS to produce the inherent characteristics
of each.

The sensitivity volume of UPS is determined by the escape depth
of the excited electrons inasmuch as the incident photons penetrate
deeply into the solid. In the ultraviolet frequency range the
escape depth is of order 5 to 20 Å. The sensitivity function $S_P(d)$
of UPS is indicated schematically as curves 1 and 2 of Fig. 39.
Two curves are shown to indicate that the sensitivity function and
hence the volume of sensitivity for UPS will vary both with photon
energy and substrate material. UPS averages over the surface monolayer and two or three layers of the substrate.

The sensitivity volume of INS is determined by factors that
differ fundamentally from those operative in UPS. The sensitivity
volume of INS is basically that region within which the up electron
in the Auger neutralization process is excited. This is the spatial
region over which the matrix element of the Auger process has
appreciable magnitude. There are good experimental reasons for
believing that this volume lies essentially entirely outside the
solid but certainly no deeper than the first monolayer. These will
be pointed out in later discussion.

The most illuminating form in which the matrix element of UPS
can be written is:

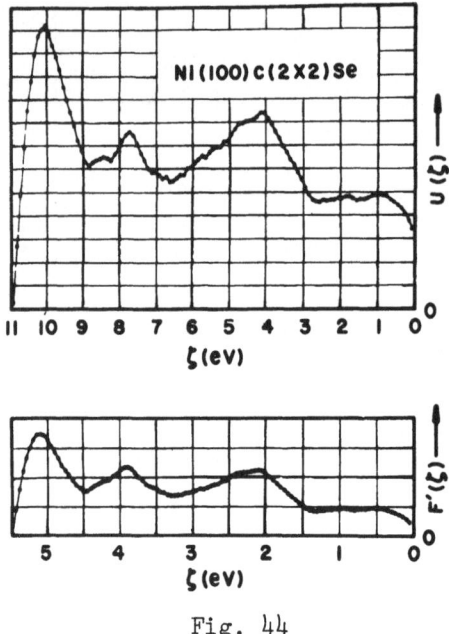

Fig. 44

$$M_P = \int u_f^*(\vec{r})(\hat{\epsilon} \cdot \text{grad}_m V) u_i(\vec{r}) d^3 r. \qquad (42)$$

Here we see that photoemission is largest where the component of
the electromagnetic field in the direction of the gradient of the
electric potential is largest. This results in d wave functions,
which lie closer to the nucleus than p or s at equal energy, having
larger matrix elements than s or p electrons at the same energy.
For localized initial state orbitals the symmetry character of the
initial state also affects the frequency range over which photoemission has appreciable probability. Photoemission from d wave
functions extends to higher photon energy than does that for s
electrons.

We may write the INS matrix element for an elemental process as

$$M_I = \iint u_g^*(1) u_b(1)(e^2/r_{12}) u_e^*(2) u_b(2) dT_1 T_2. \qquad (43)$$

Here we have grouped wave functions for a given electron, 1 or 2,
on the same side of the interaction term. The wave functions may
be identified in Fig. 39. We note that the matrix element may be
viewed as a Coulomb interaction integral between two charge clouds
$u_g^*(1) u_b(1)$ and $u_e^*(2) u_b(2)$. The first of these clouds is localized

near the ion position because of the localized character of $u_g^*(1)$.
Thus the magnitude of the matrix element depends in part at least
on the magnitude of $u_b(1)$ function at the ion position. The matrix
element for INS decreases as wave function symmetry character goes
from s to p to d at the same energy. This is the opposite behavior
to that of the UPS matrix element. INS favors initial states whose
k vector is normal to the surface also because then the magnitudes
of the u_b functions are largest near the ion.

A theoretical study of ion neutralization which illuminates
and justifies the basic assumptions underlying the INS method has
been carried out by Appelbaum and Hamann.[31] Position dependent
densities of states were calculated for Si(111) and H-covered Si(111)
for the surface monolayer (V) and at the position of the ion (W),
2 Å from the surface. Appelbaum and Hamann then compared the folds
V*V, W*W, and V*W [Eqs. (30, (31)] with the form of the experimental
INS kinetic energy distribution. They found that only V*W, that is
the cross fold of the local state density at the surface with that
at the ion, produced adequate agreement with experiment. However,
when they calculated the local state density (U) at a point halfway
between the ion and the surface, that is at a point 1 Å from the
surface, and folded this with itself to produce U*U, as good agreement with the experimental data was obtained as for V*W. Thus
unfolding $F(\zeta)$ as a self convolution does appear to produce a
function U that has the characteristics of the convolution mean of
V and W.

XVIII. ADSORBATE ELECTRONIC STRUCTURE

It was shown very early that the Auger neutralization process
is very sensitive to the adsorption of gases on the surface of a
solid. This can be seen in Fig. 45 where total yield, γ_i, from
polycrystalline W is plotted (top of figure) for He^+ and Ne^+ as a
function of exposure time to N_2 gas.[32] γ_i is seen to saturate
at the exposure where the ballistic pressure increase, Δp,
accompanying thermal desorption of N_2 on target flash also saturates.
The basic reason for the sensitivity of ion neutralization to
adsorption is seen in Figs. 46 and 47. In Fig. 46 is illustrated
the broadening and shifting of an atomic level in a distant atom as
the atom is adsorbed to the surface. The adsorbate wave function
ψ_A^2 produces a "virtual bound state" or resonance of breadth Δ_A
whose center is shifted by S_A from the atomic level. Many wave
functions pass through the region of ψ_A^2 which acts in this region
like a band pass filter for wave functions incident on the surface
from the inside of the crystal. This is illustrated in Fig. 47
where we see that the wave function magnitude at the ion in the
energy range of the virtual bound state is greater in the presence
of the adsorbed atom than in its absence.

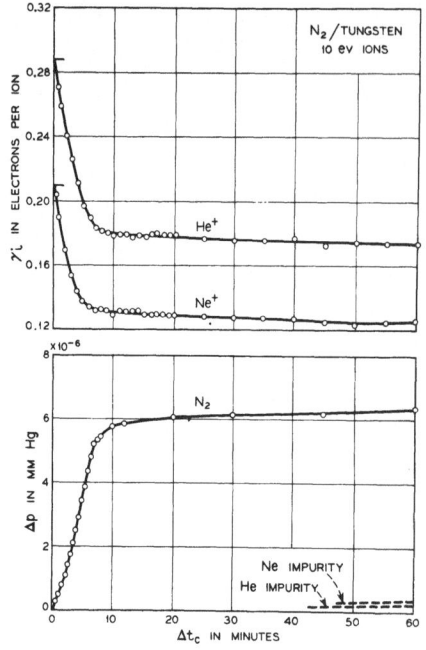

Fig. 45

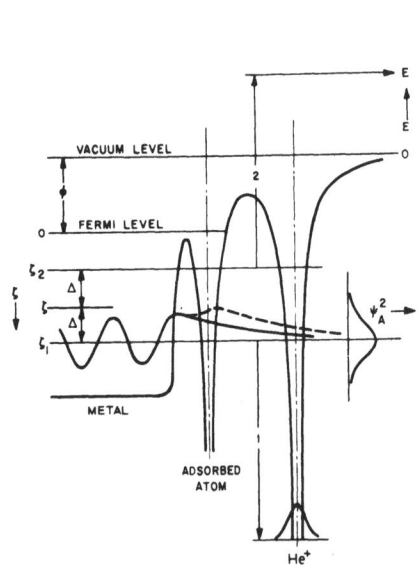

Fig. 47

Fig. 46

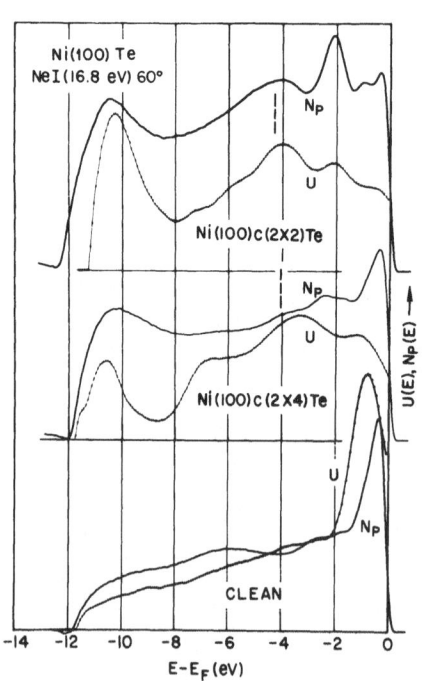

Fig. 48

In Fig. 48 are illustrated UPS spectra, $N_P(E)$, and INS unfold functions $U(E)$ for several nickel surfaces.[33,34] $N_P(E)$ is directly the measured UPS kinetic energy distribution reproduced photographically in the figure. $U(E)$ is reproduced from the computer graphic display of the digital unfold function obtained from Eqs. (39), (40), and (41). The three surfaces for which data are presented in Fig. 48 are clean Ni(100) and Ni(100) with a c(2×4), quarter monolayer, and a c(2×2), half monolayer, superstructure of Te atoms adsorbed upon it. A number of interesting results appear in this figure.

Note first that adsorption reduces the ability of each spectroscopy to see the d band of the Ni substrate. This effect is far more pronounced for INS than for UPS lending credance to our earlier conclusion that the INS sensitivity volume does not extend into the crystal beyond the surface monolayer. Secondly, we observe that adsorption of Te produces orbital peaks in the $N_P(E)$ and $U(E)$ spectra for both the c(2×4)Te and c(2×2)Te surfaces. The single broad peak for c(2×4)Te clearly seen in $U(E)$ is partially masked by d band structure still evident in $N_P(E)$. For the c(2×2)Te surface both UPS and INS show two Te associated orbital peaks at the same energy but with interestingly different intensities. It is possible that the $U(E)$ peak at $E - E_F \sim -2$ eV is weaker than the corresponding peak in $N_P(E)$ because the orbital has more d character than that at $E - E_F \sim -4$ eV or because its geometrical orientation is such that it produces less intensity at the position of the probing ion.

Perhaps the most intriguing result depicted in Fig. 48 is the fact that the Te related orbital spectra for c(2×4)Te and c(2×2)Te are completely different. As has been known for a long time in molecular chemistry this is direct, bona fide evidence that the bonding configurations at these two surfaces are completely different. As Fig. 49 indicates, in the c(2×2)Te structure [parts (a) and (b) of the figure] each Te atom has two Ni atoms to itself whereas in the c(2×4)Te structure [parts (c) and (d)] each Te has four Ni atoms to itself.[35] The local surface molecule may be regarded as Ni_2Te on the c(2×2)Te surface and Ni_4Te on the c(2×4)Te surface. Thus the local bonding on a uniformly covered surface or macromolecule is cognizant of the structure of the entire surface.

XIX.   WAVE FUNCTION VARIATION OUTSIDE A SOLID SURFACE

A unique characteristic of INS enables us to probe the variation of wave function components outside a solid surface.[36] This capability will be illustrated for the case of Hg adsorbed on Ni(100) and Si(111). Hg is particularly suitable as an adsorbate because of its nonbonding sharp $5d_{5/2}$ and $5d_{3/2}$ orbitals. Figure 50 indicates that when Hg is adsorbed on Ni(100) these orbitals are

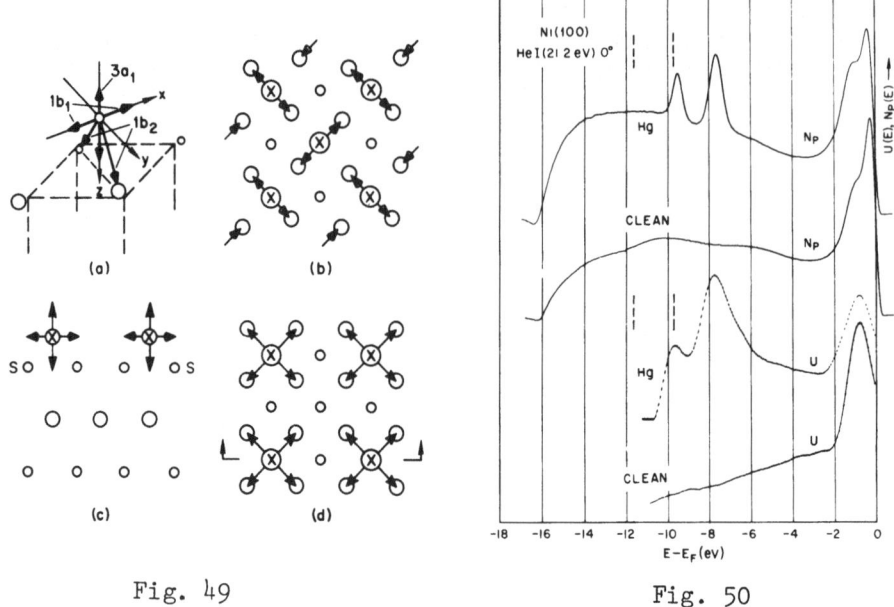

Fig. 49   Fig. 50

seen clearly at $E - E_F \sim -7.7$, -9.7 eV by both UPS and INS. This
is an interesting fact in itself since it demonstrates that there
is negligible hole-hole interaction between the two holes left in
the final state surface electronic structure after the Auger process
has occurred. Hole-hole interaction of the kind observed in some
core Auger processes could shift the orbital peaks as much as 10 eV
from the positions observed by UPS.

When Hg was adsorbed on Si(111), however, we were presented
with the startling result (Fig. 51) that UPS could "see" the Hg5d
orbitals but INS could not. It soon became evident that the basic
reason for this was that on Ni(100) the Hg5d orbitals were competing
in intensity at the ion position principally with the d orbitals of
Ni substrate. On Si(111), however, the Hg5d orbitals were essentially
completely swamped at the ion position by the p wave functions of
the Si substrate which extend much more strongly from the
substrate.[36] We also determined that the two-curve extrapolation
method (Sec. XVI) we had used in the INS data reduction in fact
discriminated against the possibility of observing a small admixture of d wave function in a predominantly p wave function.
It was for this reason that the one-curve extrapolation method was
devised.

When the one-curve method of data reduction was applied to the
Si(111)Hg surface the results of Fig. 52 were obtained. Here the
lower half reproduces the UPS data of Fig. 51. In the upper half
we see that the Hg5d orbitals are clearly seen at the correct

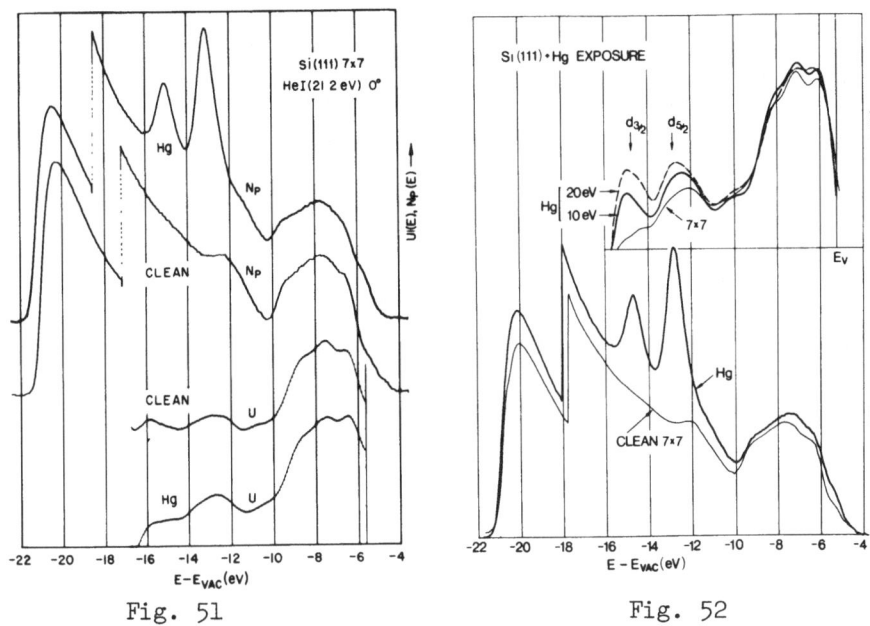

Fig. 51   Fig. 52

energies. Furthermore the Hg5d peaks are greater in magnitude
relative to the Si p peak at an incident ion energy of 20 eV than
in the U(E) spectrum for 10 eV ions. This results from the fact
that the magnitude of the Hg5d wave function component falls off
faster outside the surface than does that of the higher lying Si p
component and the fact that 20 eV ions are neutralized closer to
the surface than are 10 eV ions. Thus the relative spatial
magnitude variation of two wave function components, Hg5d and Si3p,
has been experimentally demonstrated. This could be a significant
addition to the capabilities of electron spectroscopies enabling us
to understand intensity differences of orbital resonances in chemisorption systems.

XX.   UNIQUENESSES AND LIMITATIONS OF INS

This paper is concluded with a discussion of the characteristic
advantages and disadvantages of INS. Perhaps its most distinctive
characteristic is its extreme surface sensitivity. In this respect
INS holds out the possibility that, in combination with UPS results
on the same surface, it will enable us to distinguish orbitals in
the surface monolayer from those in the selvedge or near surface
bulk. The surface sensitivity of INS is in one respect dis-
advantageous since without a companion spectroscopy like UPS
simultaneously available to give depth perception, INS results can
be confusing. The specific transition probability characteristics
of INS, particularly when used in conjunction with another electron

spectroscopy such as UPS or ELS having quite different transition probabilities, makes possible the characterization of surface orbitals beyond that possible with a single electron spectroscopy.

INS is considerably more local in character than UPS, for example, in that the incident ion interacts principally with the nearest surface atom. This has the advantage that INS is perhaps more sensitive to local surface inhomogeneities. A related characteristic has to do with the requirement that the Auger process upon which INS is based requires two electrons from the solid for its consummation. This results in the phenomenon that a solid can present to the incoming ion two distinguishable types of electron distribution, namely those that have been called folding and non-folding distributions.[20] A folding distribution can supply both electrons needed for the Auger process whereas a nonfolding distribution cannot. An example of a folding distribution is the valence band of a semiconductor, of a nonfolding distribution the surface state band. In the latter case the single surface atom with which the ion primarily interacts can supply at most one surface state electron. The second must come from a neighboring atom with considerably reduced probability. This characteristic of INS results in a relatively low probability of surface state detection despite its great surface sensitivity.

The more complicated data reduction procedure of INS vis-a-vis UPS, for example, must certainly be listed as a disadvantage. Its success, however, despite its complications, has illuminated the nature of two-electron Auger-type processes. The interplay between physics and the mathematics of ill-posed deconvolution is an interesting chapter in the development of INS.[28] Yet, there is unfinished business in this aspect of the phenomenon. The simple digital deconvolution scheme of Eqs. (39)-(41) does not work for fold and unfold functions that have low intensity near the Fermi level which is the origin for these functions. Such a situation prevails for the case of H adsorbed on Si where the s electrons of the H adsorbate lying at lower energy have considerably larger intensity at the position of the probing ion than do the higher lying p electrons of the substrate. Here a more sophisticated sequential unfolding formulation is needed that includes specific means for damping out point-to-point instability.

Finally the development of INS is a part of the broader field of the study of low energy particle-solid interactions that have occupied us during a major portion of this paper.

XXI. REFERENCES

1. H. D. Hagstrum, Phys. Rev. 96, 335 (1954).
2. S. S. Shekhter, J. Exptl. Theor. Phys. (U.S.S.R.) 7, 750, (1937).

3. H. S. W. Massey, Proc. Cambridge Phil. Soc. 26, 386 (1930).
4. A. Cobas and W. E. Lamb, Jr., Phys. Rev. 65, 327 (1944).
5. M. L. E. Oliphant and P. B. Moon, Proc. Roy. Soc. (London) A127, 388 (1930).
6. H. D. Hagstrum, Rev. Sci. Instrum. 24, 1122 (1953).
7. H. D. Hagstrum, Science 178, 275 (1972).
8. H. D. Hagstrum, J. Vac. Sci. Technol. 12, 7 (1975).
9. E. E. Chaban and H. D. Hagstrum, Rev. Sci. Instrum. 47, 828 (1976).
10. H. D. Hagstrum, J. Research, National Bureau of Standards 74A, 433 (1970).
11. H. D. Hagstrum, Phys. Rev. 150, 495 (1966).
12. H. D. Hagstrum, Phys. Rev. 104, 672 (1956).
13. H. D. Hagstrum, Y. Takeishi, and D. D. Pretzer, Phys. Rev. 139, A526 (1965).
14. H. D. Hagstrum, Phys. Rev. 122, 83 (1961).
15. W. F. Van der Weg and D. J. Bierman, Physica 44, 177 (1969).
16. W. F. Van der Weg and D. J. Bierman, Physica 44, 206 (1969).
17. C. W. White and N. H. Tolk, Phys. Rev. Lett. 26, 486 (1971).
18. W. E. Baird, M. Zivitz, J. Larsen, and E. W. Thomas, Phys. Rev. A10, 2063 (1974).
19. C. B. Kerkdijk, C. M. Smits, D. R. Olander, and F. W. Saris, Surf. Sci. 49, 45 (1975).
20. H. D. Hagstrum and G. E. Becker, Phys. Rev. B8, 1592 (1973).
21. H. D. Hagstrum and G. E. Becker, Phys. Rev. B8, 107 (1973).
22. R. L. Erickson and D. P. Smith, Phys. Rev. Lett. 34, 297 (1975).
23. T. W. Rusch and R. L. Erickson, J. Vac. Sci. Technol. 13, 374 (1976).
24. J. C. Tully, Phys. Rev. B (submitted).
25. E. Everhart, G. Stone, and R. J. Carbone, Phys. Rev. 99, 1287 (1955).
26. M. T. Robinson, Oak Ridge National Laboratory Report ORNL-3493 (1963).
27. D. P. Smith, Surf. Sci. 25, 171 (1971).
28. H. D. Hagstrum, "Techniques in Metals Research" Vol. 6, Part 1, p. 309 (1972).
29. H. D. Hagstrum and G. E. Becker, Phys. Rev. B4, 4187 (1971).
30. P. H. Van Cittert, Z. f. Physik 69, 298 (1931).
31. J. A. Appelbaum and D. R. Hamann, Phys. Rev. B12, 5590 (1975).
32. H. D. Hagstrum, Phys. Rev. 104, 1516 (1965).
33. H. D. Hagstrum and G. E. Becker, The Physical Basis for Heterogeneous Catalysis (E. Drauglis and R. I. Jaffee, eds.) Plenum Press, p. 173 (1975).
34. H. D. Hagstrum and G. E. Becker, J. Vac. Sci. Technol. 14, 369 (1977).
35. H. D. Hagstrum and G. E. Becker, J. Chem. Phys. 54, 1015 (1971).
36. H. D. Hagstrum and T. Sakurai, Phys. Rev. Lett. 37, 615 (1976).

INTRODUCTION TO SECONDARY ION MASS SPECTROMETRY (SIMS)

H.W. Werner

Philips Research Laboratories

Eindhoven, The Netherlands

1. INTRODUCTION

This contribution aims to give to the beginner in SIMS a survey of the present state of our knowledge of secondary ion emission, of basic experimental embodiment of today's SIMS instruments and of the potential of this method for research and as an analytical tool.

Let us first take a look at the history of SIMS. The first experiments with SIMS were carried out by Arnot[1] and Herzog and Vichböck[2], who described sputter ion sources for mass spectrometers. It was almost ten years later that interest on the part of a number of researchers became evident. The investigations of several workers[3] showed a widening interest in the processes of secondary ion emission. In the sixties a number of workers[4] added more knowledge about the general features of secondary ion emission. At that time emphasis was mainly on bulk analysis. A number of specialised SIMS instruments were developed for microanalysis[5]. In 1967 the application of SIMS to thin film analysis was urged[6,7] and quantitative analysis of depth profiles[6] was reported, i.e. not only was the ion current measured as a function of bombardment time, but the concentration (number of particles per cm^3) was given in terms of the depth ($\mathring{A}$) and diffusion coefficients were determined. Since then growing interest in SIMS has been noted. Analytical applications are to be found in reviews[8]. Today a dozen companies offer SIMS instruments with a diversified programme on a commercial basis.

The principle of SIMS is shown in fig. 1 : the target is bombarded with a beam of primary ions having energies of several keV. Target particles are sputtered due to this ion impact as atoms (ground state or excited state) and ions (positive or

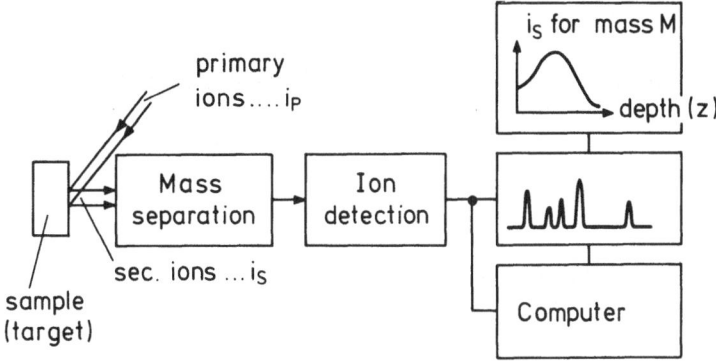

Fig. 1 : Principle of SIMS.

negative) either in the ground or the excited state. These ions
are extracted from the target region and pass a mass analyzer, where
they are separated according to their mass-to-charge ratio. These
mass-to-charge separated ions are detected by suitable means and
the information is fed to a recorder or computer.
    The information obtained from such mass spectra[6] is illustrated
in figs. 2 and 3 : in fig. 2 one can see atomic ions of the target
Al^+, Al^{++}, molecular ions ($AlOH^+$), and ions of the electropositive
elements Na^+, K^+. Note that H^+ is also seen in this spectrum, a
unique capability of SIMS. Ions of electronegative elements usually
do not appear in great abundance in the positive ion spectrum. In
the negative secondary ion spectrum they appear in high intensity,
as illustrated in fig. 3.
    Qualitative and semiquantitative survey analyses of all
elements from hydrogen to uranium can be carried out with SIMS. The
minimum detectable concentrations are element and matrix dependent.
Variations from element to element and from matrix to matrix may be
up to several orders of magnitude. The minimum detectable concentration is typically between 1 ppm and 1 ppb.
It must be realised, however, that the minimum detectable concentration is related to material consumption[9], since SIMS is a
destructive method. For example, Si bombarded with O_2^+, 5.5 keV and
a primary ion current density j_p = 4.10^{-4} A/cm^2, bombarded area
A_b = 7.10^{-4} cm^2, erosion rate $\dot{z}$ = 8 Å/sec, sputtered volume/sec
$\dot{V}$ = 6.10^{-11} cm^3/sec gives an extrapolated minimum detectable
concentration for Si of 1 ppm, on the assumption of a minimum
detectable ion current of 3.10^{-18} A.

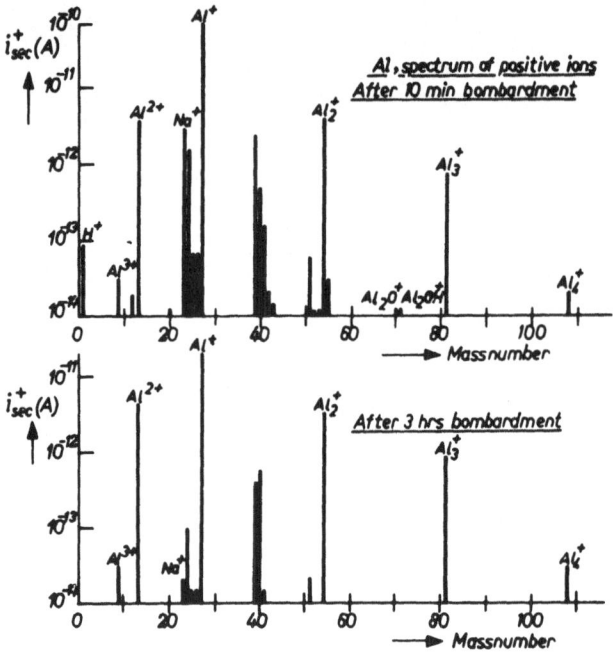

Fig. 2 : Positive secondary ion spectrum of Aluminium target
          obtained under Ar⁺ bombardment.

Sample size : in principle a sample diameter of one millimetre
is sufficient, but for ease of operation the dimensions accross the
sample should be about one centimetre. The sample thickness may
vary between 100 Å (on a substrate) and 20-30 μm. In the latter
cases the maximum obtainable erosion rate (≈ 5 μm/h) is the
limiting factor, in view of reasonable analysis times. Minimum
erosion rates as used in static SIMS[10] are about 1 Å/h. Usually
these two extreme values of the erosion rates cannot easily
be realised in a single instrument. SIMS requires no sample preparation. At the most ion etching is applied in order to remove
surface contamination.

Importance of SIMS as thin film analytical technique

SIMS can be used for the detection of surface contamination,
e.g. on technical samples, or for studying surface or catalytical
reactions. It can also be used to carry out depth analyses, as
illustrated in the lower parts of figs.2 and 3. In these cases,
comparison of the spectra obtained after ten minutes and after
three hours allows one to visualise how the composition of the
sample as a function of bombardment time (proportional to depth)
changes with depth (cf. also fig. 24).

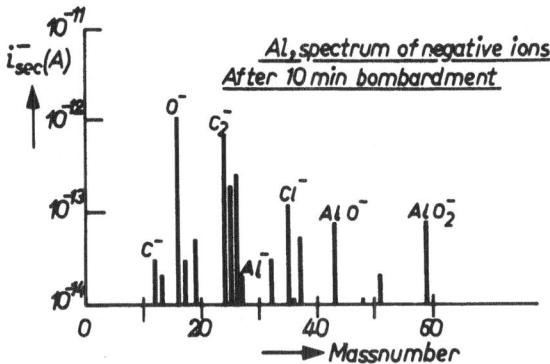

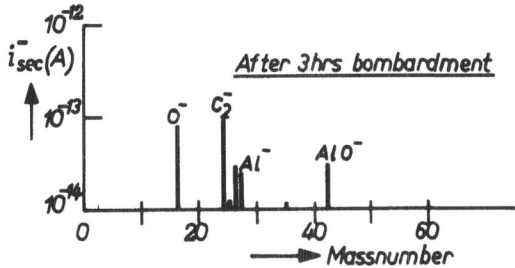

Fig. 3 : Negative secondary ion spectrum obtained under the same
conditions as in fig. 2.

Further fields of application are analysis of thin films, semiconductors, ceramic materials, glasses, implantation profiles,
diffusion profiles and local analysis of spots with dimensions of
a few microns. The position of the primary ion on the sample can be
aligned with accuracies of a few micrometres. Element mapping, i.e.
determination of elementary distributions across the surface, can
also be carried out. The samples to be analysed may consist of
organic or inorganic materials.

Organic samples have been studied by Benninghoven[11], as
illustrated in fig. 4, showing the positive ion spectrum of
phenylalanine. One can distinguish the protonated molecular ion
peak $(M+1)^+$ together with fragmentary ions such as $C_6H_5^+$, $CH_2C_6H_5^+$
etc. The silver peak is from the silver substrate on which this thin
film of phenylalanine was deposited. This figure shows that SIMS
can also be used to study the molecular structure of organic
molecules.

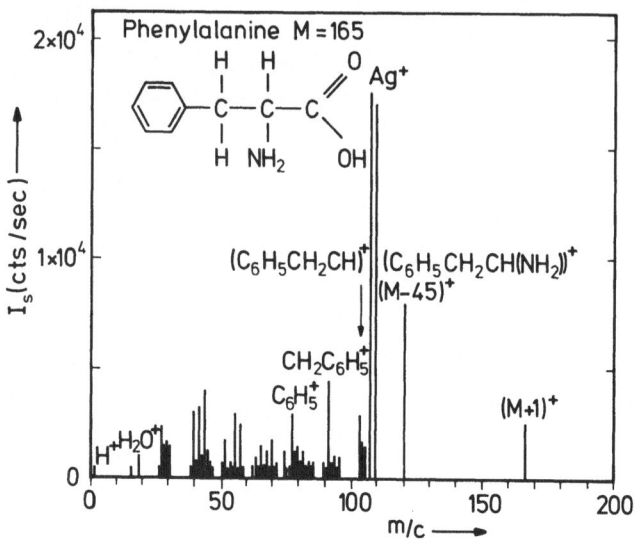

Fig. 4 : Positive secondary ion spectrum of phenyl-alanine (after
        Benninghoven [11]).

2. THEORY OF SIMS

2.1 Physical Basis

   The sputter process
   Primary ions (Ar⁺, O⁺) of keV energy penetrate into the solid
(Fig. 5), where binary collisions with target atoms occur, yielding
recoil target atoms. Primary particles will finally come to rest
after a slowing down time at the so called penetration (excitation)
depth, i.e. they are implanted into the solid. In the meanwhile
first, second, and n-th generation target recoils yield a dense
cascade, i.e. part of the target atoms are moving, others are at
rest. A special case of a cascade is a spike; this is a dense
cascade in which all particles of the volume considered are tempor-
arily in motion. The target atoms are displaced by such cascades.
If they are close enough to the surface and their velocity vectors
are pointing outside the target, the process called sputtering
occurs i.e. a target particle is emitted from the surface. Sputtering by direct knocking out of target atoms by primary ions occurs
relatively scarcely.
   The information depth (escape depth) is that depth at which
the particles which are sputtered were present prior to the
emission process.

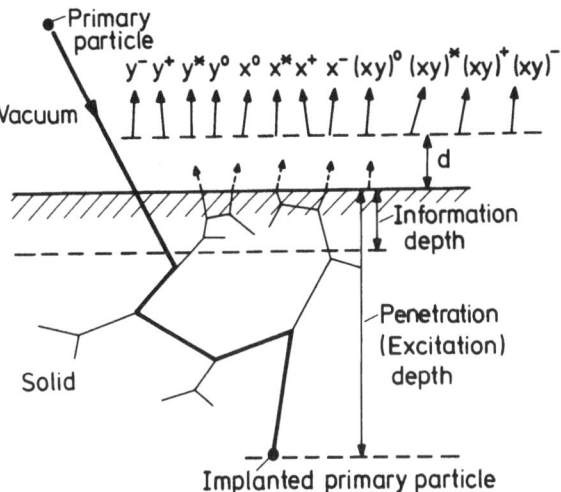

Fig. 5 : Schematic representation of an ion-solid interaction,
         leading to the emission of neutral, excited and ionized
         (+ or -) target atoms (X,Y) and of molecules (X Y).In
         surface near regions of thickness d, processes may take place
         which change the state of the particles as emitted (cf.
         table 1).

The information depth depends on the mass m_1 and energy E_1 of
the projectile ions and on the mass m_2 and energy E_2 of the recoil
matrix atoms. Obviously the greater the energy of the recoil atoms,
the greater the depth at which they were originally situated. The
relatively small number of particles with escape depths in excess
of 20 Å (fig. 6) is therefore due to the high energy recoils, while
the majority comes from zones immediately below the surface with a
mean escape depth of about 6 Å. These values are in good agreement
with estimations of the projected range of low energy particles. It
is obvious that with increasing atomic number Z and therefore with
increasing atomic radius the escape depth decreases. In particular
the escape depth of cluster ions, i.e. two or more atoms emitted as
one unit, is restricted to one or two atomic layers, since the

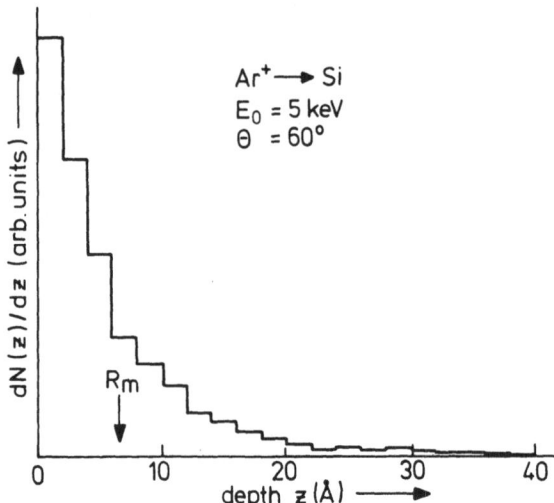

Fig. 6 : Distribution of the original depth of sputtered atoms
versus depth (after T. Ishitani and R. Shimizu[146]).

dimensions of cluster particles are large compared to the average
distance between two target atoms. This holds particularly for the
protonated parent peaks of organic molecules (fig. 4).
    The information depth is often confused with the mean free
path. There is a well defined relation between these two quantities
as well as between mean free path and the extinction coefficient[12].
    State of emitted particles
    In the theory of sputtering the ejection of target particles
irrespective of their state of excitation is considered. It is
clear, however, that in the collision cascade with its many violent
collisions, excitation of atoms and single and multiple ionization
occur. The reverse process, viz. de-excitation and neutralization,
can occur as well, together with the formation of negative ions by
electron capture. The state of these particles as generated inside
the target can be changed, however, by "reactions" in zones of at
the most 50 Å in front of the target. An excited particle, for
instance, can fall back into its ground state by emitting a light
quantum, or it can become ionized. In table 1, a number of possible

Table 1 : Some mechanisms which may take place after particle
          ejection in zones close to the surface. (State of
          particles: X^0= ground state atom, X^*= excited; X^+, X^- =
          positive or negative ground state ion).

State of ejected particle	$\approx$ Life time (sec.)	Mechanism	Final state of particle
X^*	10^{-8} 10^{-14} 10^{-14}	De-excitation by light emiss. Auger de-excitation Resonance ionization	$X^°$ $X^°$ X^+
X^{**}		Autoionization (Auger de-excitation)	X^+
X^+	10^{-14} 10^{-14} 10^{-8}	Auger-neutralization Resonance neutralization Radiative neutralization	$X^°$ $X^*, X^°$ $X^°$
$X^+, Y^°$		Charge-exchange	$X^°, Y^+$
$(X^+Y^-)^*$		Bernheim processes Dissoc. of *ionic* bond	$X^+ + Y^-$
$X^°$		Electron capture	X^-

processes which the ejected particle can undergo are listed, together
with the final state of the particles (see also fig. 7). A general
rule is that the probability P of escape of a particle from the
surface region without undergoing any of these transitions is given
by $P = \exp(-s/v)$ where s is often called the survival parameter,
P is the survival probability, and v is the velocity[13]. This
equation shows that low energy (low velocity) particles have a
small survival probability, which is obvious, because due to their
small velocity they remain longer in the reaction zone and therefore
have a greater probability of changing their state by one of the

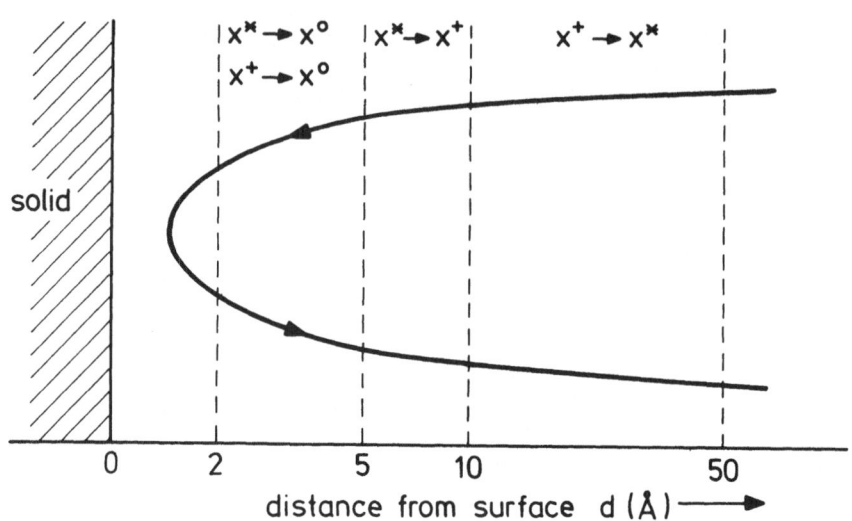

Fig. 7 : Different zones in front of the target in which the
original state of the emitted particles may change, after
Hagstrum[147].
From 0-2 Å violent collisions between incoming ions and
lattice atoms positioned at d=0 take place; d=2 - 5 Å :
Auger deexcitation or neutralization may take place;
d=5 - 10 Å : resonance ionization; d=10 - 50 Å resonance
neutralization, after Kerkdijk[148].

above mentioned reactions. It is clear that many complicated
processes play a role in the emission of secondary ions.
       There is no complete theory at the moment which accounts for
these processes from the very beginning right up to the emission
of the particles. However, a number of theories which describe
details of the above mentioned processes have been given. In many
cases these theories hold for certain experimental conditions only
(e.g. emission from pure metals under argon or oxygen bombardment)
or they relate to secondary ions of given kinetic energy.
Models for secondary ion emission :
       The theories to be discussed below (for a detailed review see
[14]) are unrelated. Probably all processes described in these
theories occur simultaneously, although one or other of these
processes may be predominant, depending on the experimental
conditions.
       Kinetic model : the kinetic model[15] refers to ions emitted,
with kinetic energies greater than 30 eV, from pure metals under
bombardment by noble gas ions. The process of secondary ion

emission in this model takes place in several steps : penetration of the primary particle into the target and emission of secondary particles via collision cascades. During the violent collisions between target particles in the dense cascade, electrons of inner shells are excited. The target atom is considered to be emitted as an excited neutral particle and deexcites outside the metal into an ion via an Auger effect. The agreement between theoretical and experimental values (for Mg) is quite good. In particular one must consider that the values were calculated without the use of any fitting parameter.

The autoionization model[16)] : This model, valid for rare gas ion bombardment, assumes inner shell excitation into an auto-ionizing state and relaxation of the latter in vacuum via an Auger process, yielding a secondary ion and an Auger electron. With a detailed knowledge of the electronic structure of the metal and the free atom, absolute values of the ion yield have been calculated to within a factor of two in respect of many elements.

Surface effect models : in contrast to the models described above, it is assumed that all processes relevant to the formation of ions take place at or near the surface of the sample. A particle while being ejected is assumed to change its electron structure from that of the bulk metal to the free atomic or ionic level. Electron transitions from these particles back to the metal may take place during the ejection processes. The ion yield is then calculated as the probability of finding an ion at a great distance from the surface i.e. ion yield is equated to the ratio of number of emitted ions/number of particles emitted in an arbitrary state. Different formulas for the calculation of this ion yield have been given by Sroubek [17)] and Schroeer[18)]. Their formulas contain a term $(E_i-\emptyset)$, where E_i is the ionization energy and $\emptyset$ is the work function. The shortcoming of this consideration is that the process of sputtering is not taken into account explicitly. This has been done in the work of Gries and Rüdenauer[19)] who have extended Schroeer's formula by taking into consideration the energy distribution of the secondary ions and the sputter yield (atom per ion) of the cascade theory of Sigmund[20)].

Cini[21)] has extended the existing considerations to the influence of Fermi electrons on the ionization process.

Antal[22)] has considered a target particle which moves through the metal as an interstitial whose local excess of positive charge is compensated by an electron cloud moving together with the particle. A plane "de Broglie" wave is assumed to represent this electron cloud. The electron cloud is partially reflected at the potential barrier of the sample surface. The ion yield is then equated to the quantum mechanic reflection coefficient of the electron wave at the surface. Calculated values show good agreement, without the use of fitting parameters, with the experimental values for several elements. An exception is provided by the noble metals, where deviations of several orders of magnitudes occur.

Bond breaking model : the elements are considered to be present as ionic compounds M^+-O^- or M^--Cs^+. The ionic bonds are considered to be broken[6],[23-26] giving M^+ and O^-, or M^- and Cs^+.

Thermodynamic model

Andersen[27] has assumed that a plasma in complete thermodynamic equilibrium is generated locally in the solid by ion bombardment (local thermal equilibrium model). The ratio of the number of ions generated in the plasma to the number of particles present in the plasma in any state is assumed to be unchanged during the emission. Neutralization processes as described above (see table 1) are neglected. The generation of positive ions is assumed to follow the equation $X^o = X^+ + e$ (analogous considerations hold for negative ions and oxide ions).

Assuming equilibrium, the mass action law can be applied to this reaction to find the ratio of the concentration (n) of ions, neutrals and electrons, yielding $n_{X^+} \cdot n_e / n_{X_O} = k$. From thermodynamics or statistical mechanics one can calculate[28] the value of k, which inserted into the above equation yields the Saha Eggert equation :

$$n_{X^+} \cdot n_e / n_{X^o} = 2\, Z^+(T)/Z^o(T) \left[(2\pi mkT)^{3/2}/h^3\right] \exp\left[-(E_i - \Delta E_i)/kT\right]$$

where Z^+, Z^o and 2 = partition functions of ions, atoms and electrons respectively, E_i = ionization energy, ΔE_i = depression of ionization energy in the plasma; T = temperature, which acts as a parameter correcting for matrix effects. Taking $n_{X^+} = I_s$ (secondary ion current), $n_X \sim$ concentration c_X of element X and plotting $\log(n_{X^+} Z^o / 2 n_{X_O} Z^+)$ against the ionization energy where I_s are the ion currents obtained from elements with known concentrations, a straight line is to be expected when the Saha Eggert equation is valid. This was first shown by Andersen[27] and later by several other authors[29-35]. Fig. 8 shows Saha Eggert plots obtained from a glass sample[30].

Application of the Saha Eggert equation for quantitative SIMS analysis. The concentration c_X of an element can be calculated from the Saha Eggert equation, assuming as discussed above that $c_X \sim n_X$, $n_{X^+} \sim I_{X^+}$, when electron density and temperature have been determined beforehand by measuring the ion currents of at least two elements with known concentration in the same matrix (internal standard method). An exponential dependance of the ion yields on the ionization energy had already been suggested neglecting, however the influence of the "temperature" on the ion yield[7]. Quantitation ran into trouble when different matrices were used (matrix effect).

Since errors can be introduced in the determination of these parameters due to fluctuations in the current measurements and/or inhomogeneous distribution of the internal standards of the sample and also because the ionization energy (defined for free atoms) may change in the plasma, Andersen[27] has developed an iterative method (Carisma) for the determination of the concentration by means of a computer programme. He has shown that semi quantitative

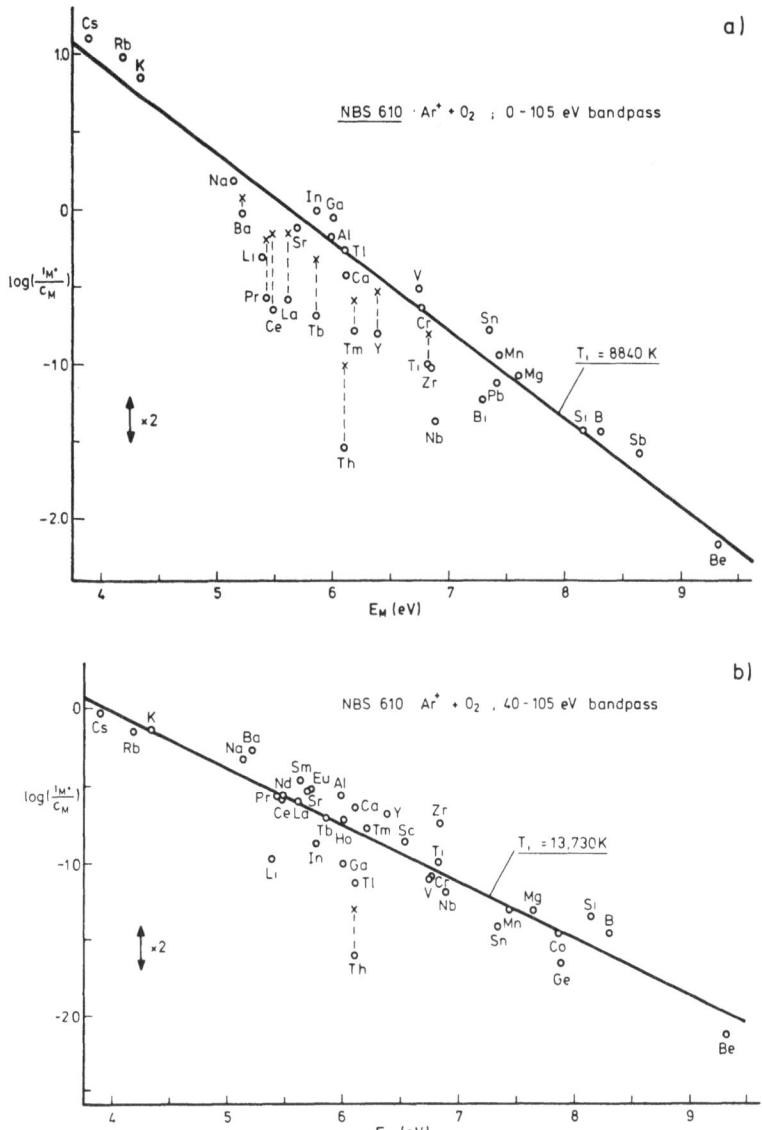

Fig. 8 : Reprinted with permission from Analytical Chemistry.
Copyright by the American Chemical Society.
Saha Eggert plot of ion currents from glass sample. Log
of reduced ion current $i_M{}^+$ divided by atomic concentration c_M vs. ionization potential E_M.
$i_M{}^+$ is the product of the measured current $I_M{}^+$, (mass
number)$^{1/2}$, and the atom-to-ion partition function ratio at T_i. Crosses result when $I_{MO}{}^+$ is added to $I_M{}^+$
(from Morgan and Werner[30]).

analysis (errors up to a factor of two and under optimum conditions
quantitative results (accuracy ± 20 %) can be obtained with this
method. The model of Andersen which has proved to be best suited
for practical quantitative analysis has been tested and further
developed by many other authors. In particular the influence of the
partition functions on the results has been tested by Evans[31] and
Morgan[30]; the distribution of the results obtained has been
studied by means of well chosen standards by Newbury[32]; the
annoying fact was studied by Shimizu[33] and Rüdenauer[34] that in
particular within one and the same sample n_e can have values
which differ up to a factor thousand,dependent on the T-value and
the choice of the element as internal standard.[32]

The collision cascade model has been developed from the LTE
model by Morgan and Werner[35]. Sputtering and ionization are
described in terms of the collision cascades initiated by
penetration of primary particles (see fig. 5). The main points of
this model are :

1) From the Saha Eggert equation as given above, it is sufficient
   that the reaction $X = X^+ + e$ is in equilibrium. The temperature
   which occurs in the correspondent equation is interpreted as
   the ionization temperature which can be different for the
   temperature derived from other processes such as generation of
   molecular ions, negative ions or the temperature T_k equivalent
   to the most probable kinetic energy of the secondary ions.
   (What was thought to be the dilemma of obtaining two different
   temperatures for one and the same sample viz. T_i from the Saha
   Eggert equation (cf fig. 8) and $T_{k'}$ is irrelevant since the
   surface binding energy determines the maximum in $N(E_k)$).

   T_i in the collision cascade model is considered as a useful
   parameter for carrying out calculations related to the
   physical essential quantity $E_c = k.T_i$ where E_c is the energy
   available for ionization in a individual cascade.
2) The influence of the experimental conditions on the value of T_i
   has been studied[30] :
   T_i depends on the matrix (insulators have higher T_i values than
   metals).
   T_i depends on the kinetic energy of the ions chosen for the
   determination of T_i. The larger the kinetic energy of the ions,
   the larger T_i. This is explained as an increase of the ionization
   cross section with energy, analogous to the situation for the
   ionization of gases by electron impact.
   T_i increases with increasing oxygen content in the sample, which
   goes parallel in general with an increase of the ion yield.
   T_i can therefore be considered as a parameter which describes
   the degree of ionization.
3) The generation of molecular ions was found to be also dependent
   on the dissociation energy. A rule to predict the ratio of
   molecular ions to atomic ions has been derived[35]. In particular

the generation of MO^+ or MOH^+ (where M indicates a metal atom) could be explained when considering the dissociation energy and the valence state of the atoms.

For the application of this model to quantitative analysis an iteration process (QUASIMS) using only one fitting parameter (T_i) and an exponential dependence on the ionization energy has been developed. This one-parameter model .has the advantage, in contrast to the two parameter model of Andersen or models derived therefrom, that it avoids the annoying result that n_e determined from one and the same sample can differ by up to a factor of thousand dependent on the choice of the internal standard. The values of T_i determined with QUASIMS are constant under constant experimental conditions and independent of the choice of the internal standard. One can therefore determine T_i for a sample with given concentration and then carry out semiquantitative analysis of samples with similar values of the major components, without the use of internal standards. When T_i has not been determined beforehand at least one internal standard is needed.

Practical results obtained with QUASIMS[30,35] give similar results as obtained with CARISMA or QUASIE. In particular the values of T_i determined with QUASIMS are independent of the choice of the internal standard.

The Andersen model and related models have been discussed here more extensively than other models because it has proved to be the best for practical quantitative analysis. Apart from small deviation of the different models in e.g. the choice of the fitting parameters and in the choice of the iteration processes, these models have all in common that they assume an exponential dependence of the ion yield on the ionization energy and use a parameter (T) which takes into account for the influence of the matrix.

In conclusion one can say that the mechanism of ion formation is not fully understood and is also not covered by one theory alone. It is hoped, however, that a unifying theory which explains the fundamental processes and also enables a simple practical application for quantitative analysis, will be available in the near future.

Other models :

Lodding[36] was one of the first to use SIMS for investigation of biological hard tissues. He studied the F^+/Ca^{++} content of apatites familiar as constitutents of, among other things, human teeth. He found that the measured F^+/Ca^{++} ratio varied, depending on the prevailing experimental conditions. To overcome this experimental influence, he plotted $\log(F^+/Ca^{++})$ against $\log(P^+/Ca^{++})$ obtained from a number of samples with known concentrations. The linear curves obtained were used as a sort of working curve for the determination of the absolute concentration of F^+ in apatite.

Lodding interpreted these linear curves as an exponential dependence of the secondary ion current on the ratio of ionization energy divided by kT, similarly to the model of Andersen. He also pointed out that the extremely high F^+ ion current indicated an

ionization energy (see Saha Eggert equation) of 10 eV rather than
the tabulated E_F = 17.5 eV. His conclusion was, in agreement with
the results of Morgan and Werner[29], that another mechanism than
that discussed above must be responsible for the F^+ production in
that case.

Sparrow[37] has proposed a formula for the determination of the
relative ion yield of compounds. His considerations, in good
agreement with the work of Morgan and Werner[38] are based on the
following.

For an atom to be emitted from the sample, ionized, and then
detected, chemical bonds must be broken, the atom M must be
ionized (ionization energy E_M) and emitted. He assumed an exponential
dependence of the ion yield on the ionization potential; the ionic
bond strength is assumed to be represented by X, the electro-
negativity of the element; the number of bonds n involved is put
equal to the valence state of the atom :

$$S^+_{rel} = \left(S^+_M / S^+_{ref}\right) \exp(-E_M/K.T)/nMX^2.$$

The model of Benninghoven and Plog[39] predicts abundances of
different molecular oxide ions in different oxidation states. The
basic idea is that the valence of a given atom in the original
lattice (lattice valence) and the "nominal" valence of a fragment
ion (fragment valence) are related to each other.

The authors have shown that ions with a given atom/oxygen
ratio are emitted in great abundance when the fragment valence is
equal to the lattice valence, in good agreement with the experimental results of Werner[40] and the considerations of Morgan and
Werner[38].

Jurela[41] has used the so-called Dobretsov equation which is said
to apply to a non-equilibrium :

$$S^+ \sim (Z^+/Z_o) \exp((\emptyset-E_c))/k.T$$

where $\emptyset$ is the work function and E_c is the ionization potential
(energy) at the critical distance for charge exchange. Comparison
of this formula with the Saha Eggert equation shows that both
include an exponential term. It has been shown[28] that the
exponential term when derived either via thermodynamics or
statistical mechanics is based on the assumption of the existence
of an equilibrium. From the work of Jurela it is not clear why the
Saha Eggert equation should then apply to equilibrium whereas the
Dobretzov equation should not.

2.2 Basic Formula For a SIMS Analysis

It is the aim of every SIMS analysis to determine the
concentration of element M from the measured secondary ion current
I_s corrected for the isotopic abundance of the element of mass M.
This can be carried out in principle from the formula

$$I_S = I_p S \beta^+ c.f = I_p S^+ c.f \tag{1}$$

where I_p is primary ion current; S is sputter yield = (number of particles sputtered N_s^k/number of impinging primary ions N_p) = (N_s^k/N_p), where k relates to neutral or charged particles, excited or in the ground state atoms, or clusters, respectively, β^+ is the degree of ionization = (number of particles which are sputtered as (positive) ions/number of particles which have been sputtered in a state k) = N_s^+/ N_s^k.

Obviously S$\beta^+ = N_s^+/N_p^+$ = (number of sputtered ions/number of impinging primary ions for cf=1)=S^+= ion yield; c is the fractional concentration = (number of atoms (summation of all isotopes) actually present per cm^3)/(number of atoms maximally possible per cm^3); f = mass spectrometer transmission = measured ion current at the collector ion current emitted from the target. The transmission can be split into two parts. Firstly there is the fraction of ejected ions that is collected in the focusing optics. This fraction increases with increasing extraction field. In magnetic sector type instrument, this field strength may be very high e.g. 1 kV/mm. Hence a large fraction may enter the ion optics, provided the aperture is not too small. In quadrupole instruments, where the injection energy must not be too high, only very weak electric fields are applied. This results in only a small fraction of the ejected ions being collected via the ion optics aperture. Acceleration-deceleration systems were reported to give improvement in this respect[37]. Secondly there are the further losses between the extraction lens and the collector, which are determined by the geometry, e.g. slit width and/or diaphragms which are kept small to keep ion optical aberrations and hence peak width low in magnetic sector instruments. In quadrupole, apart from the entrance aperture, no diaphragms need be used to ensure small peak widths.

In order to obtain some feeling for SIMS let us consider the order of magnitude of the parameters which play a role : I_p varies between 10^{-11} and 10^{-5} A,(see table 2) depending on the chosen mode. Apart from the total ion current I_p one must also consider the primary ion current density $j_p = I_p/A_b$ where A_b = effective bombarded area. It can be seen from table 2 that the values of j_p can exhibit wide differences, even for the same value of I_p if different beam diameters d_p or different scanning areas are used. In the latter case j_p is the effective ion current density. The ion current density is most important because it is proportional to the erosion rate $\dot{z}$, the layer thickness sputtered per unit of time as shown in eqn. 2. (The erosion rate is often called sputter rate but this may cause confusion with the term sputter yield.) :

$$\dot{z}(\mu m/h) = 3.6 \times 10^{-4} (M(amu)/\rho(g/cm^3))j_p(\mu A/cm^2)S \tag{2}$$

Figs. 9 and 10 enable the erosion rate to be rapidly estimated for given values of S, I_p, A_b, and j_p, respectively, taking M/ρ = 10.

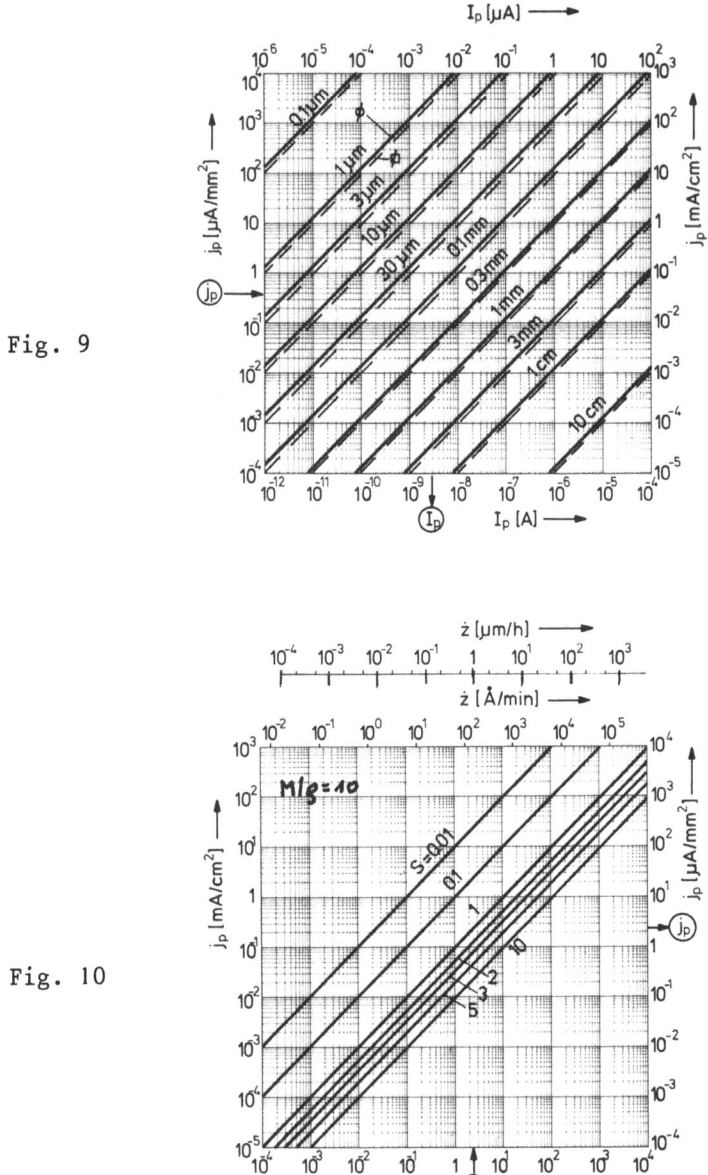

Fig. 9

Fig. 10

Fig. 9 : Nomogram for the determination of ion current density
j_p $(=I_p/A_b)$ from the primary ion current I_p vice versa,
for different beam dimensions (circular beam diameter indicated with $\emptyset$, square beam diameter indicated with $\square$).
Fig.10 : Erosion rate $\dot{z} = f(j_p)$ for different sputter yields S.

Table 2 : Typical values for primary ion current I_p, beam
         diameter d_p, primary ion current density j_p and sputter
         (erosion) rate $\dot{z}$ calculated therefrom, taking S=3.

Mode	I_p	d_p	j_p	$\dot{z}$ (S = 3)
Static	10^{-11}A	1mm	1 nA/cm^2 (= 0.01 nA/mm^2)	1 Å/h
Dynamic	3×10^{-8} A	1mm	3 μA/cm^2 (= 30 nA/mm^2)	300 Å/h
	10^{-11} A	1μm	1 mA/cm^2 (= 10 μA/mm^2)	10 μm/h
	10^{-5} A	1mm	1 mA/cm^2 (= 10 μA/mm^2)	10 μm/h

The total material consumed per unit time, $\dot{V} = \dot{z}$ A_b, is shown
also for A_b = 1 mm^2.
     The sputter yield depends on a large number of parameters[42,43]:
Mass M_1, energy E_1, and angle of incidence θ to the surface normal,
the projectile ion mass M_2, binding energy U_o of the target atoms,
and some parameters which take into account the interaction between
the two particles.
     Figs. 11 to 14 show the dependence of S on E_1, θ, Z_1, Z_2, and
on the angular (fig. 15) and energy (fig. 16) distribution of the
sputtered atoms, respectively.
β^+ and S^+ :
β^+ (equation 1) depends on the element (ionization energy), on the
matrix (ionization temperature), and on the presence of oxygen and
other electronegative elements or the presence of carbon (electropositive elements).
β^+ or $\beta^+ . S = S^+$ can be determined from equation (1) if all other
parameters, in particular f, are known.
     Benninghoven has determined S^+ for some metals and oxides
after having determined the transmission of his instrument
(Table 3). From his values we have estimated the order of
magnitude of β^+, assuming the sputter yield to have the same
value S=2 for all elements in metal or oxide form (Table 4). This
assumption is valid in many cases, as shown by Kelly[44].

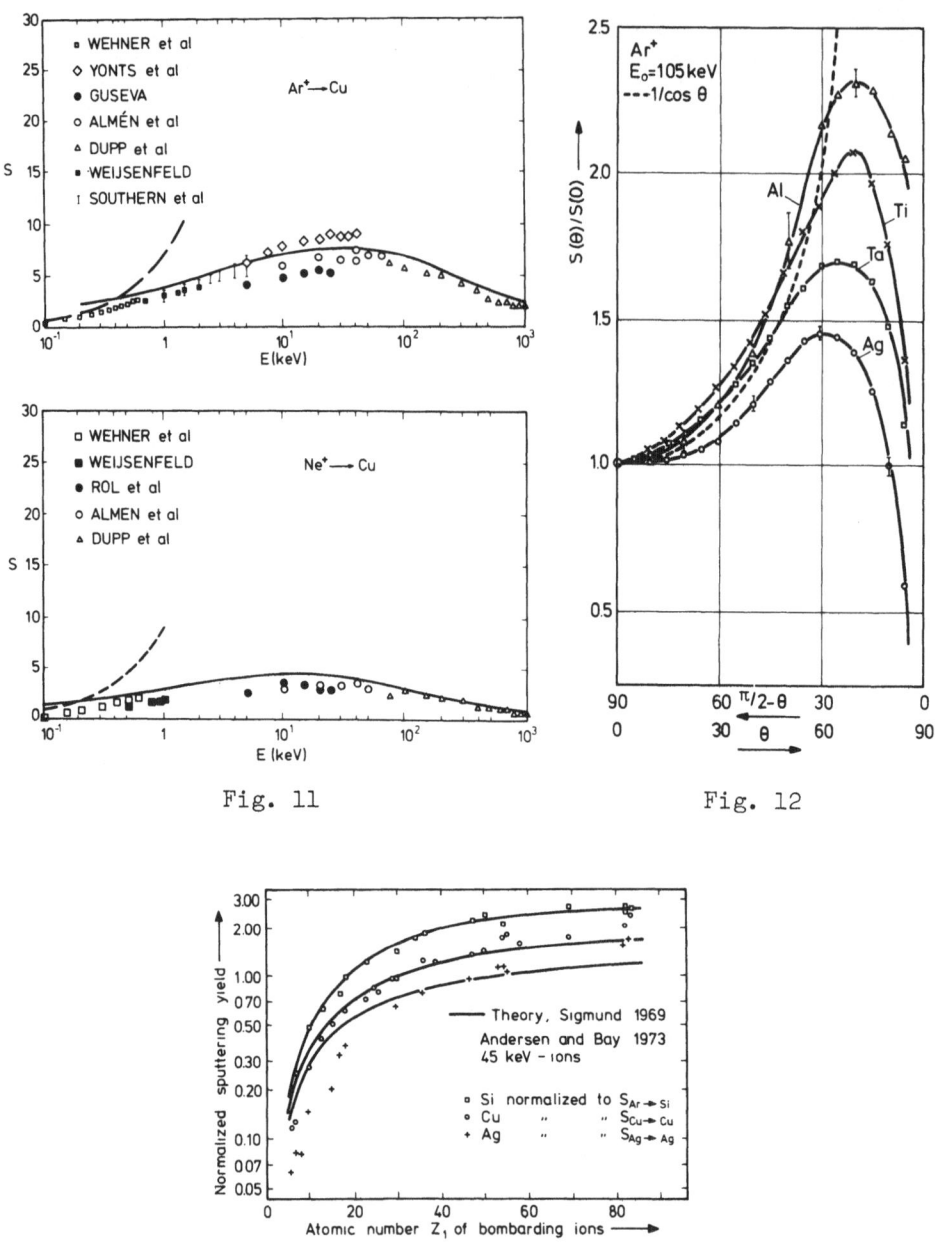

Fig. 11

Fig. 12

Fig. 13

Fig. 11 : Sputter yield S as a function of the primary energy E_p for Ar^+ and Ne^+ on Cu, after Sigmund[150].

Fig. 12 : Sputter yield as a function of incident angle, after Oechsner[151]. θ is the angle against the target normal

Fig. 13 : Sputter yield as a function of atomic number Z_1 for bombarding ions, after Oechsner[152].

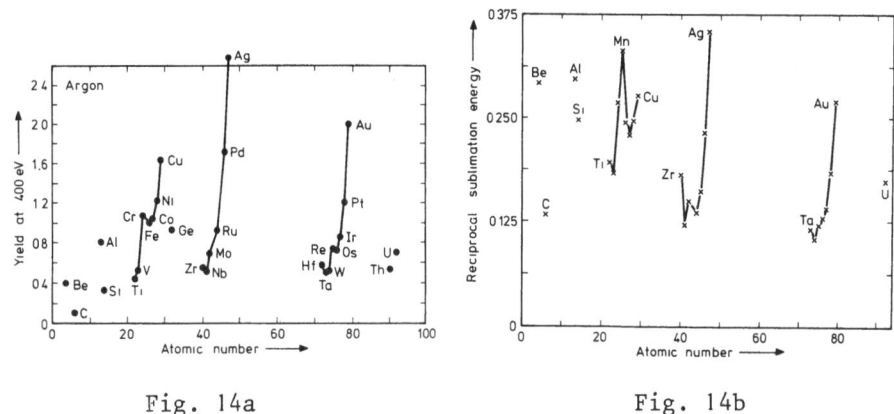

Fig. 14a   Fig. 14b

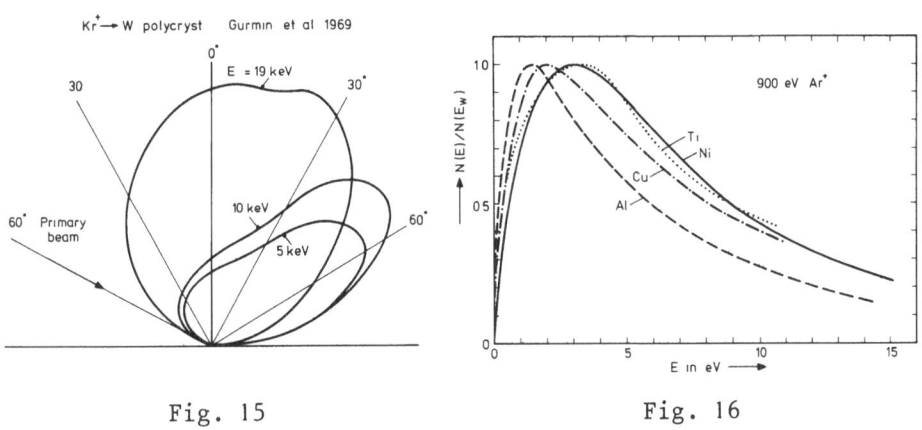

Fig. 15   Fig. 16

Fig. 14a : Sputter yield as a function of atomic number Z_2 of the
           target, after Laegreid and Wehner[153].
Fig. 14b : Reciprocal of sublimation energy as a function of atomic
           number Z_2 of target; (Rosenberg and Wehner[154]).
Fig. 15 : Angular distribution of sputtered material for poly-
           crystalline W bombarderd with oblique incident ($\theta = 60°$)
           Kr^+ ions of different energy, after Gurmin[155].
Fig. 16 : Normalized energy distribution of neutral particles sput-
           tered along the target normals from different poly-
           crystalline materials with normally incident 900 eV
           argon ions, after Oechsner[156].

Table 3 : Absolute positive ion yields for clean metals and their
respective oxides, using Ar^+ as primary ions (E_p = 3 keV,
Θ = 70°, p = 10^{-10} Torr, after Benninghoven[157]). The
degree of ionization β^+ = S^+/S has been calculated assuming that the sputter yield ($S \approx 2$; cf. Table 4) is the same
for the metal and oxides. Increase in ion yield from clean
to oxidized surfaces is shown in the last column.

Element	S^+ clean	β^+	S^+ oxide	β^+	S^+ oxide/ S^+ clean
Mg	8.5×10^{-3}	4×10^{-3}	1.6×10^{-1}	8×10^{-2}	20
Al	2×10^{-2}	1×10^{-2}	2	1	100
V	1.3×10^{-3}	7×10^{-4}	1.2	6×10^{-1}	10^3
Cr	5×10^{-3}	3×10^{-3}	1.2	6×10^{-1}	200
Fe	1×10^{-3}	5×10^{-4}	3.8×10^{-1}	2×10^{-1}	380
Ni	3×10^{-3}	2×10^{-4}	2×10^{-2}	1×10^{-2}	7
Cu	1.3×10^{-4}	7×10^{-5}	4.5×10^{-3}	2×10^{-3}	30
Sr	2×10^{-4}	1×10^{-4}	1.3×10^{-1}	7×10^{-2}	700

Table 3a: Absolute positive ion yields S^+ (Hennequin[157]), Ar^+,
Θ = 0°, 8 keV.

Element	Si	Ti	Ag	Ta	Mg
S^+	10^{-3}	1.6×10^{-3}	5×10^{-5}	5×10^{-4}	4×10^{-3}

Element	Al	Fe	Ni	Cu	
S^+	2.5×10^{-2}	10^{-3}	8×10^{-4}	6×10^{-4}	

Influence of the matrix on S^+

One must well realise, when using the S^+ values of tables 3 and 6,
and figs. 18-20, that these are valid for pure elements only[10];
in the described experimental conditions. The S^+ values of the
same elements in different matrices may differ by several orders
of magnitude. In order to correct for these "matrix effects" one
can use either empirical formulas[45] or models with a fitting
parameter, CARISMA[27], QUASIMS[30] etc., or an empirical approach
utilising calibration standards.

Transmission f : Reports on the experimental determination of
f are very scarce. Benninghoven[46] has determined f = 10^{-4} in his
sector type instrument. Vallerand[47] has calculated a transmission
of 10^{-3}. Obviously the transmission increases with increasing slit
width in sector type instruments, i.e. with decreasing resolution.

Table 4 : Sputtering yields at E = 1 keV for perpendicular ion incidence (from Oechsner[156]).

Target	He$^+$	Ne$^+$	Ar$^+$	Kr$^+$	Xe$^+$
Be	0.35	0.80	1.1	0.8	0.7
Al		1.13	1.94	1.53	
			1.94		
Si			1.0		
Ti			1.13		
Fe		0.84	1.34	1.44	
Ni	0.22	1.45	1.86	1.89	2.0
		1.24	2.18	1.73	2.22
			2.16		
Cu	0.65	2.75	3.64	3.62	2.42
		1.88	2.90	3.43	3.70
			3.2		
Ge			1.55		
Zr			1.06		
Nb			0.98		
Mo		0.43	1.14	1.41	1.63
			1.24		
Pd			3.06		
Ag	1.8	2.4	3.8	4.7	
			4.7		
Cd			11.2		
Ta			0.91		
W			1.10		
Pt	0.08	0.85	2.0	2.35	2.52
Au		1.53	3.08	3.86	
			4.02		

Transmission f in a first approximation is $f \sim 1/R$, where R is the mass resolution (Fig. 17). It should be noticed that f is also dependent on secondary ion kinetic energy and ejection angle. In quadrupoles, the transmission is determined with <u>gas</u> ion (energy width a few tenths of an electrovolt). The values for <u>secondary</u> ion transmission may be much smaller, however, since they have a larger kinetic energy. Typical values of the parameters considered are given in table 5.

Estimate of I_s:

      The secondary ion current from pure $Cu(c=1)$, made up of the sum of all isotopic ion currents, can be estimated from equation (1) and the following data; with O_2^+ as primary ion, $E_1 = 5.5$ keV, $I_p = 3$ µA, $A_b = 1.2 \times 1.2$ mm^2, so that one finds $j_p = 2.1$ µA/mm^2. With $S_{Cu} = 3$ (cf table 4) and $S^+ = 4.5 \times 10^{-3}$ (table 3), $f = 10^{-3}$ one finds $I_s = 1.35 \times 10^{-11}$ A.

Calculation of c :

      In the above case, since all parameters are known, one can calculate c_{Cu} from the measured currents. In most cases, however, f and the <u>absolute</u> yield S^+ are not known. In these cases one can calculate $\bar{c}$ by relating the measured (absolute) ion current I_M for element M to the ion current I_R measured with a sample of element R

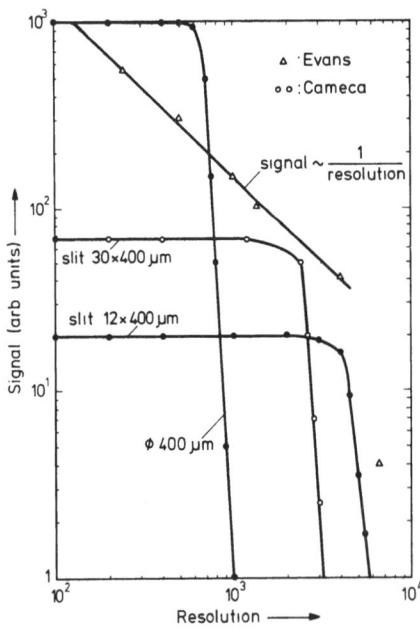

Fig. 17 : Dependence of measured ion current on resolution. From
         C.A. Evans Jr (NBS Special Publ. 427 p. 64) and
         Cameca S.A. (unpublished results).

Table 5 : Range of values for the parameters of eq. (1).

S	S^+	β^+	f
1 − 10	$10^{-5} - 1$	$10^{-5} - 1$	$10^{-5} - 10^{-1}$

in identical circumstances[6,10].

$$I_M/I_R = (I_p S_M^+ c_M f_M)/(I_p S_R^+ c_R f_n) \tag{3}$$

Taking $f_M = f_n$

$$I_M/I_R = I_{M,rel} = (S_M^+/S_R^+)(c_M/c_R) = S_{rel}^+/c_{M,rel} \tag{4}$$

where S_{rel}^+ is the relative ion yield for element M and $c_{M,rel}$ is the relative concentration. The concentration can then be calculated from the measured ion current I_M by means of the value of either S^+ or S_{rel}^+, as follows :

$$c_M = I_M/(I_p S_M^+ f) \tag{5a}$$

$$c_{M,rel} = I_{M,rel}/S_{rel}^+ \tag{5b}$$

Values of S^+ and S_{rel}^+ and also S^- determined from pure elements (c_M=1) are given in tables 3 and 6 and figs. 18-20. The relative ion yields in matrices containing more than one element as a major component can be quite different from these values[10]. For an appropriate correction one can use a semi empirical correction method[45] or one measures S^+ or S_{rel}^+ in a matrix of similar composition.

Quantities of importance in SIMS :
        Apart from the determination of the concentration it is important to have an idea on the following quantities :
1) the lowest detectable concentration c_{min} of a particular element present in the sample,
2) amount of material consumed during an analysis,
3) waste of material,
4) standard error due to statistical fluctuations in the ion current,
5) erosion rate in relation to background pressure,

1) The lowest detectable concentration (Table 6) can be derived from equation 5a as

$$c_{min} = I_{min}/(f S_M^+ I_p) \tag{6a}$$

where I_{min} is the minimum detectable ion current of the element of mass M. It can be kept low by means of an electron multiplier (see below). I_p, the primary ion current, cannot be chosen quite

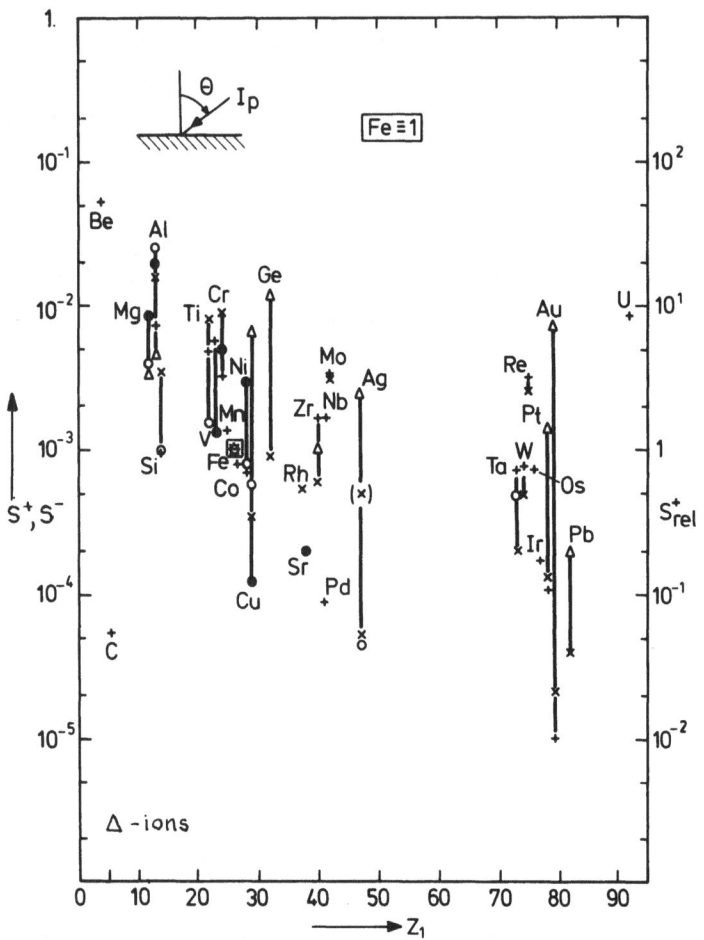

Fig. 18 : Relative positive and negative ion yields S^+, S^- for dif-
          ferent elements obtained under Argon bombardment by dif-
          ferent authors[157].
          Ar$^+$ 3 keV;   θ = 70°; Benninghoven S$^+$
          Ar$^+$ 8 keV;   θ = 0°; Hennequin   S$^+$
          Ar$^+$ 11 keV;   θ = 70°; Werner   S$^+$
          Ar$^+$ 12 keV;   θ = 45°; Beske   S$^+$
          Ar$^+$ 12 keV;   θ = 45°; Werner   S$^-$

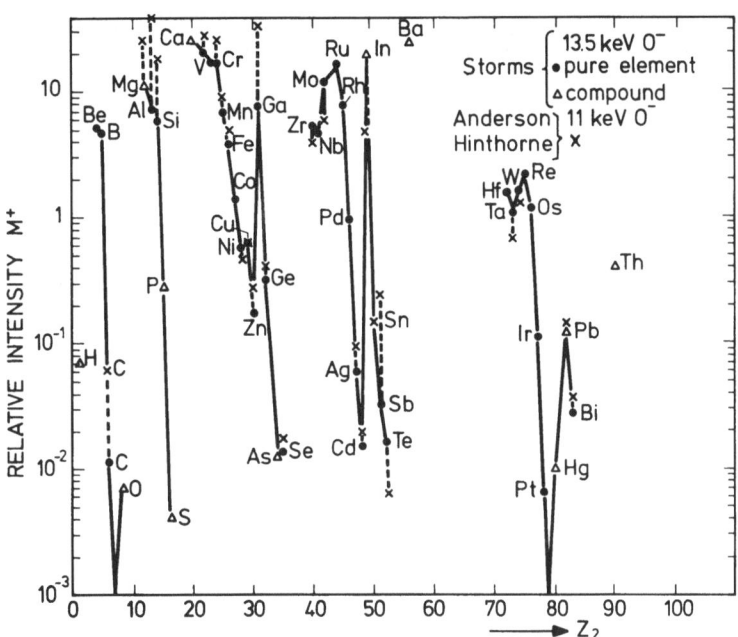

Fig. 19 : Relative positive ion yield S$^+$ for different elements
obtained under O$^-$ bombardment by different authors[158].
Copyright 1972 by the American Association for the
Advancement of Science.

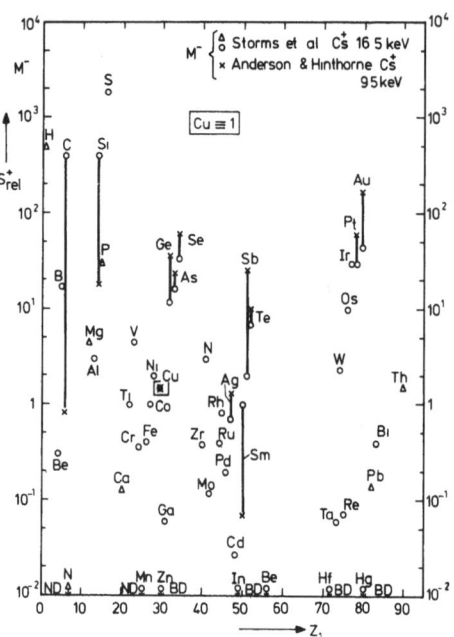

Fig. 20 : Relative negative ion yield for different elements obtained
          under Cs^+ bombardment.[158)]
          Copyright 1972 by the American Association for the
          Advancement of Science

freely,   since the volume sputtered per second $\dot{V}$ increases
with I_p and consequently the volume consumed during an
analysis, $V=\dot{V}_{ta}$, also increases (t_a is time for an analysis).
The transmission f can be optimized by a suitable choice of the
ion optics[46)]. The calculated values of c_{min} (Table 6) give the
order of magnitude under ideal circumstances. In practice,
however, c_{min} is higher, due to coincidence of lines of
different elements in particular molecular ions[10,50)]. In many
cases a lowering of c_{min} can be achieved by the use of oxide
ions[48)]. From (5b) one obtains

$$c_{min,M} = (I_{min,M}/I_p) \cdot (c_R/S^+_{rel}) \tag{6b}$$

from which c_{min} can be estimated if one has determined S^+_{rel} in
a given matrix.
2) Relation between c_{min} and material consumption
        It has been shown[48)] that

$$c_{min} = (I_{min}c_R)/(\dot{z}A_b \varkappa)_R S^+_{rel} \sim 1/\dot{V} = 1/\dot{z}A_b \text{,} \tag{7}$$

where $\varkappa$ (As/cm^3) = $10^5 [(S^+/S)(\rho/M)f]_R$,

Table 6 : Calculated values of the lowest detectable concentration
c_{min} for some elements, for various primary ion currents
I_p assuming a transmission $f = 10^{-3}$, and a minimum
detectable ion current at the ion collector $I_{min} =$
2×10^{-18} A. Values for S^+ (ion yield) have been taken
from table 3.

Elements	S^+	c min	
		(ppma)	(ppma)
Cu	1.3×10^{-4}	150	15
Ni	3×10^{-3}	7	0.7
Al	2×10^{-2}	1	0.1
Oxides	S^+	c min	
		(ppma)	(ppba)
Cu	4.5×10^{-3}	5	500
Ni	2×10^{-2}	1	100
Al	2	0.01	1
	$I_p(A) \rightarrow$	10^{-8}	10^{-7}

depends on the reference (matrix) element R. It is a measure for
the quality of an instrument and indicates how much charge has
been collected per cubic centimeter of sputtered material[10].
The larger the value of $\varkappa$, the lower that of c_{min} which can be
determined; $\dot{V}$ is the volume sputtered per unit time (cm^3/sec).
        A quantity related to $\varkappa$ is χ(number of secondary ions
collected)/(number of sputtered matrix particles) $= \beta^+ . f$.
Typical values of χ (determined on the assumption of $f = 10^{-4}$)
are found to be between 2 ions/10^7 matrix particles ($\beta^+ = 10^{-3}$)
and $2/10^4$ ($\beta^+ = 1$). A nomogram to determine an equivalent quantity,
viz. the number of collected ions per sampled volume, has
recently been given by Liebl[51].
        In order to lower c_{min} one must increase I_p (eq. 6);
at the same time the quantity of sputtered material increases
(eq . 7). Table 7 gives a relation between primary ion current,
bombarded area, erosion rate, sputtered material, layer
thickness removed per analysis, and c_{min}. As can be seen, for a
given value of the bombarded area A_b, any increase in $\dot{V}$ means
an increase of the erosion rate $\dot{z}$, since $\dot{V} = \dot{z} A_b$. Due to the
finite "thickness" of thin films, $\dot{z}$ cannot be chosen too high,
specially if one wants to carry out an in-depth analysis. The
two requirements of small values for c_{min} and small values of
the erosion rate are in fact incompatible.

Table 7 : Connection between I_p, bombarded area A_b, erosion rate $\dot{z}$, sputtered material m_{sp}, layer thickness d_{sp} or material [5] m_{sp} removed per analysis (200 sec) and minimum detectable concentration c_{min}.

$I_p(A)$	$j_p(\mu A/mm^2)$[1]	$\dot{z}$ (Å/s)[2]	(gr) m_{sp}[3]	(Å) d_{sp}	c_{min}(ppma)[4]
10^{-8}	1	2	2 ngr	400 Å	2
10^{-7}	10	20	20 ngr	0.4μm	0.2
10^{-6}	100	200	0.2μgr	4 μm	0.02

1) A_b= 100 x 100 μm = 10^{-4}cm^2;
2) For S = 3, M/ρ = 10; see also fig. 9 and 10
3) For ρ = 5 gr/cm^3;
4) For $I'_{min,M}$= 2 x 10^{-19}A, S_M^+ = 10^{-2}, f = 10^{-3};
5) Time for recording a complete spectrum, from H to U is taken to be t_a = 200 sec. Material consumption is smaller if only a part of the spectrum is recorded or if multiple collector techniques are employed.

3) <u>Efficient use of the sputtered material</u>

On scanning of the mass spectrum with only one collector for the ion currents, only one mass at a time is detected. The rest of the sputtered material is wasted. Simultaneous multi-element detection by photographic plate or electrical multi-collector detection promises a more efficient use of the sputtered material. The decision whether to use photographic detection or electrical detection depends on how much consumption of material is permissible, how many elements are to be analyzed, and what limit of detection is required, as can be derived from table 8.

Single-collector electrical detection has in its favour the lower limit of detection, as it needs only 10 ions.s^{-1} per peak, but it is limited to one element at a time. The photographic plate, needing about 10^4 ions for a just detectable line, has a higher limit of detection but measures all ions of the spectrum at the same time.

Application of multi-collector electrical detection, e.g. oblong channel plates with sufficiently small channel diameters, may shift the decision in favour of electrical detection[52].

Table 8 : Comparison of limits of detection with single-collector
electric detection (multiplier) and photographic-plate
detection for different amounts of material sputtered.

Volume V (cm^3) sputtered in $t_a = 1$ s	Number of particles in sputtered volume V	Number of ions coll./number of atoms sputtered	Limit of detection	
			Multiplier (1 line)	Photoplate (ca 100 lines)
2×10^{-12}	10^{11}	10^{-5}	10 ppm	10^4 ppm
2×10^{-9}	10^{14}	10^{-5}	10 ppb	10 ppm

4) Statistical fluctuations in the measured ion currents.

The material consumption $V(cm^3)$ should not be decreased
beyond a certain minimum value V_{min}, due to the corpuscular
character of the ion current. Slodzian[53] and Morabito[50] have shown
the relation between this minimum volume V_{min} and the admissible
statistical fluctuation, expressed as a standard error $\sigma_s(\%)$ in
the detected current :

$$V_{min}(cm^3) = \dot{V}t_a = \frac{1}{\varrho N/M} \cdot \frac{10^6}{c_M \beta^+ f} \cdot \frac{10^4}{\sigma_s^2(\%)} \qquad (8)$$

where ϱ = density (g.cm^{-3}), N = Avogadro's number, M = mass
number, and β^+ = S^+/S, the degree of ionisation; S is the
sputter yield = number of sputtered particles per primary ion;
c_M is the fractional concentration in 'ppma'.
For c_M = 100 ppm, β^+ = 10^{-3}, f = 10^{-3}, and V = 4.10^{-10} cm^3, one
can calculate from (8) a fluctuation in the measured ion
current of $\sigma = \pm 2 \%$ taking ϱ N/M = 6 x 10^{22} cm^{-3}.

5) Erosion rate $\dot{z}$ in relation to background pressure p_b.

In order to minimize the contamination of the surface, due
to adsorption of molecules from the residual gas, the erosion
rate and hence the primary ion current density j_p must be
large enough. A rule of thumb states [7,54] that this is the
case for

$$j_p(\mu A/cm^2)/p_b(Torr) \geq 10^8 \qquad (9)$$

2.3 Basic Experimental Embodiment

The validity of SIMS for the analysis of solids depends
largely on the choice of the experimental conditions. The following
parameters play an important role in the choice of these conditions.

Primary ions

For the <u>primary ions</u>, Ar^+, O_2^+,O^- N_2^+, or Cs^+ are usually chosen.
For the determination of the oxygen content of a sample, Ar^+ is
chosen as the primary ion. It has the disadvantage, however, that
the ion yield (number of secondary ions/number of primary ions) is
small, strongly element-dependent and can vary by a factor of 10^3.
When O_2^+ is used, supported by the bleeding in of oxygen gas, the
differences in the ion yield are drastically reduced and the
yields show better reproducibility. For quantitative analysis, the
small remaining differences can be taken into account by means of
correction formulas including at least the ionization energy. The
use of oxygen moreover increases the yield in respect of positive
secondary ions (cf. table 3).

Oxygen and nitrogen are also used because amorphous layers are
formed and the formation of structures (cones, ridges) on the
surface is reduced[55]. The corresponding reduction of surface
roughness is important if one wants to determine depth profiles
with good depth resolution. O^- is often used for the analysis of
insulators[56] because the charge which is brought to the target by
this negative ion bombardment may be compensated by electrons
leaving the target. The increase of S^+ with oxygen is explained
in a simple model by a breaking of ionic bonds $(M^+O^- \rightarrow M^+ + O^-)$ were
M is the metal atom. In the same way the use of Cs^+ as primary
ions gives an increase in the yield with regard to the negative
metal ion by breaking of the bond of M^-Cs^+, leading to an
increased yield of M^- ions[57] (see fig. 20). The use of cluster
ions[10] results in a decrease of c_{min}.

<u>Primary beam mass separation</u> : is used to obtain an extremely pure
ion beam[51], in particular for fundamental studies. For analytical
problems it usually suffices to use a pure gas in the ion source.
Primary ion mass separation is carried out by sector mass spectrometers as reported by Fralick, Rüdenauer, and AEI [58]   or by
a Wienfilter[59].

<u>Energy of the primary ions</u> : in order to obtain a high
sputter yield (see fig. 11) one wants to use   high primary ion
energies. The atoms of a given element can be knocked deeper into
a given sample, however, by collisions with primary ions (knock-on
effect[60]). But when primary ion energies of less than 5 keV are used,
this effect is not significant[61].   The increase of the incident
angle θ also increases the ion yield via the increase of S (Fig.12).

<u>Primary ion current density</u> :
The higher the primary ion current density, the higher the
sensitivity, but also the higher the erosion rate, i.e. the
consumption of material. A compromise must be made therfore
between the primary ion current density chosen, the desired sensitivity, and the speed at which the mass spectrum is scanned,
particularly in the analysis of thin films. The ion current density
for Ar^+ must always be chosen so that the erosion rate is higher
than the rate of adsorption of residual gas atoms. For all these

reasons, ultra high vacuum instruments are particularly recommended
for the analysis of very thin layers. For study of surface
reactions, e.g. oxidation, one uses small ion current densities and
large bombarded areas to increase the sensitivity. The corresponding
erosion rate of 1 Å/h is very small. The layer is practically
unchanged during a measurement (static SIMS[10]) cf. (table 2).
By contrast, for the determination of depth profiles, high erosion
rates are used in dynamic SIMS (table 2).

Primary beam diameter :
    With beam diameters of, typically, one millimetre it is
possible to carry out survey analyses, specially because possible
inhomogeneities average out over the sample surface. For local
microspot analysis, beam diameters of about 1 μm are used.
    For imaging of surfaces, use is made either of beams of 1 μm
diameter in the "raster mode" or of a beam of 300 μm diameter in
the "ion imaging mode" (for a review see Liebl[62]).

Secondary ions :
    Atomic ions are emitted as singly charged particles, e.g.
Fe^+, Cu^+, Al^+, or as multiply charged ions, such as Al^{++}, Si^{++},
Si^{3+}, etc. In the negative ion spectrum the rule is that only
singly charged atoms are found.

Polyatomic ions :
    Among the polyatomic ions one finds molecular ions, such as
AlO^+ etc., generated because of the use of primary oxygen ions,
$AlOH^+$, resulting from the presence of water in the system,
heterogeneous cluster ions, such as $AlCu^+$, $AlFe^+$, etc., which appear
when alloys are analysed with high primary ion current density, and
isogeneous cluster ions such as Al_2^+, Al_3^+, Al_4^+.
    The ion yield of positive and negative ions, and its dependence
on the presence of oxygen or   cesium, have been discussed already.
The energy distribution of the emitted secondary ions (Fig. 21)
shows a behaviour similar to that of the energy distribution of
the sputtered neutral particles (Fig. 16). The maximum of the
distribution, typically a few eV, depends on the bombarding
conditions, surface binding energy and on the emitted
species, molecular ions, for instance, have a lower kinetic energy
maximum than atomic ions (Fig. 21).
    Discrimination between molecular ions and atomic ions can be
carried out in two ways :

Discrimination due to the difference in initial kinetic energy
    Mass spectra obtained by Poschenrieder[63] with low and high
energy secondary ions is given in respectively figs. 22 and 23.
One can see that a large number of polyatomic ions populate the
low energy pass (fig. 22). However, when only ions with high
initial energy are accepted (fig. 23) the mass spectrum reduces
essentially to a few atomic ion peaks, the majority of the
polyatomic ions having their energies reduced to below the limit
of detection. Only small peaks of molecular cluster ions (Al_2^+
and Al_3^+) are still present.

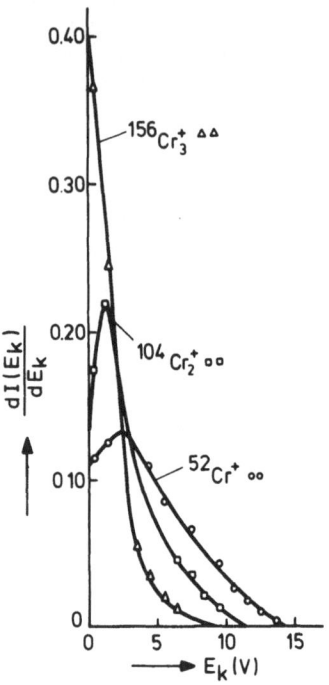

Fig. 21 : Energy distribution of sputtered atomic and poly-atomic
          ions obtained by 5.5 keV, Ar[+] bombardment on pure Cr.[159)]

        The energy selection, as discussed above, is carried out in
electrostatic designs. In electrostatic fields the trajectories
of the ions irrespective of their mass, depend only on their
kinetic energy and the applied voltage comparable to the deflection
in a parallel plate condensor; details of the energy discrimination
can be found in Section 3. Moreover the separation of molecular
ions from atomic ions has an influence on the determined ionization
temperature[30,64]. The use of an energy pass band is also
advantageous in trace analysis to remove the prepeak[65].
        The prepeak is a tail to lower mass units of all mass peaks.
It is due to ions of the same mass which, however, have been
generated in front of the target. Accordingly they have not
obtained the full accelerating voltage and therefore appear at
lower magnetic field values, i.e. mass numbers. The prepeak has
been found in sector type instruments but is also to be expected
in quadrupole mass spectrometers. In particular in the analysis
of traces this effect can often lead to disturbing results (in low
mass resolution instruments) if, for instance, a dope has to be
detected which has a mass units smaller than the mass of the main
component, e.g. Zn(M=64-70) in Ga(M=69-71).
        Three possible mechanisms leading to a prepeak have been

Fig. 22   Fig. 23

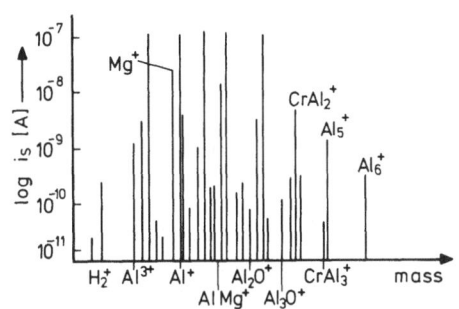   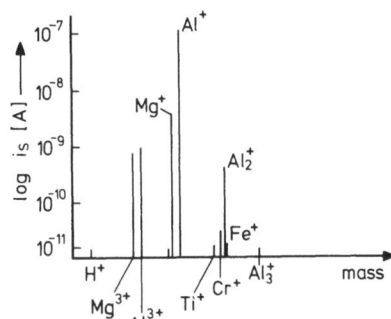

Fig. 22 : Mass spectrum of an Al - Mg alloy bombarded with 12 keV
         Ar^+ ions. Analyser adjusted to accept low energy se-
         condary ions (after Herzog et al.[63])
Fig. 23 : Mass spectrum from the same sample as in Fig. 22, however,
         with strong suppression of polyatomic ions due to shifting
         the energy acceptance of the instrument to accept high·
         energy secondary ions only.[63]

suggested[65] :
Ionization of sputtered neutral matrix particles by reflected
primary ions or by fast sputtered matrix particles, and generation
of ions from emitted metastable particles. More investigations
will still have to be carried out in respect of this topic.

Discrimination due to the mass defect
      By means of mass spectrometer of high mass resolution[66]
(about 5000)the disturbing interferences between some atomic and
polyatomic ions can be eliminated (Table 9). One must consider,
however, that the sensitivity decreases with increasing
resolution (Fig. 17).
Fingerprint spectra
      It has already been pointed out[67] that it is most ad-
vantageous in chemical analysis to use atomic ions. For studies
of chemical compounds, on the other hand, polyatomic ions, which
constitute the fingerprint spectrum of a given compound, can be
used with advantage[68,69]. The occurrence of molecular ions and
cluster ions has another implication : the mass range in a mass
spectrometer must extend to mass one thousand, in particular when
heavy ions combine to clusters or molecular ions[70,71], e.g.
U_2O^+, U_3O^{++} or Bi_3^+, Bi_3O^+.
Organic molecules
      The study of organic molecules[11] by SIMS may also require
extention ·to cover high masses (for determination of the parent
peak) and possibly also a high mass resolution.

Table 9 : Required resolution to resolve analytical ions (atomic ions) from interfering poly-atomic ions (after Evans[66]).

Interference type	Interfering ion	Analyt. ion	Required resolution
Multiply charged	$^{28}Si^{2+}$	$^{14}N^+$	950
matrix ion	$^{62}Ni^{2+}$	$^{31}P^+$	3200
Matrix selfpolymers	$^{16}O_2^+$	$^{32}S^+$	1800
ions	$^{28}Si_2^+$	$^{56}Fe^+$	2950
Prim. ion-matrix	Cu_3O^+	$^{207}Pb^+$	1050
molecular ions	Si_2O^+	$^{75}As^+$	3250
	AlO_2^+	$^{59}Co^+$	1500
Hydride ions	$^{30}SiH^+$	$^{31}P^+$	4000
	FeH^+	$^{55}Mn^+$	3300
	SnH^+	$^{121}Sb^+$	19500
Hydrocarbon ions	$C_2H_3^+$	$^{27}Al^+$	650
	$C_5H_3^+$	$^{63}Cu^+$	650
	$C_2H_2^+$	CN^+	2000

2.4 Type of Information Obtainable

    2.4.1 Qualitative or quantitative survey analysis. All
elements from hydrogen to uranium and combination of these
elements (polyatomic ions or cluster ions) can be detected. The
atomic ions can be used for quantitative chemical analysis. The
polyatomic ions can be used for the identification of chemical
compounds or for the study of chemical structures. Applications :
detection of impurities on the surface and determination of their
depth distribution (see e.g. figs. 2 and 3).

    2.4.2 Surface analysis. Since the information depth in SIMS
is very small (see Section 2.1) it is possible to analyse only the
uppermost atomic layers if the current density of the primary
beam is kept so low that only about one monolayer per hour is
sputtered away (static SIMS). An extremely good vacuum (p $\approx$ 10^{-10} Torr)
must be maintained. Application : study of surface reactions[72,73]
(catalysis, oxidation, reduction).

    2.4.3 Depth analysis. If the sputter rate is chosen high
(see Table 2) an analysis of the distribution of elements at
depths between 50 Å and several μm is achieved in typical times
from 20 minutes up to 1 hour (dynamic SIMS). This mode is used for
the investigation of diffusion processes and for the determination
of diffusion coefficients[6], for the study of implantation
profiles[74] in the semiconductor industry (B, As, P, F implanted
in silicon or other materials[50]), for the study of thin film
sandwich layers[69] (Fig. 24), for the study of electrodes for
devices (Co, Si) as well as for a better understanding of soldering
problems, adhesion problems, ceramic-to-metal adhesion, etc.
    Principle of depth analysis : ion bombardment continuously
removes the surface of the sample "layer by layer". Simultaneously
the secondary ion current of one or more elements is detected as a
function of the bombardment time. At constant erosion rates the
bombardment time is proportional to the depth[6]. The signal can be
converted into absolute concentration/cm^3. A method to convert
measured ion current vs. bombardment time obtained from
implantation profiles into concentration (cm^{-3}) vs. depth (Å) has
been described recently[75].
    The absolute depth resolution is that layer thickness which
must be sputtered away before the signal obtained from an element,
distributed in depth according to a step function, has increased by
a given amount[76], e.g. from 16 % of its maximum value to about 84%.
The depth resolution obtained depends on the concentration profile
in the sample and on instrument factors such as information depth,
the basic sputter process, the original and final surface roughness,
knock-on effects, ionic mixing and atomic transport[77,78]. Processes
as illustrated in fig. 25 can lead to a deviation of the measured
from the actual profile. This can partly be overcome by electrical
and also mechanical gating, i.e. only from a small portion A$'_e$ of the

DETERMINATION OF Al-PROFILE IN GaAs/GaAlAs-LAYERS

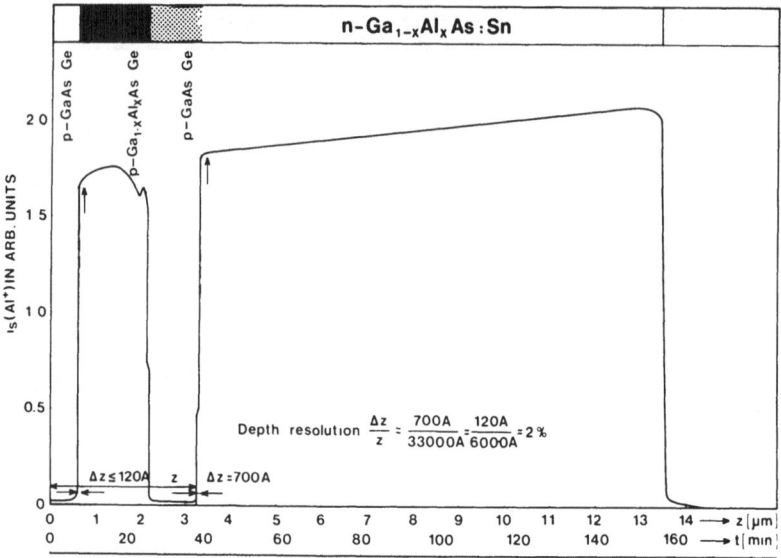

Fig. 24 : Concentration depth profile (concentration versus depth) of Al in Al Ga As sandwich layers.[69]

bombarded surface can ions enter the mass spectrometer. "Crater effects" and effects derived therefrom are suppressed[75]. Recent studies have shown that for not too deep profiles (≈ 200 Å) factors determined by the sample such as generation of surface topography on an atomic scale by the sputtering process[78] and information depth determine the resolution, while at greater depths instrument-determined factors, such as beam inhomogeneity, play an important role, leading to a constant value for the relative depth resolution[75,79] $\Delta z/z$ of typically 1–2 % (cf. figs. 24 and 26).

It can be shown easily that $\Delta z/z$, the relative depth re-solution, is constant when beam inhomogeneity is the dominant disturbing factor[76]. Obviously at any time t in the middle of the crater ($r=r_0$) a sputter depth z_0 will then be obtained, whereas in the peripheral zone (distance r from the centre) a depth z_1 is obtained.
Since
$$z_0 \sim j(r_0)St \text{ and } z_1 \sim j(r_1)St, \text{ it follows that}$$
$$\Delta z = z_1 - z_0 = (j(r_1) - j(r_0))S \cdot t$$

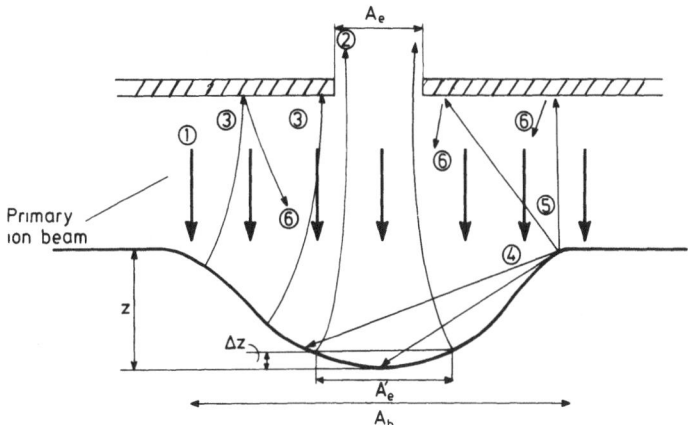

Fig. 25 : Illustration of different effects (schematic) causing
distortion from the ideal relation between measured and
true profile.
(1) Primary ion beam (considered to be in a plane outside
of the paper); bombarded area A_b.
(2) Secondary ions which can pass the extraction
electrode (A_e).
(3) Secondary ions starting outside the tolerated extraction region, intercepted by the extraction
electrode.
(4) Sputtered neutrals from crater edge, deposited in
central crater region.
(5) Sputtered neutrals from crater edge deposited on the
extraction electrode opposite the target.
(6) Particles sputterdeposited from the target onto the
extraction electrode, being resputtered.

and the relative depth resolution
$$\frac{\Delta z}{z} = \frac{j(r_1)-j(r_0)}{j(r)} = \text{const.}$$, irrespective of the depth, if the
primary current density distribution and S as a function of r are
constant. Experimental proof of these considerations can be derived
from fig. 24 and the work of Doi et al.[79] (Fig. 26).

2.4.4 <u>Element mapping</u> i.e. the determination of the distribution of elements across the surface, is obtained in two
modes :
<u>Ion microprobe</u>[80]
Elemental images are obtained in a similar way as in an
electron microprobe. The mass spectrometer is tuned to a given

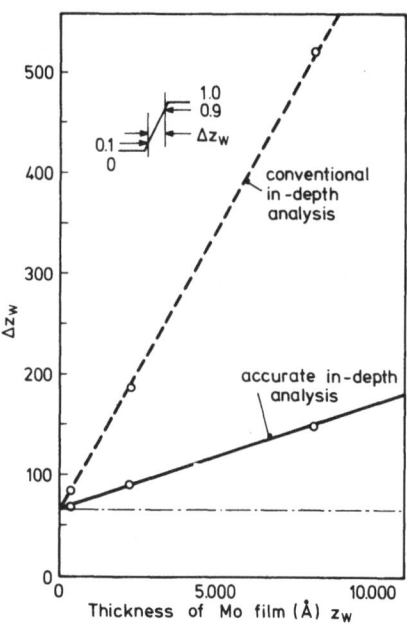

Fig. 26 : Relation between absolute depth resolution Δz and sputtered layer thickness z showing that $\Delta z \sim z$ if knock-on effects etc. are neglected (after Doi)[79]: $\Delta z = \alpha + \beta z$

mass number and a fine ion beam (1 μm diameter) scans the surface. The secondary ion current of mass M_1 measured at the detector is displayed as a deviation in the y-direction on a cathode ray tube, whose time base is in synchronism with the scanning unit of the beam (y-modulation). The lateral resolution is limited here by the beam diameter and is typical 1-2 μm.

Ion microscope :

An ion optical image of the surface in the "light" of the secondary ions is obtained via the ion optics of the mass spectrometer, which is tuned to a particular mass. The image is made visible on, for example, a fluorescent screen or on a photographic plate (intensity modulation). The lateral resolution is here determined by aberrations in the ion optics and is typical 1 μm[81]. Narrow slits are needed for small aberrations, leading to loss of sensitivity. The brightness with which a surface arrangement is imaged depends on the primary ion current density, the concentration of element M, the ionic yield, and the acceptance of the mass spectrometer. The necessary contrast to obtain an image of the surface at a homogeneous primary ion current density is caused by the following processes[81-86]:

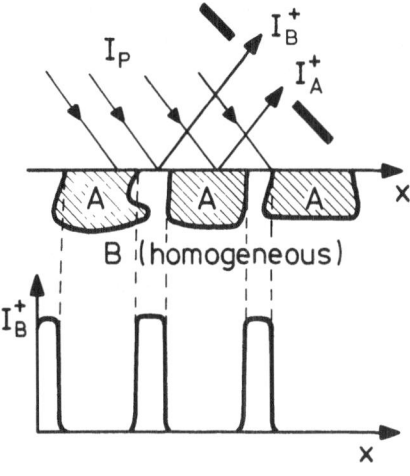

Fig. 27

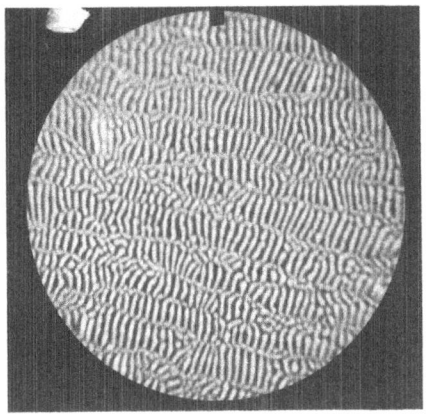

Fig. 28

Fig. 27 : Origin of concentration contrast (schematic).
Fig. 28 : Illustration of concentration contrast : Ca^+ image of
Ca-Al alloy, obtained with secondary ion microscope
(IMS 300,CAMECA), forming Ca or Al rich zones of 1 μ
thickness. (viewing field 250 μm).

1) Material (concentration) contrast
   A signal is only obtained from zones where element M is present
   (Fig. 27 and 28). But the ion current actually measured is not
   a direct measure of the concentration of M, because other
   processes (see below) also contribute to the secondary ion
   current.
2) Orientation contrast
   This occurs because S^+ depends on the orientation of the
   crystallographic planes with respect to the primary beam
   (Fig. 29).
   On rough surfaces the following effects occur in addition to
   these two effects, caused by the acceptance of the mass spectrometer :
3) Topographic contrast
   Because of the inclination of the incident beam, no secondary
   ions are released from areas shadowed by a "hill" on the surface
   (Fig. 30-32).
4) Chromatic (energy) contrast[86]
   Due to different initial energies, ions of element B (low
   energy) are collected more than element A ions (fig. 33) (high
   energy) and will enter the acceptance optics of the mass

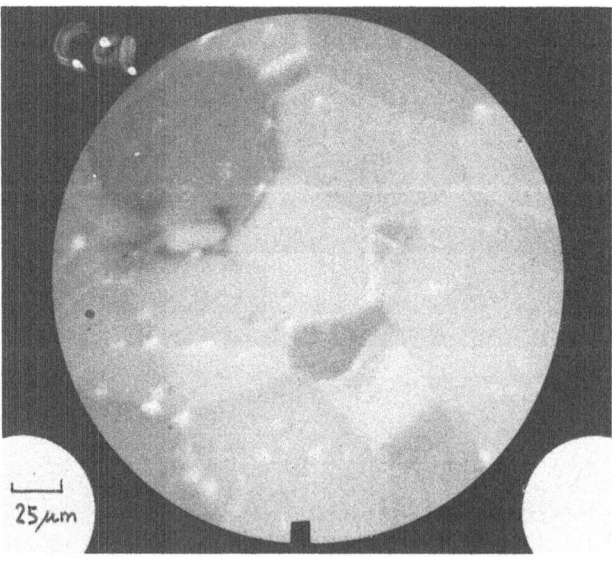

Fig. 29 : Illustration of orientation contrast : image from ferrite
          revealing crystalites with different orientation. (Ca-
          image).

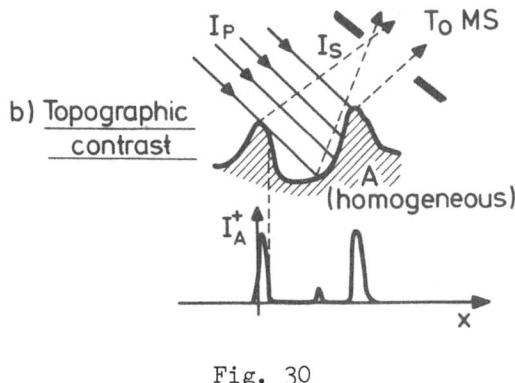

Fig. 30

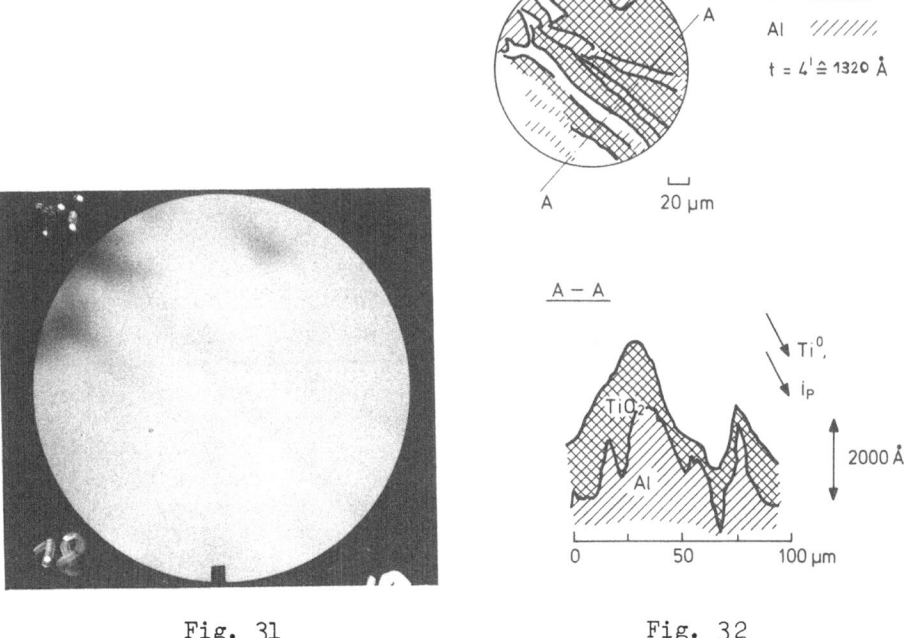

Fig. 31   Fig. 32

Fig. 30 : Origin of topographic contrast.
Fig. 31 : Illustration of topographic contrast : Ti^+ image of Al-layer covered with TiO_2 (viewing field 250 μm).
Fig. 32 : Explanation of mechanism leading to figure 31.

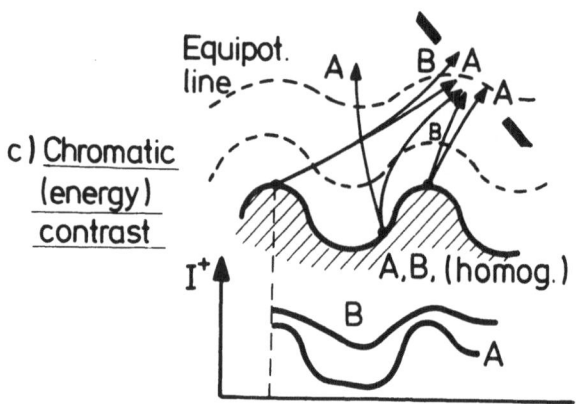

Fig. 33 : Origin of chromatic (energy) contrast.

spectrometer due to the action of the electrostatic field.
Therefore at certain spots they give a relatively higher signal
than element A ions.

2.4.4.1 Microspot analysis, line scan, 3D analysis :

Microspot analysis, i.e. the analysis of a small, pre-
determined zone with dimension of a few micrometre can be carried
out by means of primary ion beams of small (μm) diameter and an
appropriate design which allows of adjusting the beam with
sufficient precision (typical a few microns) to desired spots.
During the bombardment the beam position can be controlled by an
optical direct-viewing microscope. Other possibilities for control
are simultaneous imaging of secondary electron or absorbed ion
currents and secondary ion current.

Line scan : A line scan is obtained if the position of the
beam relative to the sample is changed along a straight line. In
most cases the ion beam moves across the sample and the intensity
of the secondary ions as a function of the beam position is
detected. Sometimes also the sample is moved with respect to the
beam, the beam being kept in its original position. The ion image

in y-modulation consists of many of these line scans.

    3D-analysis : A combination of element mapping and depth
analysis occasionally carried out in several laboratories[87],
provides three-dimensional information about the element distribution.

2.4.5 Structural analysis

2.4.5.1 Crystal structures

    The short range structural order of the atoms in surface-
adjacent regions or layers (information depth) in crystals has
been determined with SIMS. The resulting secondary ion spectrum
was shown to be related to an atomic arrangement on cleavage planes
of crystals[88,89]. From the crystal structure of CaF_2, for instance,
(Fig. 34) one can expect to find the following secondary ion
clusters to be emitted : CaF_2 and its fractional derivatives CaF;
if F is substituted by oxygen (surface oxidation) one would expect
ions such as Ca_2OF, Ca_2O, Ca_2F. All these ions have been found in
secondary ion spectra[88] (Table 10). Other clusters to be
expected from the structure of CaF_2 oxidized on the surface
(substitution of one or more F atoms by oxygen or OH groups) are
Ca_3OF_3, Ca_3OF_2, Ca_3O_2F, Ca_2OF_2 and, from the non-oxidized parts
Ca_2F_4. The relation between the chemical structure of the emitted
secondary ions and the structural formula is evident from Table 10.

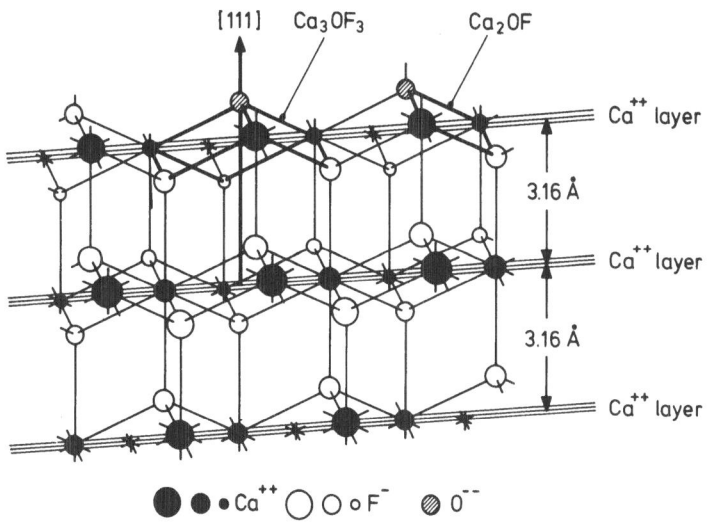

Fig. 34 : The (111) surface of CaF_2. Several groups (Ca_2OF, Ca_3OF_3)
         in the surface layer are emphasized where coordination
         sites are occupied by O and/or OH instead of by F
         (after Bühl and Preisinger[88]).

Table 10 : Positive secondary poly-atomic ions obtained from a
CaF$_2$ crystal. Only the most common isotopes ^{40}Ca, ^{19}F,
^{16}O are considered; after Bühl and Preisinger[88].

Mass	Ion composition	Structural formula
96	Ca_2O^+	Ca–O–Ca
97	$Ca_2(OH)^+$	Ca–O(H)–Ca
99	Ca_2F^+	Ca–F–Ca
115	Ca_2OF^+	Ca(–O–)(–F–)Ca
134	$Ca_2OF_2^+$	Ca(–O–)(–F–)Ca–F
156	$Ca_2F_4^+$	F–Ca(–F–)(–F–)Ca–F
171	$Ca_3O_2F^+$	Ca(–O–Ca)(–O–)(–Ca)F ring structure
174	$Ca_3OF_2^+$	Ca(–F–Ca)(–O–)(–Ca)F ring structure
193	$Ca_3OF_3^+$	Ca(–F–Ca)(F–)(–O–)(–F)(–Ca) ring structure

2.4.5.2 <u>Compound identification by means of fingerprint spectra</u>

From polyatomic ion spectra (fingerprint[90] spectra) assumed
to be representative of every compound[91] it is possible to conclude under given circumstances that different chemical compounds
(phases) must be present in one and the same sample. The principle
of this method of fingerprint spectra is as follows : first one
must show,by means of pure samples, i.e. samples consisting of only
one compound, that different compounds, e.g. CoSi, Co2Si,Si, have
a characteristic abundance of individual polyatomic ions (Fig. 35).
If several of these compounds are present in a given sample
simultaneously, they may contribute to one and the same mass line.
The measured secondary ion mass spectrum is therefore a superposition of the fingerprint spectra of the individual compounds
(Fig. 36). The ion current measured at the individual peaks 1, 2,

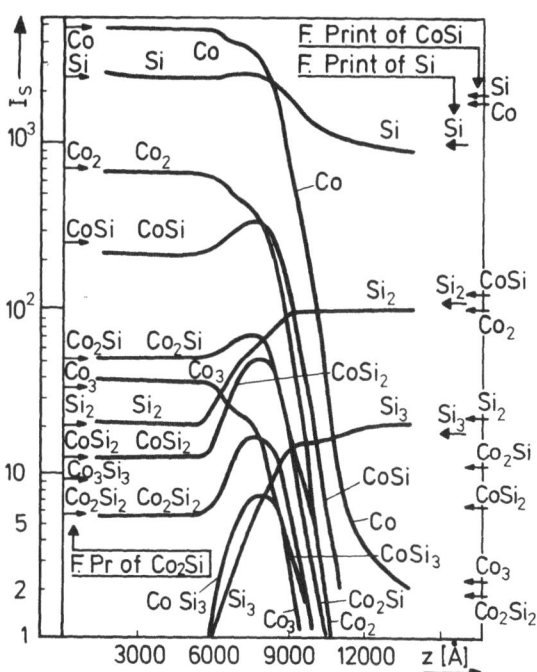

Fig. 35 : Ion current as function of the depth from sandwich layer
with variable composition in Si and Co. Arrows on the
right and left side indicate fingerprint spectra of
Co2Si, CoSi,Si obtained from pure calibration samples of
the respective materials[160].

3, ... i as a function of bombardment time t, proportional to depth z, then becomes :

$$I_1(t) = I_{I1} \; c_I(t) + I_{II1} \; c_{II}(t) + \ldots I_{III1} \; c_{III}(t) + \ldots$$

$$I_i(t) = I_{Ii} \; c_I(t) + I_{IIi} \; c_{II}(t) + \ldots I_{IIIi} \; c_{III}(t) + \ldots$$

On the assumption that, for example, there are only two compounds present in the sample, i.e. $c_I + c_{II} = 1$, the concentrations of phases I and II can be determined from two of the measured ion currents of the set given above. The fingerprint spectra of phase

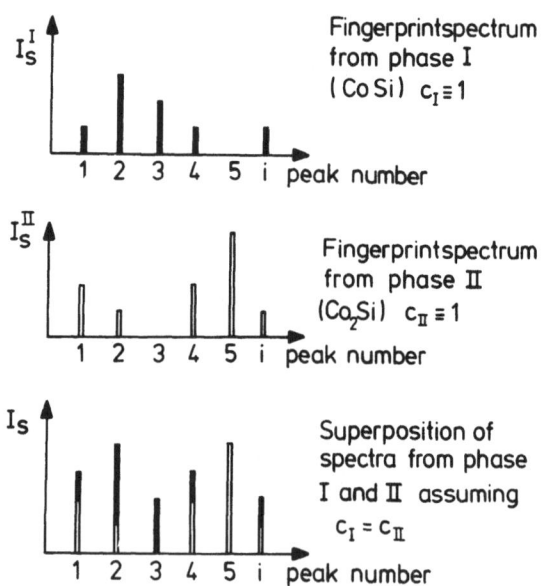

Fingerprintspectrum from phase I (CoSi) $c_I \equiv 1$

Fingerprintspectrum from phase II (Co$_2$Si) $c_{II} \equiv 1$

Superposition of spectra from phase I and II assuming $c_I = c_{II}$

Fig. 36 : Illustration of the principle of fingerprint spectra of two different phases I and II and the resulting spectrum obtained by their superposition from a sample containing both phase I and phase II [160].

I and phase II $\{I_{I_i}\}$, $\{I_{II_i}\}$, have to be determined beforehand
from calibration samples. The sign $\{\ \}$ here indicates the first parts
of the first and second columns in the above equations. The
fingerprint spectra are determined from calibration samples of
phase I with $c_I=1$ and $c_{II}=0$ and one obtains $\{I_i(t)\} \equiv \{I_{I_i}\}$, i.e.
the fingerprint spectrum of phase I. Analogous considerations hold
for phase II. Note that the fingerprint spectra of phases I and II
give no direct information about the original atomic short range
structure in phases I and II, because due to the high energy (keV)
of the ions and due to the high dose needed for a depth analysis,
the sample may have changed. One can nevertheless determine the
concentrations of the two phases I and II in an unknown sample, on
the assumption that the destruction in the calibration samples and
in the unknown target are the same. fig. 37 shows the concentration
of Co_xSi_y phases determined from fig. 35.

2.4.5.3 Determination of the chemical structure and the
molecular weight of organic compounds
Here one uses a low dose (static or semistatic SIMS), so that
the original structure is not disturbed[11,69). fig. 38 shows the
positive secondary ion spectrum of a Teflon sample, which reflects

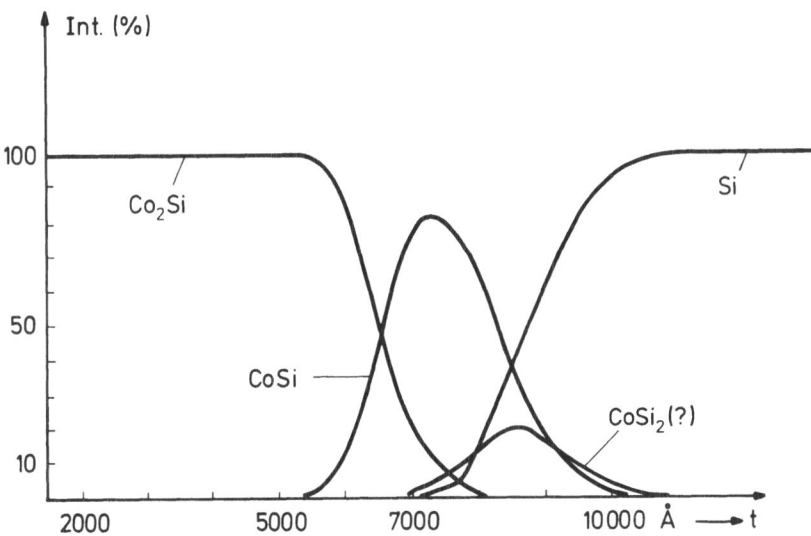

Fig. 37 : Reduced data obtained from fig. 35 by applying the
fingerprint spectra superposition theorem. As a result
the concentration of Co_2Si, $CoSi$, Si and of $CoSi_2$ as a
function of depth have been obtained [160).

the arrangement of C and F in Teflon, i.e. the arrangement of
carbon with respect to F. The chain character of Teflon can be
derived from the presence of ions of the series CF_2, C_2F_2, C_3F_2,
C_4F_2, C_5F_2. Besides, the degree of branching in this linear chain
can also be determined from the height of the CF_3 peaks by means
of calibration standards having a known degree of branching.
<u>Molecular weight determination</u> is possible by means of the
protonated parent peak $(M+1)^+$ or the parent peak having lost one
hydrogen atom $(M-1)^-$ in the negative secondary ion spectrum[11].

2.5 Derived and Related Techniques

2.5.1 <u>Neutral particle bombardment</u> : Instead of positively or
negatively charged ions one has used also neutral particles of keV
energy to obtain emission of ions from the target by sputtering.
The neutral particles[92] N_2^0 or Ar^0 are obtained by charge
exchange of a beam of fast (keV) ions in collision with a molecular
beam crossing its path.

The kinetic energy of the original ions is not essentially
changed by these charge exchanges and one obtains a beam of atoms
with high (keV) energy. The sputter and the ion yield are not
changed very much either, because sputtering and ionization are
mainly due to effects of the collision cascades and only to minor
degree to direct impact of projected particles on target atoms.
The advantage of using neutral particles (in that case it is
better to speak of sputter ion mass spectrometry instead of

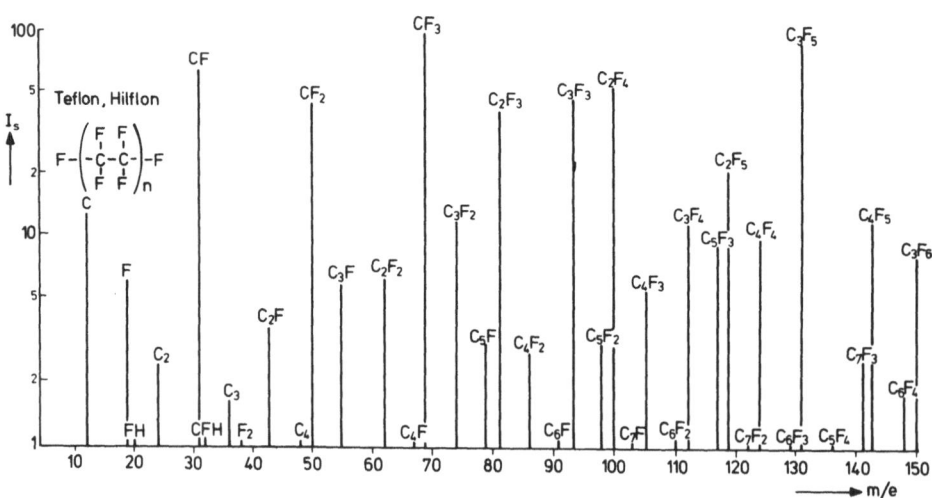

Fig. 38 : Teflon : positive secondary ion spectrum[160].

secondary ion mass spectrometry) becomes apparent in the analysis
of insulators[93]. In that case, charging is only caused by bombardment-induced electron emission ("secondary electrons") because
the primary particles, in contrast to bombardment with ions, do not
bring any charge to the insulator surface.

The disadvantage of this method is that focusing to a fine
beam (µm diameter) or rastering of the neutral beam across the
surface for imaging or for microspot analysis is in principle not
possible. One could, however, think of rastering the original beam,
but due to the scattering process and the following charge exchange
it is to be expected that the beam diameter may be somewhat disturbed.

2.5.2 Post-ionization of sputtered neutral particles : It has
already been mentioned that the ionic yield $S^{\pm} = \beta^{\pm} S$, of different
elements can differ up to a factor of 1000. The large differences
in the yield are due mainly to differences in the degree of ionization. The differences in sputter yield are much smaller and also
less dependent on the matrix. It seems reasonable, therefore, to
analyse the sputtered neutral particles by means of post-ionization
and not to use, as described up to now, the ions which are directly
emitted without any further ionization. Two approaches for such a
post-ionization have been reported :

2.5.2.1 Post ionization with a transverse electron beam as
usual in gas mass spectrometry[6,94], was found to give an insufficient increase in the secondary ion current, compared to the ion
current already present from particles emitted directly as ions.
The reason is that due to the high velocity of the secondary ions
(10-20 eV) the effective density in the ion beam was too small and
therefore ionization was not sufficient.

2.5.2.2 Sputtered neutral mass spectrometry : Oechsner[95]
generates a plasma in a gas of 10^{-5} Torr pressure. The electron
temperature in this plasma is about 10^5 K. Due to this relatively
high temperature the energy of the electrons is high enough for
ionization by electron impact. The ionization cross-section of
different elements shows small differences.

2.5.2.3 Glow-discharge mass spectrometry (GDMS) : Coburn
and Kay[96] generates a plasma in a gas (Ar) of 10^{-2} Torr. The
electron temperature is rather low (5000 K) and therefore the
(thermal) electron energy is not sufficient to ionize by electron
impact. Ionization of the gas molecules is obtained via metastable
argon atoms (Penning ionization) :

$$X + Ar^* \rightarrow X^+ + Ar + e$$

Here, too, the ionization cross-section of different elements shows
small differences. It is expected that both these methods of post-
ionization will improve quantitative SIMS. When these methods are
used the following points have to be considered :

(1) Contamination of the target surface by the residual gas
in front of the target. The system of Oechsner will give less

contamination, due to the smaller gas pressure.

(2) Careful shielding is necessary to avoid contamination of the target surface by sputtered particles from the probe holder.

(3) The method, at the moment, is limited to large area analysis.

Both approaches are very interesting also for fundamental research. The work of Oechsner[97] makes it possible to compare the yields of ionic molecular ions and neutral molecular ions. From the measurements of Coburn and Kay[98] one can obtain information about the lifetime of metastable atoms.

2.5.3 Bombardment induced light emission[99] (BLE) : A technique related closely to SIMS is BLE also called SCANIIR[100]. We have already mentioned that, under ion bombardment, excited atoms and ions are also emitted. These excited particles can, after being ejected from the target, fall back into the ground state by the emission of light. By means of an analysis of this emitted light with a suitable spectrograph, qualitative and quantitative analysis of the sputtered particles is also possible.

Advantage : detection of light can be carried out with a simple spectrometer but also with light filters, although in the latter case[101] one must be careful that the continuum emitted from the target is not measured as a spurious signal.

The wavelength of the emitted light does not change in consequence of charging of an insulator surface. (In SIMS, charging of the surface causes a shift of the secondary ion peaks, often up to several mass units [93]. BLE has therefore been succesfully used for the analysis of glasses, although with special designs SIMS has also been used succesfully for even quantitative analysis of glasses[102].

The sensitivity of the BLE is about the same as that of SIMS. In principle it should be even better than that of SIMS, since the cross-section for the emission of light is much greater. However, since the collection of light (emitted in all directions) requires, focusing, complicated mirror arrangements have to be used.

Disadvantage : quantitative analysis gives the same difficulties as originally found with SIMS. The yield for elements is matrix dependent. Attempts to use a Saha-Eggert type of approach for BLE gave difficulties because forbidden lines occurred resulting in a wrong determination of the excitation temperature.

MacDonald[103] has calculated the excitation temperature from the distribution of the line densities without the use of internal standards. Application to fundamental studies : the emission of particles responsible for SIMS and BLE signals is due to the same collision cascade which causes particles to be emitted in the excited state or as ions in competing processes. A more fundamental insight is therfore to be expected from a study of these two competing techniques[104].

2.5.4 <u>Laser induced ion emission</u> : Pulsed laser beams are
directed to the sample (in most cases organic materials). The
material, evaporated, among other things, in the form of ions, is
detected by means of a time-of-flight mass spectrometer. The limit
of detection is very low, due to the large transmission of these
instruments (f=50 %). The mass resolution, however, is small (100).
Quantitative analysis has still to be investigated[105].

3. PRACTICE OF TECHNIQUE

3.1 Types of Instruments

Every instrument has a vacuum chamber, primary ion source,
a sample holder, an analyser and a detector for the secondary ions,
usually followed by equipment for automatic data acquisition and
processing, and in future probably also for instrument control. Ion
guns and detectors do not differ essentially from instrument to
instrument and are therefore unserviceable distinguishing the types
of mass spectrometers. The main distinction of different SIMS types
is the use of quadrupole or sector type analysers (see table 11).

3.1.1 Primary ion sources
In most cases a gas (Ar, O_2, or N_2) is introduced into the
ionization chamber at pressures of typically 10^{-3} Torr (fig. 39)
and is ionized by means of a gas discharge. Occasionally
ionization by electron impact in a modified ionization gauge has
been reported[106].

Recently surface ionization sources have been used to produce
Cs^+ ions. In that case a salt containing cesium is put on a
filament and heated. Cs evaporates to a large extent as positive
ions[47].

The ions generated in the ion source are extracted from the
ionization region by an electric field. The stronger the electric
field F =-dU/dx in front of the ionization space, the higher the
ion current which is extracted; the higher also , if volume
ionization is applied, is the energy spread $\Delta U = \Delta x \frac{dU}{dx}$, where Δx
is the region from which the extraction takes place.

After optional mass separation, focusing of the beam to the
desired diameter on the target is achieved by a focusing optic,
used also to obtain a high, homogeneous current density of the beam,
and a small angular aperture, i.e. the ion beam must have a great
"brightness". For the alignment in, for example, the x-direction
(in the plane of the paper of fig. 39) two pairs of deflection
plates are necessary to define the place and the angle of incidence
of the beam on the target. The same considerations hold for alignment in the y-direction (perpendicular to the plane of the paper in
fig. 39). These deflection plates can also be used for raster
scanning of the beam across the sample.

Well designed ion sources use differential pumping : a
diaphragm D with a hole not much larger than the beam diameter at
that place is situated between ion source chamber and target.

Table 11 : Some criteria used to classify the different types of
secondary ion mass spectrometers.

Primary ions	
Current density :	high (Dynamic SIMS) low (Static SIMS)
Beam diameter :	~ mm (macroprobe SIMS) ~ μm (microprobe SIMS)
Secondary ions	
Mass separation :	quadrupole or sectortype (occasionally time of flight (TOF))
Energy selection :	electrostatic designs
Mass resolution :	low (300), with single focussing instruments; high (up to 10000), with double focussing instruments
Element mapping :	by scanning beam (μm diameter): Ion Microprobe by ion optical imaging: Ion Microscope
Vacuum system :	UHV (bakeable): $p = 10^{-10}$ Torr or HV $10^{-7} - 10^{-8}$ Torr.
Modular design :	compatible with other thin film analytical methods

Immediately behind that diaphragm the gas which is not ionized is
pumped away. Due to the flow resistance of the diaphragm a pressure
drop between ion gun volume and the space on the target side of the
diaphragm is therefore achieved.
Special types of volume ionization
    The vacuum extraction gauge has a high current density and
works without auxialiary magnetic field. All other sources, to be
discussed hereafter, use auxiliary magnets to quench the discharge
in a small space and to obtain great ion current densities of small

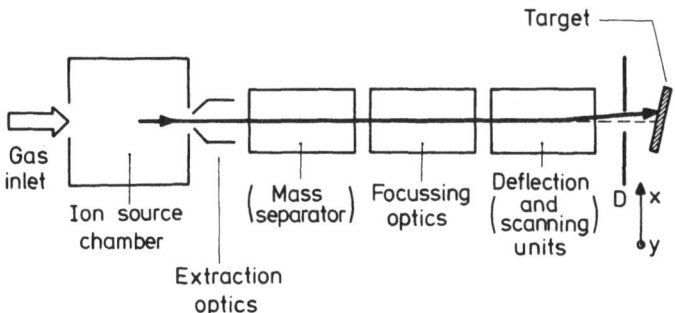

Fig. 39 : Schematic of primary ion source gun. Options (mass
          separation and scanning unit) are given in parentheses.

lateral dimensions[107].
Penning ionization gauge. Here, in a magnetic field the gas discharge
is initiated by spurious electrons and/or field emission.
Radio frequency source (RF source), where the discharge is initiated
by high electric RF fields.
    These two latter types, however, give unstable discharges, and
the life of such a source is short because the ions are extracted
with high (keV) energies. Therefore rapid erosion of the diaphragm
takes place because of sputtering, and also shortcuts occur due to
sputter deposition of material on the insulating rods.
    The duoplasmatron. (Fig. 40) which has been developed by
Ardenne[108] was first used in SIMS by Liebl and Herzog[109]. It is
a source with high brightness (high current density in small
angular aperture and low energy spread[51]). This source is called
duoplasmatron because a double squeezing of the discharge takes
place, firstly by the geometry, due to the use of a cone shaped
extraction electrode, and secondly by the use of a inhomogeneous
magnetic field in the axis of the source, by means of which the
discharge is squeezed in a very small space on the beam axis.
Inside the source, close to the extraction opening, a high concentration of ions (and electrons) is obtained and therefore also
a local "potential mountain" is generated, of about 20 eV potential
with respect to the surroundings. The ions "roll" as it were from
this mountain and need not be extracted by a (disturbing) extraction
field. Consequently the average energy of the ions is about 20 eV
and a high current density with relatively small energy spread
emerges from this source.

    3.1.2 Mass analysers

    3.1.2.1 Magnetic sector (single focusing) mass analyser : if
ions of charge e have been subjected to a voltage difference U they

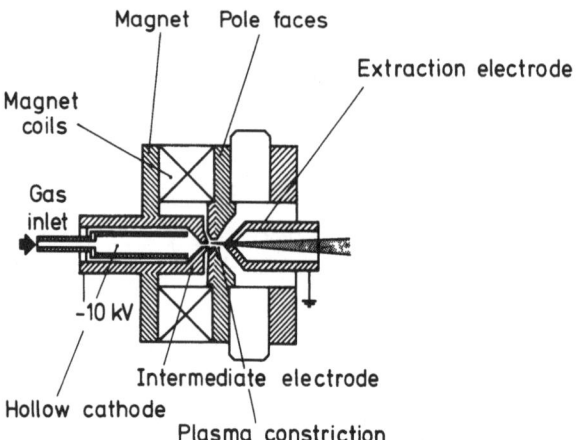

Fig. 40 : Schematic of Duoplasmatron with hollow cathode :   the
          plasma is constricted twice (duoplasmatron)
          1) by the shape of the intermediate electrode
          2) by the inhomogeneous magnetic field near the inter-
             mediate electrode.

have obtained a kinetic energy

$$E_k = mv^2/2 = eU \qquad\qquad (1)$$

where m = mass of the ions, and v = velocity.
When these ions enter a magnetic field B (normal to the plane of the
paper in fig. 41) they describe circular trajectories and the
Lorentz force $F_L = evB$ and the centripetal force $F_c = \frac{mv^2}{r}$ are in
equilibrium. The ions then move on circles with radius

$$r = mv/eB \qquad\qquad (2)$$

Ions with different momentum mv move on circles with different
radii (Fig. 41). Depending on the position of the exit slit S_1 and
the collector slit S_2, only a given mass meets the requirements of
equation (2) for a given value of B and can pass the collector
slit. If B is varied as a function of time, one mass after the
other can pass the collector slit and by recording the collector
current as a function of time one obtains a mass spectrum. In-
sertion of (1) in (2) gives

$$r(cm) = \frac{144\sqrt{(M/n)U(volts)}}{B(G)} \qquad\qquad (3)$$

where M is the mass in amu; B is the magnetic field in (G)
($1G=10^{-4}Vs/m^2 = 10^{-4}T$) and n=1,2... counts the charge state of the ion.
Since r is constant for all masses, it follows that $M \sim B^2(t)$ if B
is varied as a function of time, i.e. we have a quadratic mass
scale. By logarithmic differentiation of equation (3) one obtains

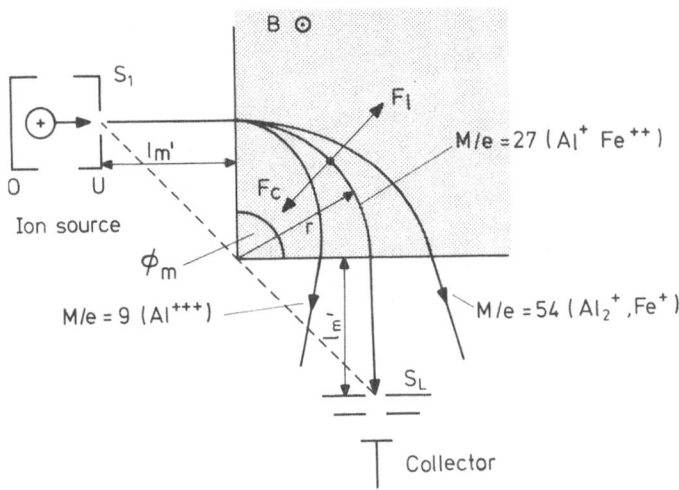

Fig. 41 : Principle of directional focusing in a sector type mass
          spectrometer with angle of magnetic sector ϕ_m=90°.
          Entrance and exit slit (S$_1$ and S$_2$) at a distance
          $l_m' = l_m'' = r$, respectively; radius of curvature of the
          central beam = r.

$$\Delta r/r = \frac{1}{2}\Delta M/M + \frac{1}{2}\Delta U/U \tag{4}$$

i.e. the magnetic sector field gives a separation according to mass
and energy if the ions exhibit also an energy spread ΔU. Good mass
resolution requires $\Delta M/M \gg \Delta U/U$.
     Mass dispersion is the distance y_1-y_2, on the recorder of two
masses M and M+ΔM having different trajectories according to
equation (3)

$$y_2-y_1 (cm) = r(\Delta M/M) \tag{5}$$

Note that for ΔM = constant, r = constant and the dispersion
$y_1-y_2 \sim \frac{1}{M}$, i.e. the dispersion decreases with increasing mass.
Peak width : in practice the ion beam has an angular aperture 2α
(Fig. 42) and moreover the ions do not all have the same energy E_k.
After deflection by 180°, as illustrated in fig . 42 the beam is
focused on the width Δy, where

$$\Delta y = \sum s + r(a_{01}\alpha + a_{10}\beta + a_{11}\alpha^2 + a_{12}\alpha\beta + a_{22}\beta^2 + ...) \tag{6}$$

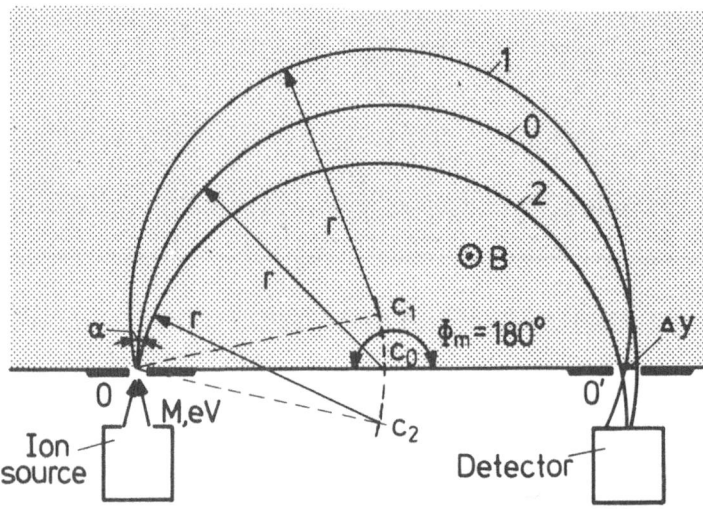

Fig. 42 : Principle of mass dispersion in magnetic sector mass
spectrometer with angle of magnetic sector ϕ_m = 180°,
showing different ion trajectories 0, 1, 2, due to the
divergence (half angle α) of the ion beam. Entrance and
exit slit denoted by 0 and 0' respectively. Magnetic
field B normal to the plane of the paper. Energy of the
ions at the entrance to the mass spectrometer : eU.

in which $\beta = \frac{\Delta v}{v} = \frac{1}{2}\frac{\Delta U}{U}$; v is velocity of the ions, $\sum s$ = sum of the
collector slit and exit slit, a_{ik} are functions of the geometry
of the arrangement, in particular of the sector angle ϕ_m of the
magnetic sector. The distances between the field boundary and ion
source and collector are respectively l_m' and l_m''. (For the case
of ϕ = 180°,fig. 42, l_m' = l_m'' = 0; for the case of $\phi = \frac{\pi}{2}$, fig. 41,
l_m' = l_m'' = r.)
By an appropriate choice of the geometry[110] one can make many of
the coefficients a_{ik} equal to zero. One speaks of single

(directional) focusing if $a_{01} = 0$, of double focusing (focusing
according to direction and energy), if $a_{01} = a_{10} = 0$.
The distance of the ion source $l_m = l_m = l_m$ in the symmetrical case
is given by $l_m = r/ctg(\emptyset_m/2)$.

Estimation of the peak width

For single focusing instruments (homogeneous magnetic field,
normal beam incidence) $a_{01} = 0$, $a_{10} = 2$, and $a_{11} = -1(\emptyset = \pi/2, \pi/3)$.
For $\emptyset = \pi$, in addition $a_{12} = a_{22} = 0$. If α, β are sufficiently small,
the quadratic terms can be neglected and one obtains

$$\Delta y = r.\frac{\Delta U}{U} + S_1 + S_2 = constant \tag{7}$$

for all masses, where Δy denotes the peak width.

Mass resolution

Two peaks with mass difference ΔM are considered to be resolved
if the mass dispersion (equation (5)) is equal to the peak width
(Figure 43c). Different criteria are used to describe the mass
resolution :

peak width resolution at x per cent peak height : if the peak
width of two peaks ,assumed to be of equal height and of symmetrical shape at, x% (10 % in fig. 43c) is
equal to the distance between the two peaks.

valley resolution : for the same physical situation one finds
20 % valley resolution, because the resulting valley is 20 %
of the peak height.

Contribution resolution : if the contribution of a peak to its
neighbour at the centre amounts to x % (1 % in fig. 43c).

Estimate of the mass resolution : from its definition a formula
for the resolution can be derived; for example, let the distance
of two mass peaks ,y_2-y_1,of masses M and M+ΔM be equal to the peak
width Δy : $r\Delta M/M = (r.\Delta U/U) + S_1 + S_2$ from which it follows that

$$M/\Delta M = r/(r\Delta U/U + S_1 + S_2) \tag{8}$$

Arbitrarily, by definition, M/ΔM is called the mass resolution
because, if r is kept constant, it is independent of the mass and
therefore a measure of the mass resolution obtained with a given
(sector) instrument. Example : for r = 15 cm, S_1+S_2 = 0.01 cm,
ΔU = 100 eV (cf. fig. 21), U = 5000 eV one would obtain a resolution of 45. This is not sufficient and one must therefore try
to decrease the energy width (see also double focusing instruments),
for example, down to ΔU = 5 eV. One would then obtain a mass resolution of 600. Note that this calculated value gives only the
order of magnitude, because Δy is an overall value and not
related to a peak width at a given height. For exact calculation
of the mass resolution one would have to use a formula analog onto
(6), but this gives Δy at different heights.

The peak widths shown in fig. 43, calculated as in the above
example for ΔU = 5 eV, have the same value Δy = 25 μm for all
masses. The distance of the peaks (mass dispersion) calculated for
ΔM=1, y_2-y_1=rΔM/M was found to be 300, 150 and 30 μm for

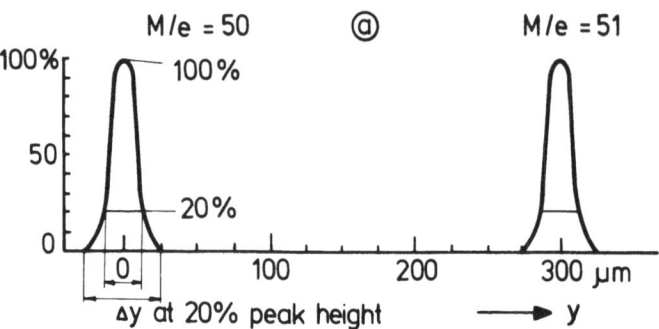

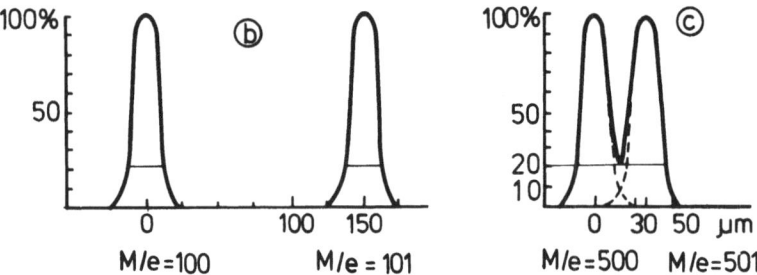

Fig. 43 : Calculated peak widths Δy and dispersion of two peak
with mass difference $\Delta M=1$ = constant, for different
mass numbers and mass resolution in a magnetic sector
type instrument with electric single collector detection
and magnetic field scanning i.e. radius of curvature
r = constant.

respective masses of 50, 100 and 500.

Energy selection.

We have already seen that due to the large energy spread of the secondary ions one has to cut out a narrow band from the energy distribution of the secondary ions in order to obtain good mass resolution. Whereas the magnetic sector gives rise to dispersion with respect to mv (momentum), an electric sector or any other purely electrostatic device exhibits dispersion with respect to energy only. The most simple case to illustrate the energy dispersion (=distance of peaks on the recorder for a relative energy difference $\Delta U/U$) is the electric cylindrical capacitor (electric sector). It consists of a sector (Fig.44a) made from two concentric cylinders at different potentials. Then ions with energy $E_k = \frac{mv^2}{2} = eU$ can only move on circular trajectories in the cylindrical capacitor if the electric force $F_e = eE_o$ is equal to the centrifugal force $F_c = mv^2/r_e$ (Fig. 44a), from which it follows that $r_e = 2U/E_o$ (9), i.e. the electric field leads to energy dispersion ($U \sim E_k$ = initial energy): $\Delta r/r = \Delta U/U$ (10)
Just as in the magnetic sector field, focusing of a beam with not too large an angular aperture occurs if the distance between ion source and field boundary is : $1'=1_e=(r_e/\sqrt{2})ctg\sqrt{2}(\phi_e/2)$. One can see that for $\phi_e = \pi/\sqrt{2}$, just as is the case of $\phi_m = \pi$, focusing occurs if the ion source is positioned at the boundary of the electric field.
The use of electric cylindrical capacitors has been reported[111].

Other electrostatic designs were also used : electric spherical sector[112] (sector consisting of two concentric spheres at different potentials) electric toroidal sector[113]; parallel plate mirror[113,114]; single plate mirror[115].
All these designs are based on the classical ion optical work of Hintenberger[110], and others[116-118]. Toroidal capacitors[119] can also be used for double focusing SIMS.
The principle of double focusing is shown in fig. 44b. Here one can use, in contrast to fig. 44a, a slit with large opening A_e, so that only little loss of intensity occurs, because the energy dispersion in the electric field is equal, but of opposite sign, to the energy dispersion in the magnetic field.
In the design of Vastel[120], in which an electrostatic spherical sector was used behind the magnetic sector, both an increase in resolution (from 200 to 300) and at the same time an increase in sensitivity by about a factor of fifty was obtained.
Energy selection is also necessary when quadrupoles are used. On the basis of the considerations of Paul[121] one can see that the velocity spread has an influence on the resolution also of quadrupoles. Compensation of the energy dispersion in a quadrupole by an electrostatic separation is not possible, because the energy dispersion in the quadrupole is not based on ion optical

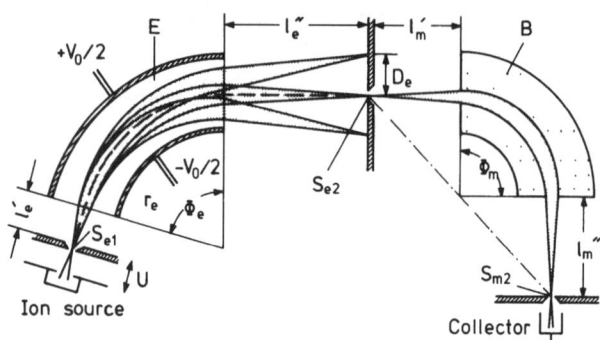

Fig. 44a

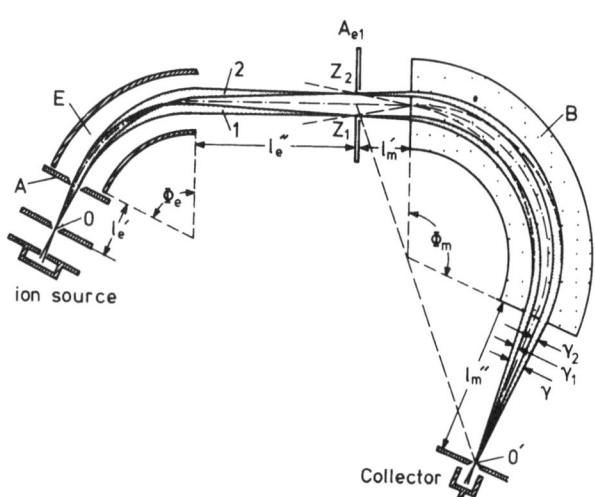

Fig. 44b

Fig. 44 : Designs for high mass resolution sector type instruments.
Fig. 44a : Magnetic mass spectrometer with high mass resolution
           obtained by use of energy filter with slits Se_1 and Se_2.
Fig. 44b : Principle of double focussing in the electric sector
           field (field strength E). Ions of different masses but
           with the same energy are selected. Only ion trajecto-
           ries of one particular mass are shown for simplicity.
           The energy dispersion of these ions in the magnetic sec-
           tor is compensated by the dispersion in the electric sec-
           tor, which is equal to but of opposite sign to the one
           in the magnetic sector. Slit Ae can therefore be ta-
           ken relatively large which gives a relatively large in-
           tensity at high mass resolution. From "Massenspektrome-
           trie" by C. Brunée and H. Voshage. Verlag Karl Thiemig
           1964.

principles. In quadrupoles double focusing in the ion optical mea-
ning of the term is not possible (Fig. 44b). One has rather to cut
out a portion of the energy band (Fig. 44a). Nevertheless the sensitivity of quadrupole instruments is about the same as that of sector
type instruments of the double focusing or single focusing type, because the transmission of quadrupoles is some orders of magnitude
better than that of comparable sector type instruments of the same
dimensions[122], among other reasons, because the use of very narrow
defining slits (necessary in sector types to obtain good mass resolution, see equation (8)) is not necessary in quadrupole instruments.

   3.1.2.2. Quadrupole mass spectrometers : Principle : for quadrupole mass spectrometers[121,123,124] no magnetic field is necessary for mass separation. Since, therefore, no magnetic stray field
is present, which could influence trajectories of charged particles, quadrupoles are ideally suited for use in apparatuses in
which SIMS is combined with, for example, AES and ESCA (see chapter 4). The mass separation in a quadrupole occurs in a system
consisting of 4 hyperbolic rods (Fig. 45) to which a voltage is
applied as shown in the figure. The potential obtained between the
rods is the result of applying simultaneously an electrostatic DC
voltage (U) – not to be confused with the accelerating voltage U
in the preceding paragraph – and a high frequency voltage (zero to
peak value V): $\emptyset_o$ = U–V cosωt, giving :

$$\emptyset(x,y,t)=\left[(U-V\cos\omega\, t)/2r_o^2\right](x^2-y^2)=(\emptyset_o/2r_o^2)(x^2-y^2) \tag{11}$$

We shall in this article use the notation of Dawson[124] which is
different from the one of Paul[121]. Ions injected in the z-direc-
tion normal to the plane of fig. 45 move along oscillatory trajectories through the rod system. Depending on the value of U/V, only
ions in a given mass-interval have amplitudes of oscillations in the
x-y plane which do not become infinitely large, i.e. they move on
"stable trajectories". These ions can therefore pass the rod system
and be collected at a collector at the end of the rods.
   The amplitudes of the other ions grow rapidly when passing the
rod system, the ions are finally intercepted by the rods. From the
equation of motion described by the so-called Mathieu equation one
finds (Fig. 46) those values of a, q for which ions in a given
mass interval have stable trajectories. The a and q values for
stable trajectories (working points) are related to the values of
U, V, r_o, and ω by the relations

$$a = 4\ eU/mr_o^2\omega^2 \tag{12}$$

$$q = 2\ eV/mr_o^2\omega^2 \tag{13}$$

from which it follows immediately that a/q = 2U/V is independent of
the mass.
   For given values of U/V = constant, the working points of ions

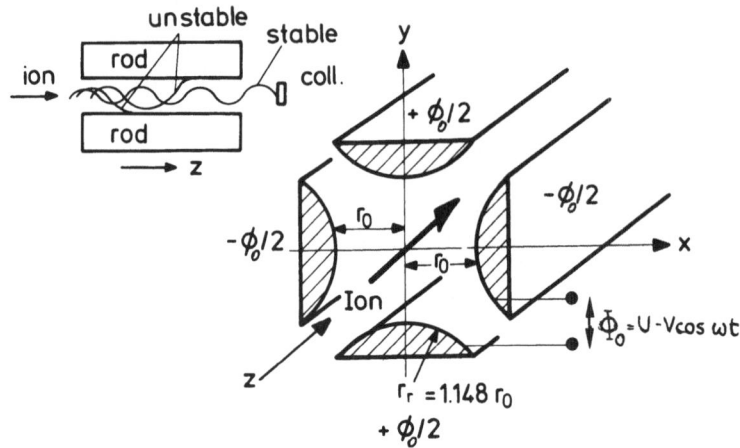

(Cross section through y-z plane)

Fig. 45 : Quadrupole mass spectrometer (schematic) and definition
of parameters. $\emptyset$ = potential applied between opposing set
of electrodes; r_r = rod radius = 1.148 r_0; where r_0 =
field radius = half distance along X- or Y-direction
between two rods. (ϕ_o =U-Vcosωt); notation of Dawson[23]:
$V=V_D$ = zero to peak RF-voltage.

with masses m lie on a straight line through the origin with slope
a/q and limited by the values q_1 and q_2, where

$$m_1 = 2 \ eV/r_o^2 \ \omega^2 q_1 < m < 2 \ eV/r_o^2 \ \omega^2 q_2 = m_2 \qquad (14)$$

On increasing of the ratio U/V the stability regions (q_1,q_2) becomes
smaller, until in the limiting case

$$U_1/V_1 = \frac{1}{2} \frac{a_1}{q_1} = \frac{1}{2} \cdot \frac{0.23699}{0.70600} = 0.1678 \qquad (15)$$

Only ions in a very small mass range (M,M+ΔM) are then transmitted.
    A mass spectrum is obtained in most cases by variation of U and V
with time, with U/V kept constant at all times. From equations (12)
and (13) one can see that m~U and m~V respectively. By using the
appropriate units one finds

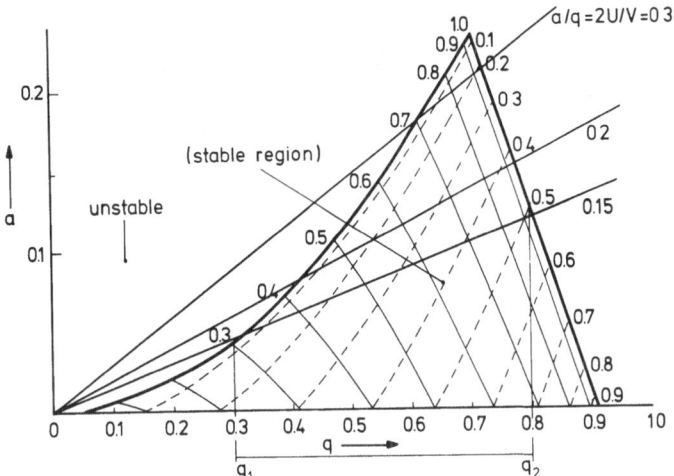

Fig. 46 : Stability diagram : stable ion trajectories are obtained
for (a, q)-values within the trianguloid section.
Masses with corresponding q values $q_1 < q < q_2$ move on
"stable" trajectories i.e. are not intercepted by the
rods. The same holds for the a-values.

$$M(\text{amu}) = \frac{0.409 U_D(\text{volts})}{f^2(\text{MHz}) r_o^2(\text{cm})} \qquad (16)$$

$$\text{or} \quad M(\text{amu}) = \frac{0.0678 V_D(\text{volts})}{f^2(\text{MHz}) r_o^2(\text{cm})} \qquad (17)$$

where U_D and V_D are the DC-voltage and (zero to top) RF-voltage
resp., applied between two adjacent rods. The index D indicates
that we have used the notation of Dawson (see fig. 45). When using
the notation of Paul one must realise, that if V_p is the (zero to
top) RF-voltage as quoted, the RF-voltage to be inserted in eq.(17)
is $2V_p$, since the applied voltage between adjacent rods is
$2\emptyset_o = 2(U + V\cos\omega t)$ (Analogue for U in eq. 16), i.e. one must sub-
stitute : $U_D = 2U_p$, $V_D = 2V_p$ in the above equations. Obviously
equations (12) and (13) are changed also when Paul's notation is
used. Table 12 gives a survey on the interrelations between the
different parameters. For $f = 0.6$ MHz, $r_o = 13.79$ mm ($r_{rod} = 16$ mm)
one needs $V_D = 10$ V for M = 1 and $V_D = 5000$ V for M = 500. Further
examples are given in table 13.
Mass scale :
        Equations (16) and (17) show that quadrupoles in the U(t),
V(t)-mode have a linear mass scale. From these equations one
derives further :

Table 12 : Interrelation between different parameters in quadrupole
mass spectrometers (U_D and V_D are DC- and (zero to top)
RF-voltage resp., applied between two adjacent rods).

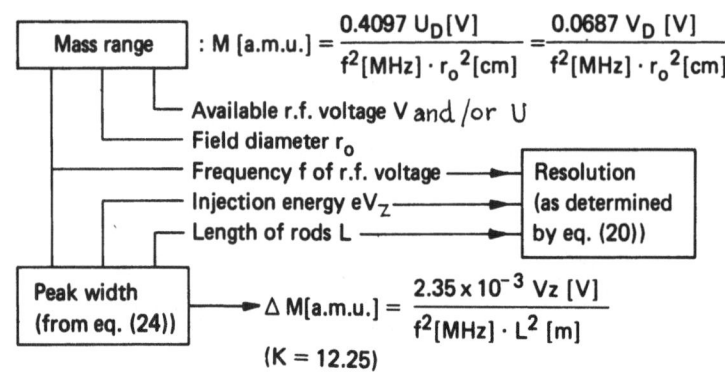

$$\Delta M = K_1 \Delta V; \quad \Delta M = K_2 \Delta V, \tag{18}$$

which means that two masses with the same mass difference ΔM on
the V or U-scale (mass spectrum) are spaced at equal distances,
irrespective of the mass M; this is in contrast to sector type
instruments, where this distance becomes smaller with increasing
mass. In other words, the mass dispersion of quadrupole instruments is mass independent and the masses are equidistantly
arranged along the spectrum.

(Theoretical) Mass Resolution

Paul[121] has derived that the theoretical mass resolution
$M/\Delta M$ is given by equations (19a) and (19b) :

$$(\text{I}) \quad \frac{M}{\Delta M} = \frac{0.126}{0.1678 - U/V} \tag{19a}$$ Low mass resolution
(trapezoidal peak shape)

$$(\text{II}) \quad \frac{M}{\Delta M} = \frac{0.2528}{0.1678 - U/V} \tag{19b}$$ High mass resolution
(triangular peak shape)

where M is the mass number and ΔM here denotes the peak width at
half the height, expressed in atomic mass units (in our earlier
notation for sector type instruments the peak width was denoted by
Δy(amu) in order to avoid confusion with the real mass difference).
These equations show that the (theoretical) value of $M/\Delta M$ depends
only on U/V (i.e. the resolution is mass independent); mass
resolution can be varied, as discussed above, by variation of
U/V. In particular for U/V = 0.1678, one would obtain infinite
mass resolution; but because of $T \sim 1/R$[121] the signal would have
decreased to zero at the same time.

Equations (19) have been derived from a detailed study of the

Table 13 : Interrelation between maximum (M_{max})–minimum (M_{min})
transmitted mass number, for different values of RF-
frequency f, RF-voltage V and field diameter r_o.
V is (zero to top) RF-voltage applied between two
adjacent rods.

V = 4000 Volt			
Frequency f (MHz)	r_o = 2.76 mm (r_r = 3.17 mm)	r_o = 15 mm (r_r = 17 mm)	r_o = 19 mm (r_r = 21.8 mm)
1 3	M = 3610 M = 401	M = 122 M = 13.6	M = 76 M = 8.5
for M_{max} = 500			
1 3	V = 554 V V = 4986 V	V = 16.4 keV V = 147 keV	V = 26.3 keV V = 236 keV
for M_{max} = 350			
1 3	V = 388 V V = 3490 V	V = 11.5 keV V = 103 keV	V = 18.4 keV V = 165 keV
for M_{min} = 1			
1 3	V = 1.1 V V = 9.9 V	V = 32.7 V V = 294 V	V = 52.5 V V = 473 V

stability diagram on the assumption that the working lines (mass-
scan lines) are parallel, i.e. they do not all go through the
point of origin. Further the assumption was made that the rod
system was infinitely long and moreover that the amplitudes of the
ions in the x, y plane were always smaller than the field radius
r_o. In order that equations (19) be valid, several conditions
must therefore be fulfilled in a practical design. We then arrive
at the following set of conditional formulas for the
Practical mass resolution :

1) $V_z \ll (1/K)f^2 L^2 (\Delta M/M).M$ (20)

where eV_z is the maximum longitudinal (axial) energy = kinetic
energy of the ions at injection. K is a constant which depends on
how the mass resolution $M/\Delta M$ has been defined. Values of K according to different authors are : K_p=12.25 (Paul); K_A = 20 for peak
width at 10 % height (Austin and Leck); K_E = 25 (Extra Nuclear
Labs.)

ΔM is the mass difference between two peaks which are just resolved,
i.e. it is assumed that the distance of 2 mass peaks ΔM(amu) is
equal to the peak width Δy(amu) as already discussed in sector -
type instruments. ΔM corresponds to our earlier Δy(amu) and must
not be confused with the actual mass difference.
Taking K = 12.25 one finds for the maximum tolerable injection
energy to yield a mass peakwidth ΔM :

$$V_z(\text{volt}) = 4.26 \times 10^2 f^2 (\text{MHz}) L^2 (m) \Delta M (\text{amu}) \qquad (20a)$$

where ΔM = peak width at half the height, expressed in atomic mass
units.
2) For ions injected in parallel to the z-axis the diameter of the
injection aperture D may not exceed the value

$$D \approx r_o / (M/\Delta M)^{\frac{1}{2}} \qquad (21)$$

3) Ions with radial energy eV_r, entering the field along the z-axis,
will only be sufficiently transmitted if

$$V_r(\text{volt}) < V(\text{volt}) / 30 (M/\Delta M) \qquad (22a)$$

or

$$V_r \sim M / (M/\Delta M) \qquad (22b)$$

since (eq. 16) $V \sim M$, where V = zero to maximum amplitude of the
RF-voltage (equation (11)).
4) In order to obtain a given resolution, the ions must stay in
the RF field during N cycles. It has been found experimentally
that[121,123,125,126])

$$M/\Delta M = (1/K)N^2 \qquad (23)$$

where K is the constant (see equation 20) which depends on the
definition of the resolution, viz. peak width at x % height or x %
valley definition, etc.
It should always be kept in mind that different authors may use
different K-values, as discussed above.
Obviously the number of cycles N during which an ion stays in the
RF field depends on its longitudinal velocity v. In fact equation
(20) has been derived from (23) by means of the relation

$$eV_z = 1/2 (mv^2), \quad t_1 = L/v = L\sqrt{m}/\sqrt{2eV_z} \; ;$$

where t_1 = residence time of ions in the field. The actual mass m(kg) has been replaced in equation (23) by $m = m_p M$, where m_p = proton mass, M is mass number. From equation (20a) and K = 12.25 one finds for the peak width ΔM :

$$\Delta M(amu) = \frac{2.35 \times 10^{-3} V_z (volt)}{f^2 (MHz) L^2 (m)}$$ (24)

With f = 2 MHz, L = 0.2 m, V_z = 5 V one obtains a peak width $\Delta M \cong \Delta y \approx 0.07$ a.m.u. From equation (24) one might conclude that the peak width $\Delta M = \Delta y$ was independent of the mass; this would only be true if equations (20) to (23) were the only ones which determined the peak width. In fact, however, the peak width is determined by the ion trajectories, i.e. by the Mathieu equation and the stability diagram derived from its solution, and it therefore depends also on equation (19). The same conclusions are valid regarding of the resolution which could be drawn from equations (21) and (23).

Operating Modes

From equation (19) one can see that the quadrupole can be operated in different modes. This is achieved in practice by making

$$U = \gamma V - \delta$$ (25)

$$U/V = \gamma - \delta/V$$ (25a)

where γ is the constant which controls the theoretical resolution of the quadrupole; δ is a fixed DC-potential. Hence, from (19b) :

$$\frac{M}{\Delta M} = \frac{0.2528}{(0.1678 - \gamma) + \delta/V}$$ (26)

a) For $\delta = 0$ it follows that $M/\Delta M$ = constant, irrespective of mass, i.e. the peak width $\Delta M \sim M$, hence the transmission T (here denoted by T, instead of f, in order to avoid confusion with the frequency f of the RF-voltage) is inversely proportional to the resolution is constant :
$T \sim 1/Res = \Delta M/M$. Then, because $\Delta M \sim M$, T is constant.

The spectrum which can be expected to be obtained with this mode is shown schematically in fig. 47 (left-hand side).

b) For γ = 0.1678 and δ = constant $\neq 0$ one finds from equation (26): $M/\Delta M \sim (V/\delta)$ and with equation (16) :

$$M/\Delta M \sim M$$ (27)

i.e. the resolution is mass-dependent.

From equation (27) one finds ΔM = constant, i.e. the peak width is mass-independent. The resulting spectrum is shown schematically in fig. 47 (right hand side). The transmission T in this case, $T \sim \Delta M/M \sim 1/M$ is mass-dependent.

When operation in a constant $M/\Delta M$ mode is chosen, equation (21) shows that the maximum aperture D for entrance of the ions is

mass-independent; equation (22b) at the same time shows that the
maximum transverse energy of the ions to be transmitted increases
proportionally with mass. In this mode one could expect that heavy
ions with a given amount of radial energy eV_r would be transmitted,
whereas lighter ions with the same amount of radial energy might
not be transmitted to the same extent.The mass filter would then
show a mass-dependent transmission, with discrimination against
the lighter ions. The same holds for equation (20).

When operating in the ΔM = constant mode, i.e. $M/\Delta M \sim M$, all
ions with given radial energy eV_r or longitudinal energy eV_z would
be transmitted equally well, as is evident from respectively
equations (22) and (20). In this mode, however, equation (21) shows
that the maximum radius of injection of ions that are to be transmitted will decrease with increasing ionic mass. The signal which
will be transmitted is the product of ion current density D_2,
$D^2 \sim r_o^2/M$. In this case there is discrimination against high mass
ions.

In practice, therefore, one will work in a hybrid mode for
survey analysis over a large mass range, or one will discriminate
against high mass ions, for example when one wants to detect B in Si,
or against low mass ions, for example when detecting Sn in GaAs.

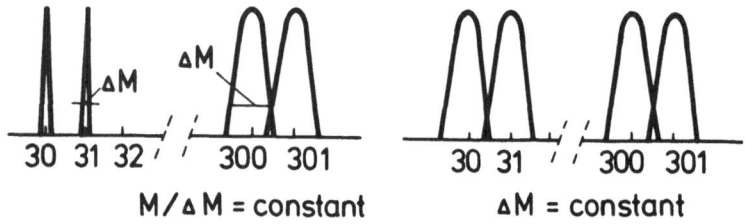

Fig. 47 : Illustration of the influence of different modes on the
mass resolution of a quadrupole. Left : $M/\Delta M$ = constant.
Since peak width $\Delta M \sim M$, the peak width at mass 300 is ten
times as large as at mass 30 and transmission is constant.
Right : ΔM = constant, resulting in equal peak width at
masses 30 and 300; mass resolution : $M/\Delta M \sim M$. Transmission
$T \sim 1/M$.
The distance between peaks with mass differences $\Delta M = 1$ is
equal for all masses in both cases.

Energy selection

Equation (24) shows that in order to have not too broad a
peak width and hence good mass resolution, the longitudinal energy
eV_z must be limited. For this reason one uses an electrostatic
device which cuts out a small band from the energy distribution of
the secondary ions. Fig. 48 shows several designs for energy
selection. Besides the designs illustrated in this figure one could
in principle also use cylindrical sectors or toroidal sectors. The
two latter and also spherical analysers, parallel plate mirrors,
and cylindric mirror analysers (CMA) give a focusing of the
secondary ion beam. A double focusing in the sense of ion optics
as used in sector type instruments is not possible, however. But
the loss in transmission with respect to the situation in sector
type instruments is compensated by the greater transmission of
quadrupoles.

The use of energy selecting units has the further advantage
that the target is removed from the direct line of sight to the
analyser and the detector. With a direct line of sight arrangement
between target and multiplier, sputtered neutrals and metastables,
backscattered fast primaries, photons, and ions reflected from the
analyser rods would be detected in the multiplier detector. A
solely out of sight arrangement of the multiplier, as shown
(Fig. 48f), is not sufficient to suppress the background. Similar
arguments hold for a primary mass selection by a sector type mass
spectrometer; in that case the neutral component emerging from the
primary ion source cannot hit the target and give difficulties
during depth profiling.

3.1.2.3 Time-of-flight mass spectrometers have been proposed
by Liebl[51] because of their great transmission (nearly 100 %).
Here in principle one must generate short pulses (ns) of secondary
ions, which are extracted by an accelerating voltage. The mass
separation is due to the flight time of the ions. Ions of the same
kinetic energy $eV = \frac{1}{2}m_1v_1^2 = \frac{1}{2}m_2v_2^2$ have different flight times
across the same distance due to their different velocities.
Amplifiers and recorders with 0.1-1 GHz bandwidth are necessary
to detect in turn one kind of ions after another and to record them
with a fast response C.R.T. (cathode ray tube).

3.1.3 Ion detection

Detection of ion currents between $10^{-8} \div 10^{-14}$ A is achieved
with a Faraday cup followed by a DC-amplifier (Fig. 49). The
current I collected on the Faraday cup generates a voltage drop
$U_R = I.R...(1)$ at the input resistor R of the DC-amplifier. With
$R=10^{11}\Omega$ and a minimum detectable voltage of $U_{min}=1mV$, the minimum
detectable current I becomes 10^{-14} A. Then a peak can still be
detected if its corresponding voltage U_R is about 2 to 3 times
greater than the rms-noise voltage of the amplifier. In well
designed DC-amplifiers this noise voltage is given by the (thermal)

(a)

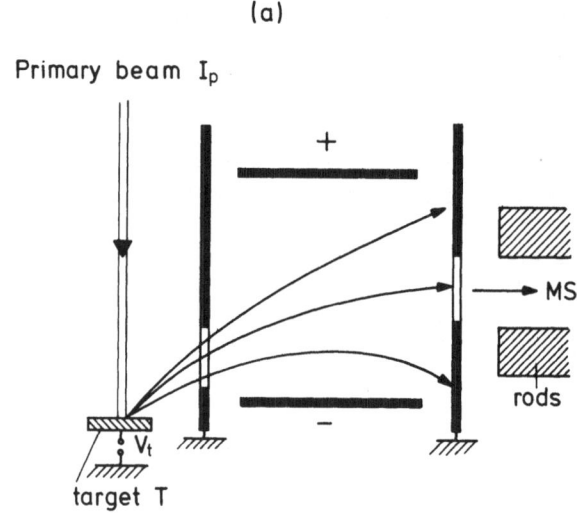

(b)

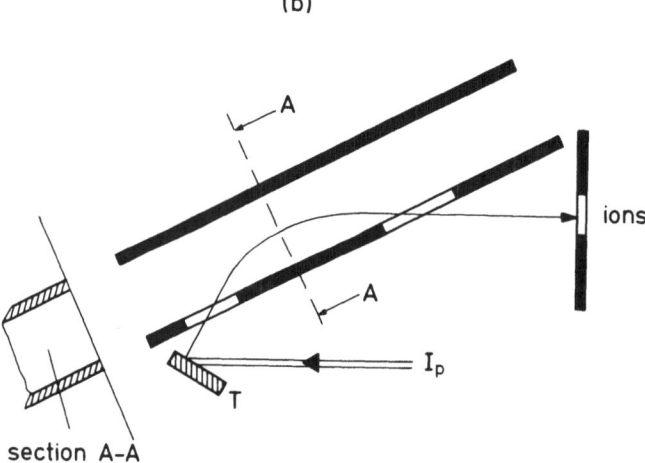

Fig. 48 : Different geometries of electrostatic designs for
          energy selection as used in quadrupole SIMS :
          a) parallel plate capacitor[127)
          b) parallel plate mirror[128) .

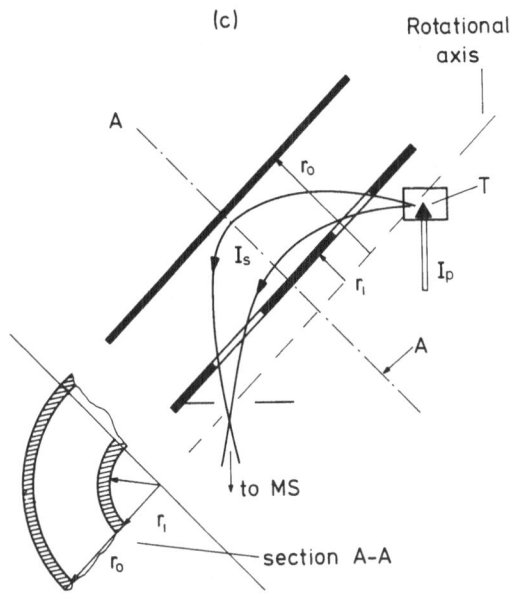

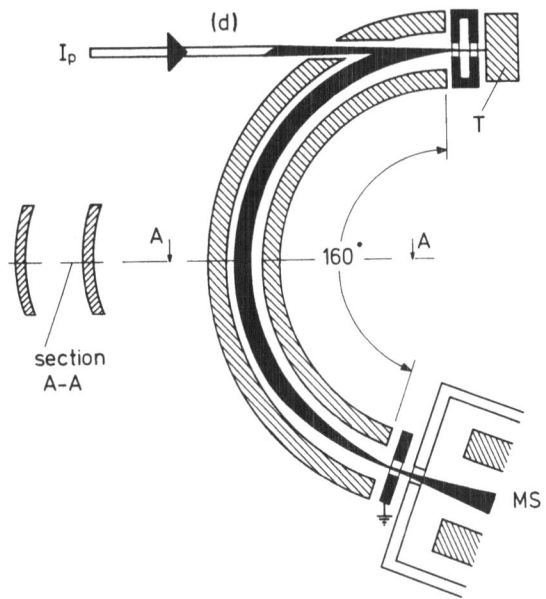

Fig. 48 : c) cylindrical mirror analyser[129].
Fig. 48 : d) spherical analyser[130].

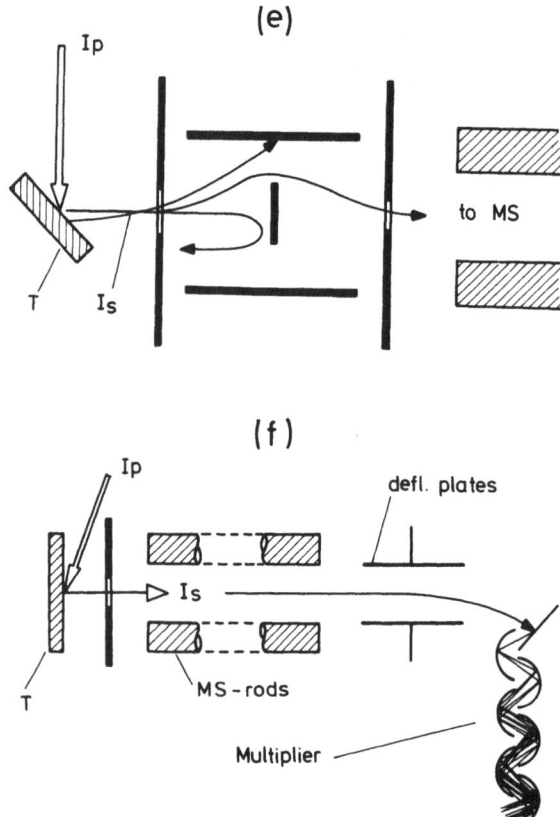

Fig. 48 : e) straight axisymmetric type with axial beam stop[131]
          (=Bessel Box).
Fig. 48 : f) no energy analyser : multiplier in off-axis position.
          Deflection of the ions behind the quadrupole by pa-
          rallel plate capacitor[132].

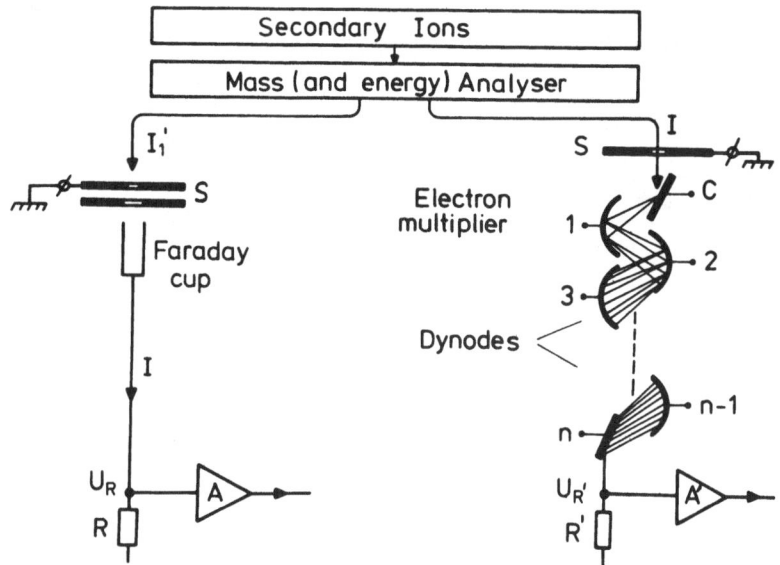

Fig. 49 : Schematic of ion detection with Faraday cup (left) and
          open (Allen type) electron multiplier (right) : S is
          collector shielding slit on ground potential. Multiplier
          dynode voltage V_c= -5000 V, V_n = 0 V. Total voltage across
          multiplier is divided equally among the dynodes.

noise of the resistance R :

$$U_N = 7.4 \times 10^{-12} \sqrt{TR\Delta f} \qquad\qquad (2)$$

where T is the absolute temperature in degrees K, R is the
resistance in Ω, Δf is the bandwidth (Hz) of the amplifier. This
noise voltage is due to thermal movement of the free electrons in
the resistor R. The amplifier cannot distinguish this noise from
the signal obtained at the input resistor, because it is generated
at the same spot. With $R=10^{11}\Omega$, Δf = 10 Hz, and T = 300 K, the
noise voltage becomes 1.3×10^{-4} V, corresponding to $I_N=1.3 \times 10^{-15}$ A.
From equations (1) and (2) the signal-to-noise ratio is found to
be $U_R/U_N \sim \sqrt{R}$ i.e. it increases with increasing R. The value of
this input resistor must not be chosen too high, however, because

the time constant of the amplifier τ_o = R.C. is already 0.5 second
for R = $10^{11}\Omega$ and C, the stray capacity parallel to the input is
5 pF.

Bandwidth consideration

The time constant τ (sec) of the amplifier determines whether
a signal (mass peak I(t)) with a duration of t_p(sec) at the base is
passed unchanged in form and height;(similar considerations hold
for any amplifier in the electronics of the ion detection system
and in particular, for the recorder). The time constant of the
amplifier is related to the bandwidth Δf (limiting or cut-off
frequency) via the relation

$$\Delta f = 1/2\pi\tau \qquad\qquad\qquad (3)$$

which indicates how quickly the amplifier can follow a sudden change
of the input signal (Fig. 50). If one wants to record a spectrum
in a reasonable time, considering that SIMS is a material consuming
method, the typical analysis time for a spectrum from mass 2 to
200 becomes 200 seconds. Then a peak width at the base of about
t_p = 0.1-0.2 sec is necessary. The above mentioned rise time τ_o
can be obtained by decreasing the value of R, but then, according
to equation (1), also the signal would decrease. Another way is to
use frequency dependent feedback of the signal[133]. In that case
one will find

$$\tau = RC/A , \qquad\qquad\qquad (4)$$

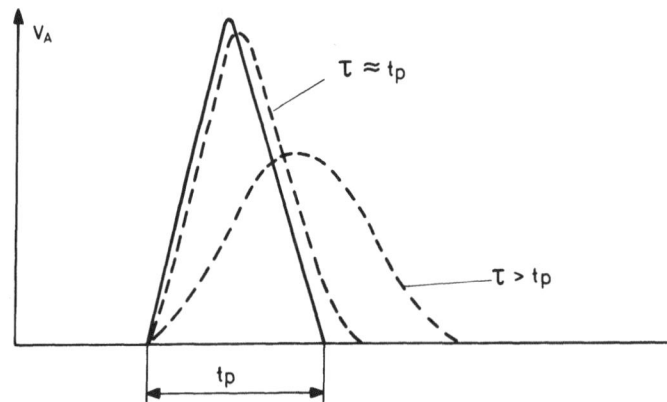

Fig. 50 : Original (solid curve) and transmitted (dashed curve)
          peak shape for different values of risetime τ of the
          amplifier or recorder. Typical values for t_p = 0,2 sec.
          For $\tau \ll t_p$, the peak shape and peak height remain unchanged.

where A is the (frequency dependent) feedback factor of about 10^4.
This gives an input time constant $\tau = 5.10^{-5}$ sec. This small time
constant is obtained because, due to the feedback, only a small
voltage IR/A builds up at the input resistor. The stray capacitor
C then need be charged to only a small voltage.

The corresponding bandwidth (equation (3)) of the amplifier
becomes $\Delta f = 3$ kHz. It cannot be fully used, however, because
(see equation (2)) the noise voltage would become too high. As a
compromise one therefore chooses, for example, $\Delta f = 30$ Hz
($\tau = 5.10^{-3}$ s $\ll t_p$).

Vibrating-reed electrometer amplifiers

Here the DC signal is transformed into an AC signal for detection by demodulation techniques. The limit of detection is
then about 10^{-16} A. The time constant (≈ 0.5-5 sec) is several
orders of magnitudes larger than with DC in the feedback mode.

Electron multipliers

Nowadays electron multipliers are commonly used for ion
current detection between about 10^{-19} A and 10^{-12} A. The higher
upper limit of detection with the multiplier ensures an overlap
with signals obtained from a DC amplifier.

Principle of the electron multiplier (Fig. 49, right hand side):
the ions to be detected strike the conversion dynode C of the
multiplier and release γ secondary electrons, ($\gamma \approx 2$-3 electrons/
ion). These electrons in turn strike the following electrode (1,
2, ... n) which is at a positive potential with respect to C and
release $\delta \approx 2$ secondary electrons. The total current amplification
G of the multiplier is $G = \gamma \delta^n$, (typical values of G are 10^4-10^8).
The input resistor R' of the amplifier can therefore be G times
smaller and still detect the same ion current as with a DC amplifier.
The noise of the input resistor (equation (2)) is then of no more
importance. The minimum detectable current is determined by background pulses[134], corresponding to about 10^{-20}-10^{21}A.

Channeltron multipliers

In contrast to the multipliers discussed in Fig. 49 with
discretely distributed dynodes (typical diameter of these
multipliers 3 cm, length 15 cm) there are designs of multipliers
with small dimensions (diameter typical 1 mm, length $\approx$ 1 cm) using
continuously distributed dynodes. These channel multipliers
(channeltrons) are used often also in 2-dimensional arrays
(channel electron multiplier array (CEMA)). Channeltrons consist
of a thin glass tube coated internally with a thin metallic layer
(continuous electrode) to provide secondary electrons. Between the
ends of the tube a voltage of several kilovolts is applied. Ions
which strike the (cone-shaped) entrance generate secondary electrons,
which give rise to an electron avalanche along the tube in the
direction of the tube anode. Current amplification between 10^4 and
10^8 is achieved.

Any electron multiplier can be used in 2 different modes.

A) DC mode :

In this case the multiplier is used as a low noise current amplifier of high gain G between 10^4 and 10^8. The minimum detectable current is 10^{-18} A in this mode.

Detection of positive ions in the DC mode in general gives rise to no problems. Usually one wants to have the DC amplifier A' at ground potential (Fig. 49). Therefore also the anode n of the multiplier must be at ground potential, and hence, in order to achieve acceleration of the electrons towards the anode N, the voltage of the first dynode V_C must be at a negative potential ($-V_{MP}$) with respect to ground (Fig. 51). The voltage $-V_{MP}$ at the same time gives an extra acceleration (in addition to an acceleration voltage V_a = difference between source potential and the average analyser potential, at ground potential) to the positive ions to be detected.

Detection of negative ions in the DC mode is obviously not simple and needs a special design, e.g. raising of the DC amplifier potential to V_{MP}, i.e. making $V_C = 0$, or using a blocking capacitor and accumulating the pulses with this blocking capacitor.

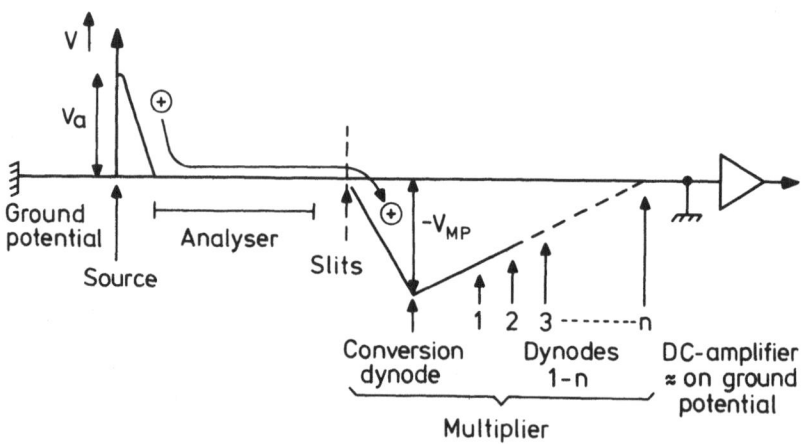

Fig. 51 : Potential distribution V(x) along mass spectrometer arrangement for positive ion detection in DC-mode.

B) <u>Pulse counting mode</u>
     The impinging ions are counted individually via voltage pulses
generated at the anode of the multiplier. The limit of detection in
this mode is between 1 ct/sec and 0.1 ct/sec $(1.6 \times 10^{-19} \mathrm{A} \triangleq 1.6 \times 10^{-20} \mathrm{A})$.
     Detection of positive and negative ions in the pulse counting
mode :
     This gives no problems. In the positive ion counting mode the
potential distribution is in principle the same as shown in Fig. 51.
For the counting of negative ions the potential of the last dynode
n is at a voltage $V_a' + V_{MP}$, where V_a' is an optional additional
accelerating voltage for the negative ions (Fig. 52). The input of
the DC amplifier, however, is at ground potential, because it is
blocked with respect to DC voltages by the capacitor C. Yet the
pulses generated at the multiplier can pass the capacitor C, because

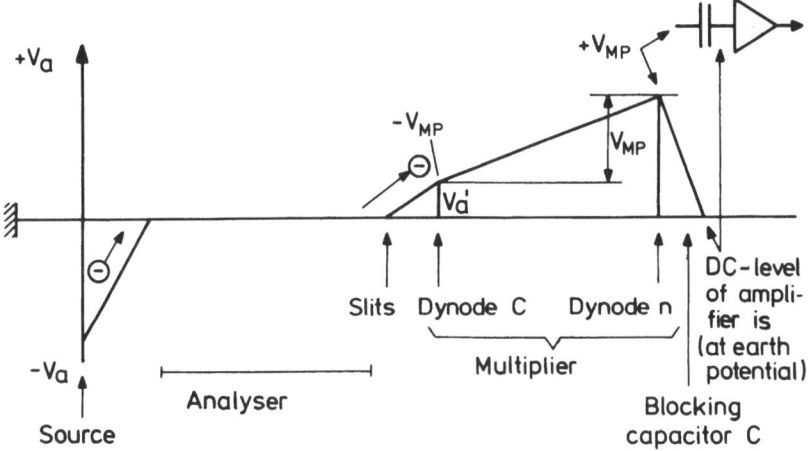

Fig. 52 : Potential distribution V(x) along mass spectrometer
          arrangement for negative ion detection in pulse counting
          mode. The input of the amplifier is at DC-ground potential,
          because it is blocked by the capacitor C (high voltage
          proof).

they contain a high frequency AC component (as can be shown by
Fourier analysis of the pulses). For these correspondingly high
frequencies f the capacitor C is a short-cut with $R_c \approx 0$
($|R_c| = 1/2\pi fC$).

Counting losses

In general one uses pulse counting modes also for higher ion
currents (up to 10^{-12} A) because the design is simpler, in
particular for the detection of negative ions.
This gives difficulties, however, because counting losses occur
at high counting rates, resulting in the measured ion current $\dot{n} < \dot{N}$
(counts/sec). This can be understood as follows : if the average
incoming pulse rate is $\dot{N}$, then one pulse may last only a little
shorter than $1/\dot{N}$. If the pulse length becomes longer then $1/\dot{N}$,
overlapping of the pulses occurs and one would count fewer pulses
then are actually coming in per unit time. Assuming constant length
of the pulses one can derive the relation : $\dot{n} = \dot{N}(1-\dot{n}\tau)$. Nowadays
pulse lengths between 1 μs and 1 ns can be obtained, corresponding
to measured count rates $\dot{n}$ as illustrated in fig. 53.

Accuracy of the measurement

The error in the ion current measurement is finally determined
by statistical fluctuations in the ion current, if all other sources
of error have been eliminated. If the average number of ions in a
measured current is N, the relative error of the current measurement
is :

$$\Delta N/N = 1/\sqrt{N} = 1/\sqrt{\dot{N}\tau} = 1/\sqrt{(I/e)\tau}$$

where $e = 1.6 \times 10^{-19}$ Cb is the elementary charge.
The relative error in the current measurement does not depend on
whether one uses a DC mode (current measurement) or, as usually,
a counting mode. Example : for $\dot{N} = 10/\text{sec} \triangleq 1.6 \times 10^{-18}$A and a
measuring time $\tau = 10$s, the relative error becomes $1/\sqrt{100} = 1/10 =$
10 %. The smaller the ion current, the larger the measuring time
must be to obtain the same accuracy.

Recorders of different designs and time constants are in use.

Pen recorders : rise time $\tau = 0.2 - 1$s. Full scale deflection :
15-25 cm. One or several tracks can be used.

U.V. recorders with natural frequencies up to a few kHz and
several tracks of different sensitivity can be used. Fig. 54 illustrates a UV recorder-plot with 3 tracks with sensitivities in the
ratio 1 : 10 : 100 . Therefore a wide dynamic range of the signals
can be handled simultaneously.

Analog Magnetic Tape Recorder (frequency modulated) : bandwidth up
to 100 kHz.

Cathode Ray Tubes (CRT) have very small rise times because of
electron beam deflection. The corresponding bandwidth is typically
100 kHz up to several MHz; expensive cathode ray tubes have band-
widths of up to 1 GHz.

Essential in the choice of a recorder is that the recorder must
have a time constant which is smaller than the peak width in the

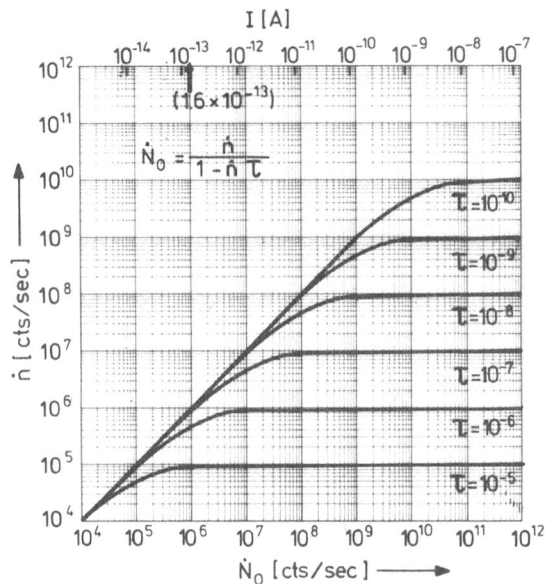

Fig. 53 : Counting losses in pulse counting mode as a function of
dead time τ_0: measured pulse rate $\dot{n}$ (in counts per sec.)
as a function of actual pulse count rate $\dot{N}$(counts per sec.).

mass spectrum, so that the mass peaks are recorded without distortion in shape and in peak height (Fig. 50).

Designs for single element detection

Faraday cup (Fig. 49) limited to the detection of currents $\geq 10^{-14}$A,
can be used for the detection of both positive and negative ions.

Allen type multipliers (Fig. 49) are mostly used in the pulse
counting mode, including pulse height discrimination in order to
obtain a low background level ($<10^{-20}$A) and because the mass dependent ion-to-electron conversion at the first dynode is eliminated
in this mode[134]. Also changes in the gain of the multiplier, a
disadvantage of the multipliers with respect to the Faraday cup,
become less important.

Richard and Hays detector[135]

The ions are accelerated onto a scintilator (Fig. 55) in the mass
spectrometer. Light flashes generated by these incident ions
release photo-electrons from the cathode of a (closed) photo

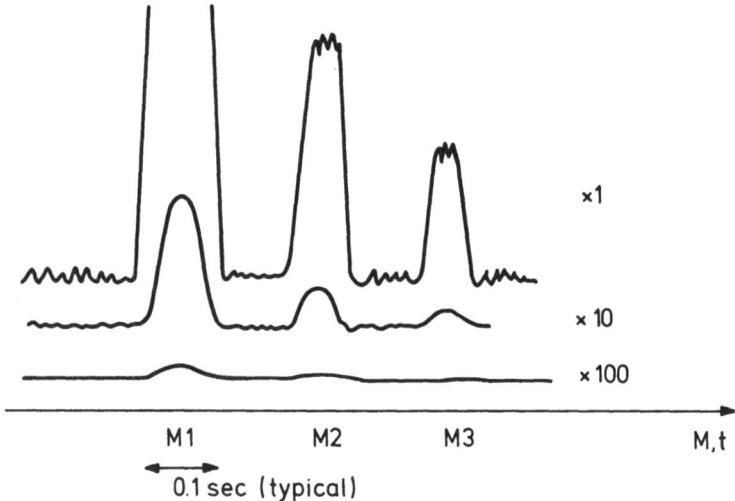

Fig. 54 : Multichannel recording of spectrum, showing part of the
spectrum (schematic) obtained with a UV-recorder using
three channels with attenuation of x1 x10 x100. (Risetime
≈ natural frequency of galvanometers is 1/500 s)

multiplier pressed to the rear of the scintilator. Disadvantage is
a rapid deterioration of the scintilator by sputtering[134].
<u>Daly detector</u> (Fig. 56) : the deterioration of the scintillator by
sputtering has been overcome by first converting the ions[136,137] at
a conversion electrode into electrons and then accelerating these
electrons on to a scintallator facing the conversion electrode. The
scintillator is at earth potential and the conversion electrode at
negative potential. Light flashes which are generated by the impact
of the electrons are transferred, just as in the Richard and Hays
detector, on to a photomultiplier cathode, amplified, and detected.
Both pulse counting and DC mode are now in use. The relatively
high background due to the dark current of the photocathode
($\approx 10^{-16} A/cm^2$) can be suppressed by coincidence circuits (splitting
of the light current and using   two photomultipliers in
coincidence) or by pulse height discrimination. Negative ions can
also be detected in this mode without difficulties, because separation of the DC potential between conversion electrode and the
potential of the multiplier entrance (photocathode) is ensured.

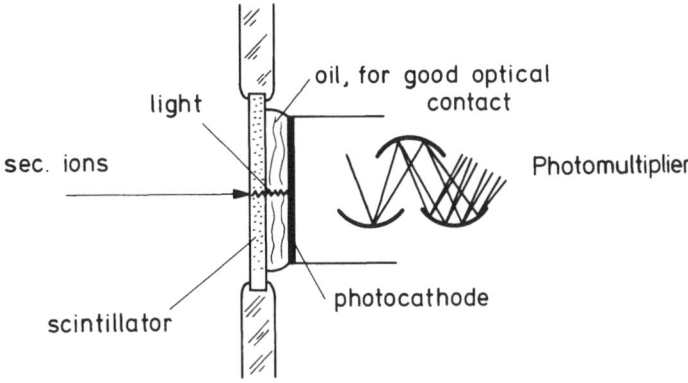

Fig. 55 : Richard and Hays detector[135]) (schematic) for positive
         and negative ion detection.

Design for multi-element detection
        As already mentioned, sample consumption plays a role in SIMS
and therefore multi-element detection can be of advantage in many
cases (cf. table 7). Since photographic plate detection, which
for many years was the only possibility for multi-element detection,
is not very practical in particular for quantitative analysis, one
has tried to use other designs :
        Channel plate[138]) : The ions are converted (Fig. 57).
into electrons at a conversion electrode arranged along the image
plane of a Mattauch-Herzog mass spectrometer[139]) The released
electrons are focused by means of the stray magnetic field[140]) on
a CEMA array (25 x 100 mm^2). As collectors behind these channel
plates, 60 wires have been used which feed the electron signals,
magnified in the channel plate array by a factor of $10^6 \div 10^7$, into
a commercial 60 channel on-line data collection unit. The mass
spectrum is electrically scanned to cover portions between the
collecting wires. In effect this arrangement is therefore equivalent
to a mass spectrometer with 60 slits. Such systems may play an
important role in future SIMS designs.
Electro-optical device for multi-element detection or image display
        This design uses a channel plate to amplify the ion currents
striking different places of the image plane. The currents I_1, I_2,
I_3 (Fig. 58) represent a) different masses (as obtained in a mass
spectrograph) or b) currents of one mass M at different spots of
the image plane (image of mass M in an ion microscope).
        The ion currents directly strike a channel plate array, which
may deteriorate rapidly by sputtering, in contrast to the arrangement shown in fig. 57. The electrons (current IxG) at the output

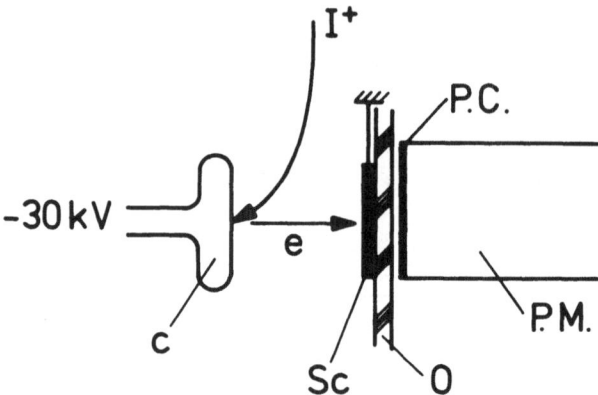

Fig. 56 : Daly detector,[136] shown for the detection of positive
          ions : I$^+$ = positive secondary ion current, C = conversion
          electrode (for ions into electrons), e = emitted
          secondary electrons, Sc = scintillator, O = glass window,
          separating the vacuum of the mass spectrometer from
          atmospheric pressure, PC = photocathode of the
          multiplier PM.

of the channel plate array are accelerated (5 keV) on to a phosphor
screen. The generated light is transferred via a fibre optics to a
Vidicon arrangement for reading out the resulting image. The
vidicon is followed by a multichannel optical analyser. Displaying
takes place on a cathode ray tube.

A similar device for image display in an ion microscope has
been described by Maurice[141]. This design uses an ion-to-electron
conversion electrode from which the released electrons are accelerated on to a phosphor screen. Behind the phosphor screen the same
arrangement as shown in fig. 58 is used. The use of a conversion
grid C in front of the channel plate array has also been
proposed.

Time-of-flight multi-element detection has been proposed by Liebl[51]:
the combination of a pulsed primary beam with time of flight, suited
for on-line data collection with extremely low waste of material.

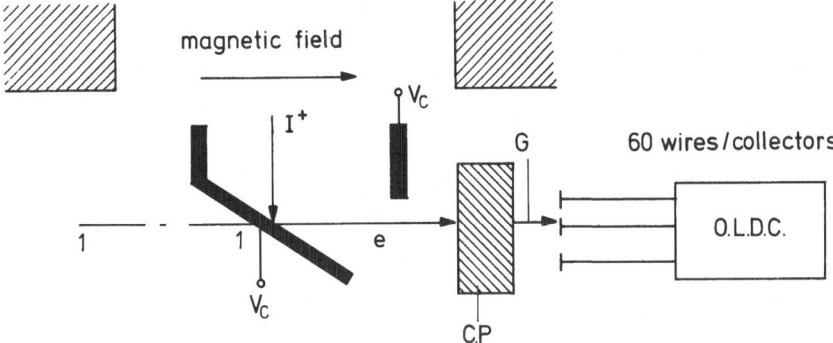

Fig. 57 : Proposed channel plate multi-element detector : I^+ =
secondary ion current, (1-1) = image plane; e = electrons,
CP = Channel Plate with Gain (G) = $10^6 \div 10^7$, followed
by 60 wires used as individual collectors; OLDC = on
line data collection units, after Beske[139]).

3.1.4 Use of computers in SIMS instrumentation
    3.1.4.1 Data handling and processing (data acquisition) :
Data acquisition consists of data handling and processing. For data
processing to be carried out the information must be offered to the
input of the computer in a form which can be "understood" by that
machine : the mass spectrum I(t) as measured (raw data)is converted
into a voltage

$$V_1(t) = I(t)/R$$ (eq.(1), par. 3.1.3).

It contains a large number of information bits in analog form i.e.
as they can be seen in the spectra on the UV-recorder or a cathode
ray screen. From these spectra the skilled operator can obtain
information on peak height and via the peak position he can
calculate the mass number. The computer requires all information
to be supplied in digital form, since it cannot "read" or "understand"
a curve (spectrum or peak). It must be given the information, e.g.
about the shape and position of the peak, in the form of currents
over many intervals, expressed directly in numbers (digits) to as
many significant places as necessary in a given interval (sampling
step) (cf. fig. 61).
Data reduction
    The digital information which has been read by the computer
can be "reduced" further. For example, the information in a mass
spectrum $V_1(I(t))$ together with the information about the magnetic
field $V_2(B(t))$ obtained at the same time (or $V_3(V_Q(t))$ obtained in

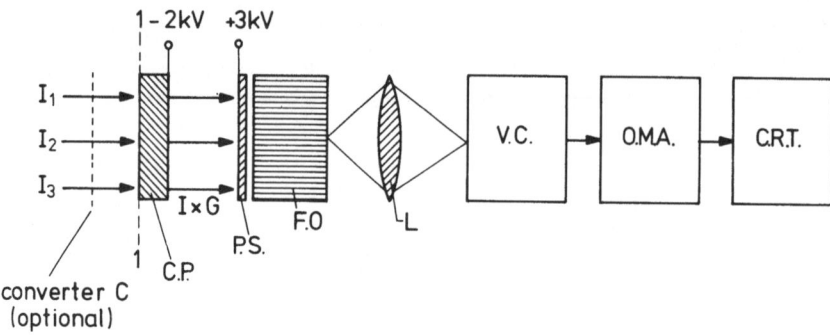

Fig. 58 : Electro-optical device for simultaneous multi-element
          detection or secondary ion image display, after
          Tuithof [161]. I_1, I_2, I_3 ion currents on different spots
          of the image plane 1-1, CP = channel plate (channel
          electron multiplier array CEMA) with channel diameters
          of 15 μm at repetition distances of 19 μm, gain≈10^3,
          PS = phosphor screen, FO = fibre optics, L = objective
          lens, VC = vidicon camera, OMA = optical multichannel
          analyser; CRT = cathode ray tube display. Option :
          grid acting as an ion to electron converter.

quadrupoles) can be reduced by a suitable computer programme to
a table containing only I as a function of M. The mass scale is
obtained from a calibration which must be known to the computer :
V_2 = f(M(B)) or V_3 = KM(V_Q). In the former case (sector type
instruments) V_2 is not a linear function of M,   i.e. the mass
lines are not positioned at equal distances   because   then
Hall-voltage (V_{Hall} = $V_2 \sim M^2$) is proportional to the square of M.
Moreover the hysteresis can modify the ideal quadratic relationship
between V_2 and M. In a quadrupole, as discussed above, V_3 is
directly proportional to M.

Off-line data processing

      Fig. 59 shows schematically how an analog magnetic tape
recorder (frequency modulated) can be used.
First the signals are recorded on the recorder. Next the recorder
is decoupled from the mass spectrometer and the information is
transferred to a computer with the speed v_{out}. The first stage

of the computer contains an ADC (an analog-to-digital convertor)
for the (analog) signals. The output velocity v_{out} can be chosen
arbitrarily because one is not bound to any time (the read-out time
does not influence material consumption of the sample). If one
choses $v_{out} < v_{in}$, one can in certain cases obtain a favourable
signal-to-noise ratio because the bandwidth can be kept low and
therefore little additional noise of the tape is introduced. We
have in this mode obtained accuracies of about 1 % several years
ago. Nowadays instead of such an analog recorder one would use a
digital tape recorder. In that case the signal is transformed from
analog to digital already in the recorder and the ADC of the
computer can be bypassed.

On-line data processing

        Today in most cases data acquisition is carried out on-line,
i.e. the information flow is fed without interruption into a small
dedicated computer. This computer with small memory capacity, e.g.
16 k, is entirely "dedicated" to data reduction from the instrument (Fig. 60). The reduced data thus obtained can be displayed
or stored in a digital store, whence further treatment can be
carried out with a big (memory-capacity) computer.
As indicated in fig. 60 the analog mass spectrometer signals V_1, V_2,

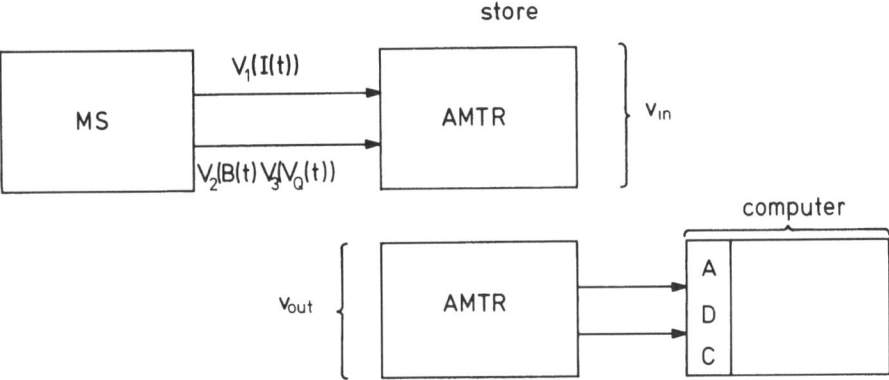

Fig.59 : Off-line data handling and processing : the signals
          $V_1(I(t))$ = mass spectrum and the mass indication
          [$V_2(B(t))$ = magnetic field indication or the RF voltage
          V_Q of the quadrupole $V_3(V_Q(t))$] are fed with a speed v_{in}
          into an analog magnetic tape recorder (AMTR).

V_3, etc. pass an ADC which is DC (galvanically) separated from the computer via opto-insulators. An opto-insulator consists of a LED (light emitting diode). The information transfer takes place in digital form via light flashes generated in the LED by the incoming digital information (pulses). Digital signals from the mass spectrometer (ion currents measured in the counting mode, V_1') are directly transmitted to the computer also by an opto-insulator. Digital data transfer is less sensitive to disturbances than analog data transfer.

Data acquisition and reduction in a computer is carried out in real time i.e. simultaneously with the incoming information flow. Display of the reduced data (e.g. M versus I) is carried out with some delay because the (mechanical) printing equipment is not fast enough to follow the offered information flow i.e. the computer may, for example, be on mass 180 while the printer is still busy at mass 50.

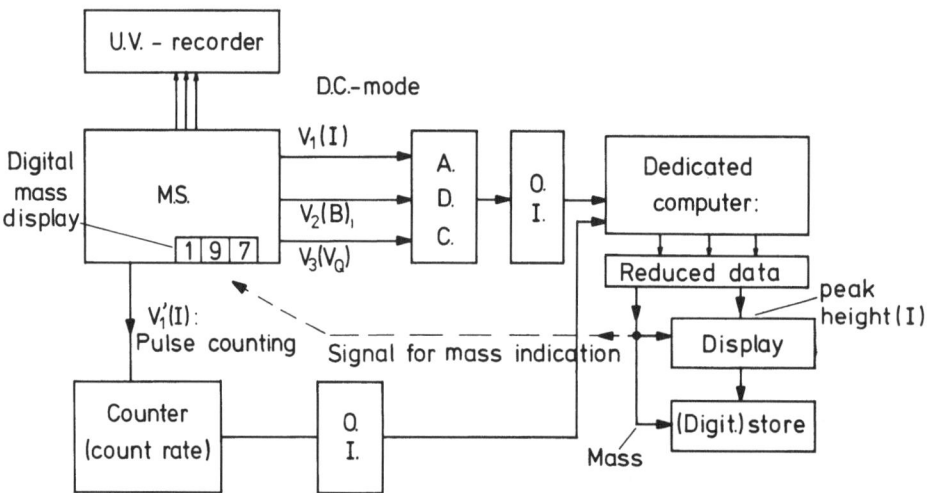

Fig. 60 : On line data processing : the analog signals $V_1(I)$, $V_2(B)$ (in quadrupoles $V_3(V_Q(t))$ are converted into digital form in an ADC. The latter is isolated from the computer via opto-isolators (OI) but digital information (pulses) can be transferred to a computer. Digital signals e.g. ion currents in the counting mode V_1' are directly fed to the computer via an opto-insulator.

<u>Store of reduced data</u> : by reduction of the information of the mass
spectrum to a mass-versus-ion current, the number of bits required
is reduced by about a factor of 10 and the reduced spectrum can thus
be stored in a digital store of not too large capacity.
<u>Some general remarks on data acquisition</u>
        Since the computer can only read digital (numerical) information,
the analog spectrum must be "resolved" into small pieces on the time
or mass axis (Fig. 61). The current integral over the sampling
period is stored (physical smoothing). Further smoothing of the
peaks is carried out digitally, i.e. mathematically, by the
computer[142]. The resulting peak shape (intensity in digits per
sampling periode) is used to find the peak position.
In order to distinguish mass peaks from spurious peaks (noise,
voltage-breakdown peaks) one uses additional criteria to define a
mass peak :
a) A peak height must be higher by a given factor than the r.m.s.
   level, e.g. 3 x r.m.s. level.
b) The peak must have a width of, for example, at least 0.1 sec at
   that level. Due to the latter criterium the spike at sample
   no. 15 in Fig. 61 will not be considered as peak. For the samples
   numbered 18-20 none of the 2 criteria is fulfilled.

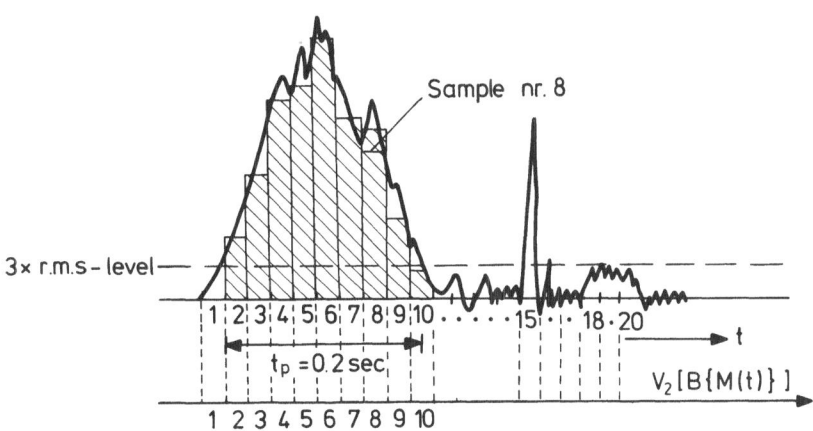

Fig. 61 : Sampling a peak : every peak is divided in e.g. 10
          sampling intervals of about 20 milliseconds. Average
          across every sampling interval is measured. Further
          smoothing of the peak shape is done digitally i.e. via
          mathematics in the computer.

The peak position considered to be determined by the position
of the maximum is defined as lying in that sample where the values of,
for example, 3 samples to the right and to the left are smaller by
a given amount. In Fig. 61 sample number 6 is the maximum; (sample
no. 8 would not be considered the maximum according to this
definition). Peak maxima found in that way or integrals across the
whole peak area are taken as a measure of the intensity I of the
given mass peak.

Relation between peak maximum position and mass number

Once the sample number which corresponds to the peak maximum
has been determined one uses the corresponding voltage $V_2(B(M(t)))$
to calculate via the computer the corresponding mass number, using
the calibration function between mass number and voltage V_2.
Intensity as a function of mass can then be displayed or stored
(cf. fig. 60).

Signal for mass indication

The calculated mass can be displayed in real time also on the
mass spectrometer panel, as indicated in fig. 60.

Other data handling procedures

Determination of a depth profile : the peak height is plotted
as a function of the depth, various smoothing procedures, calibration procedures with standards (I versus absolute fractional concentration), conversion of bombardment time into depth etc. can
also be carried out.

Quantitative analysis : calculation of absolute fractional
concentration of all peaks in the mass spectrometer by means of a
suitable correction programme.

Treatment of element mapping with the computer, i.e. recording
of lines with equal concentration, obtained in ion microscopes.

Detection of Preselected peaks : the need to trace only a few
peaks, for an efficient use of the sputtered material, has already
been discussed. In this mode the mass-defining physical quantity
(magnetic field or RF voltage, V_Q), is switched successively to
different values in order to select and detect different masses in
turn. (Switching of the magnetic field is difficult due to
hysteresis, but in practice one can obtain good results if the peaks
are switched in always the same sequence.)
The handling of these different ion currents $I_i \sim V_i (i = 1,2,3,...n)$
can take place by means of an n-channel U.V.-recorder (analog
recording) or, better, by means of a computer. Two possibilities are
discussed hereafter which are also suited to the handling of any
signal (e.g. E_p, I_p, B, V_Q) and not merely of peak heights.

1) Use of solid state switches (Fig. 62a).

An array of n solid-state, analog switches (or if only a low
switching speed is wanted, reed relais) are activated by the
computer which gives the "command" to close in turn switches
1, 2, 3, 4, ... n. Then the signals $V_1 ... V_n$ can in turn be
transmitted, converted into digital form (ADC), and fed to a
computer. The advantage of this design is, that only one ADC

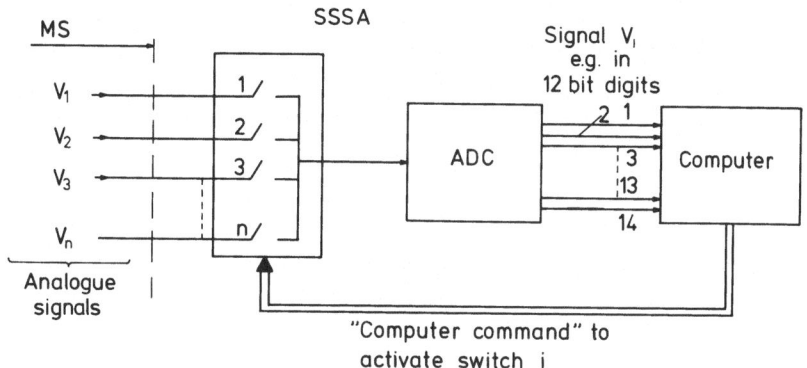

Fig. 62a

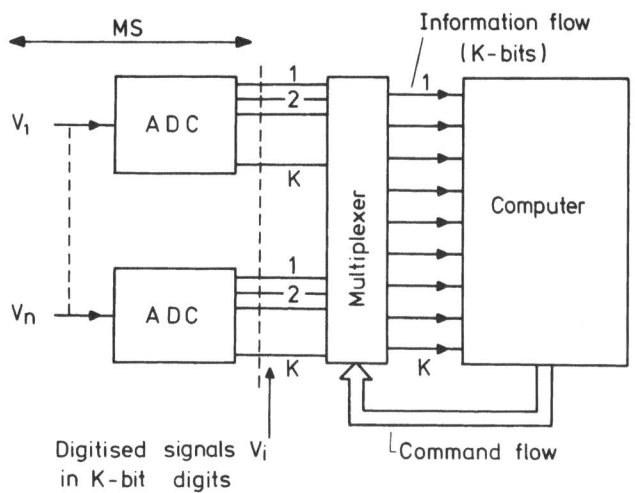

Fig. 62b

Fig. 62 : Handling of n preselected signals (e.g. mass peaks) by
         a computer.

Fig. 62a: Use of analog solid state switch array : $(V_1...V_n)$ =
         analog input signal from the mass spectrometer;
         SSSA = solid state switch array, consisting of n
         switches 1, 2, ... n; information flow is indicated by
         solid arrow; flow of computer command is indicated by
         open arrow.

Fig. 62b: Use of a multiplexor (digital solid state switch array)
         and a computer with only <u>one</u> command line.

unit is needed. This is made possible by the use of <u>many</u>
relatively cheap solid state switches.
2) Use of a <u>multiplexer</u> (Fig. 62b).
        In this case every signal V_i (i = 1, 2, ... n) has its own
ADC which gives a k-bit signal. The computer gives commands to
read out the different items of information, via the <u>command</u>
<u>flow</u>. The multiplexer consists, just as the solid switch array,
of a large number of switches, but in this case <u>digital</u> switches,
which in turn are activated by computer commands.
3) With modern three state output ADC, the use of multiplexers becomes
obsolete (Fig. 62c).

    3.1.4.2 Instrument control
        We have already mentioned (Fig. 60) an information flow from
the computer back to the mass spectrometer, viz. the signal for mass
indication. In that example, however, there was no interaction
between this information flow and the way in which the mass spectra
were obtained, i.e. the mass determining parameters (B, V_Q) were
changed independently of the computer. In modern instrumentation,
however, the mass spectrometer parameters are controlled by a
computer, i.e. the computer gives the command to vary e.g. $V_2(B)$ or
$V_3(V_Q)$ in a given interval, in other words to scan a given mass

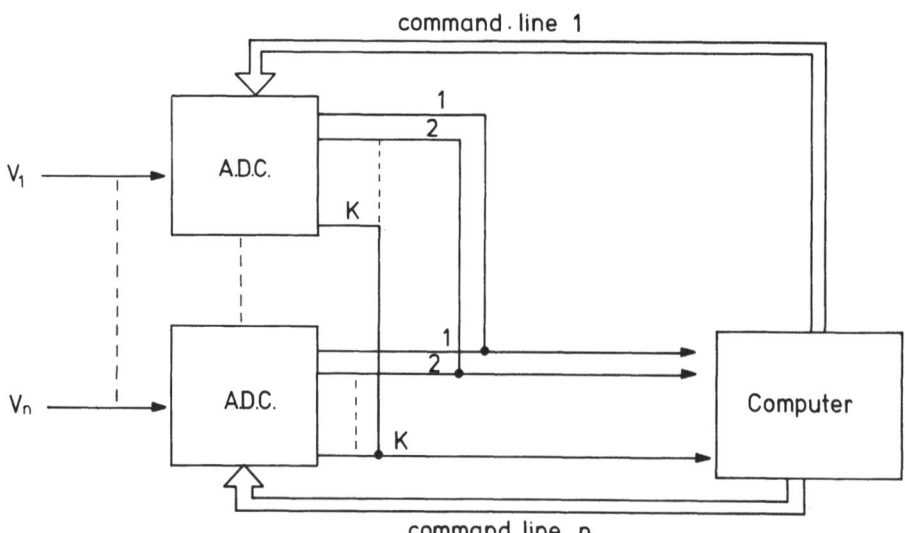

Fig. 62c : Modern handling of n preselected signals with n ADC's,
           one computer and n command lines. Note that every ADC has
           its own command line.

interval of the spectrum. Now the reverse of the process discussed
earlier under data handling occurs, viz.the signals of the
computer (digital) must be converted into analog signals (Fig. 63).

To achieve this, a digital-to-analog converter (DAC) is needed.
In order to check whether the command from the computer has been
carried out correctly (e.g. hysteresis of the magnetic field in the
magnetic sector instrument may "disturb" the correct execution of
the command) a check of the physical quantities must always be
carried out. This is done by computer <u>software</u> (Fig. 64).

Note that in this case it is the computer which, via the
software (fig. 64), finally brings V_2' to the desired value V_2.
The voltages V_2 and V_2' are galvanically (= electronically)
separated, in contrast to an electronic control with differential
amplifier (inset in Fig. 64) where the two voltages V_2 and V_2' are
compared at the input of one and the same differential amplifier
which gives a signal at the output $V_0 = (V_2-V_2')A$. As long as $V_0 \neq 0$
the current through the magnet field-coil is increased. If $V_0 = 0$
the current through the coil remains at the given value.

Instrument control is nowadays used in mass spectrometers
not only for commanding a mass spectrum but also for checking many
other parameters, e.g. primary ion accelerating voltages, ion

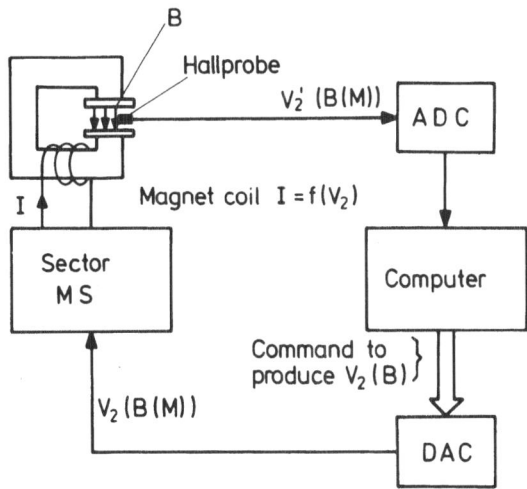

Fig. 63 : Hardware for computer instrument control.

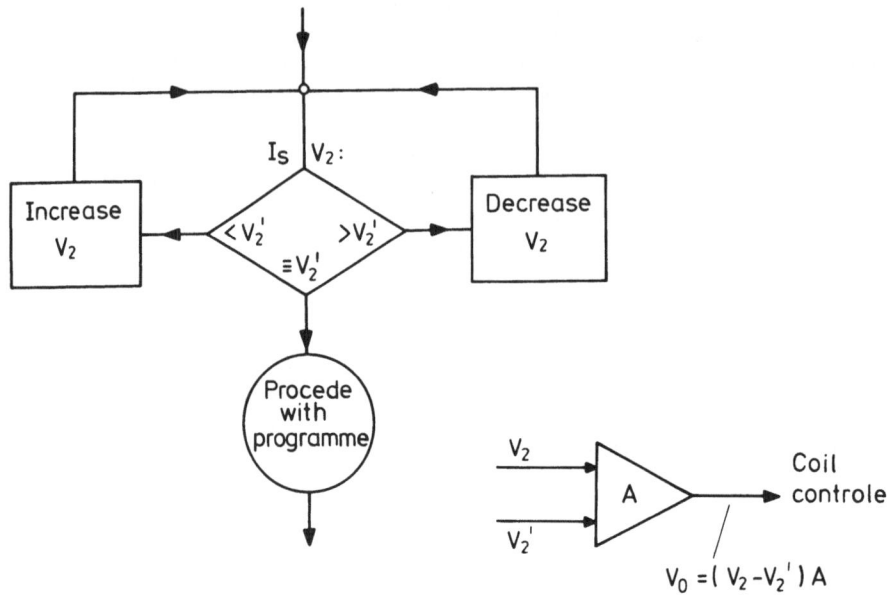

Electronic   controle

Fig. 64 : Software for instrument control as incorporated in a
          computer. Inset : principle of electronic instrument
          control.

currents, or to check whether some of the essential voltages are
excessively high or low, e.g. extreme rise of pressure, leakage
currents etc. Also parameters such as beam diameter or beam position
can be controlled. For this latter it is necessary, however, to have
the optics calibrated first, so that one knows for which voltages
in the ion optics the ion beam has the desired diameter or position
on the sample.

     The saving of time and the ease of operation which can be
obtained with   instrument control will enable the operator to
dedicate himself better to the real analytical problem without the
need to watch trivialities while controlling the instrument.

4. COMPARISON OF SIMS WITH OTHER THIN FILM ANALYTICAL TECHNIQUES

     In the chemical characterization of materials, instrumental
analytical methods are more and more being used as a complement to
classical (wet) chemical methods. The latter are element specific

and give absolute quantitative analysis with good accuracy.

Instrumental methods give qualitative survey analyses, i.e. multi-element detection, although often with widely different sensitivities; with a few exceptions, calibration by wet chemical methods is still necessary for a quantitative analysis.

Classical instrumental methods are spectrochemistry, spark source mass spectrometry, and X-ray fluorescence analysis. Discharge excitation by spark or arc as used in spark source mass spectrometry and spectrochemistry is characterized by the fact that neither the position nor the extent of the excited zone on the sample are exactly predictable.
Therefore this mode has in general been used only for bulk analysis in a special design. Analysis of the outer layers of a bulk ($\approx\mu$m thickness) has been obtained with spark source mass spectrometry.

Glow discharge excitation has also been used for depth profiling recently.

X-ray fluorescence analysis with beam excitation gives information about the composition of the topmost bulk layer. For beam excitation, beams with well defined properties (type of particular energy, beam diameter, etc.) are used. Beams with diameters down to 1 μm (ions), 0.1 μm (electrons), 1 mm (X-rays,UV) are used. By suitable positioning of the beam, a well defined area on any desired spot of the surface can be analysed. Depending on the exciting beam, the sample is more or less perturbed either by sputtering, by bombardment-induced adsorption or desorption, or by generation of defects. Elemental mapping/imaging (determination of distributions across the surface) can be obtained in many cases. Depth profiles of layers with thickness between some tens of Ångstroms and several micrometres can be determined, with the analysed surface area having dimensions of only several microns in optimal cases. A combination of depth profiling and elemental mapping gives localized three dimensional, elemental microspot analysis.

In the following we shall give a brief comparison of SIMS with other thin film analytical techniques, using beam excitation (for more details and references see review articles[143]). Table 14 gives a survey of the different particles used for excitation, emission, and the respective methods.

For an evaluation of some of these methods we shall use the characteristics given in table 15. The properties of the different techniques given there are extreme values, which up to now have only been obtained in specially designed instruments. It is not always possible to obtain these extreme specifications in one single instrument. How far these specifications can be obtained in a practical laboratory or commercial instrument will depend on the design of the instrument.

Table 14 :

Excitation and Emission as used in Thin Film Analytical Techniques

Excitation	Technique	Emission
Photons	XRF, LOES / XPS, (AES)	Photons
Electrons	EMP / AES, EM	Electrons
Ions	BXE, BLE, (AES) / SIMS, LEIS, MEIS, HEIS, (GDMS, SNMS)	Ions

XRF	:	X-ray fluorescence spectroscopy
LOES	:	Laser optical emission spectroscopy
XPS	:	X-ray induced photo-electron spectroscopy = ESCA : electron spectroscopy for chemical analysis
UPS	:	Ultraviolet photo-electron spectroscopy = PES : photo-electron spectroscopy
AES	:	Auger electron spectroscopy
EMP	:	Electron microprobe analysis
EM	:	Electron microscopy
H(L)EED	:	High (low) energy electron diffraction
BXE	:	Bombardment induced X-ray emission spectroscopy = IEX : ion excited X-ray spectroscopy
BLE	:	Bombardment induced light emission = SCANIIR : surface composition by analysis of neutral and ion impact radiation = IBSCA : ion beam spectrochemical analysis
SIMS	:	Secondary (sputter) ion mass spectrometry = IMMA : ion microprobe mass analysis
LEIS	:	Low energy ($\leq$10 keV) ion scattering spectroscopy = ISS : ion scattering spectroscopy = NIRMS : noble-gas-ion reflection mass spectroscopy = NODUS : non-destructive and ultra-sensitive single atomic layer surface spectroscopy
MEIS	:	Medium energy (20-500 keV) ion scattering spectroscopy
HEIS	:	High energy (0.5-3 MeV) ion back scattering = RBS : Rutherford back scattering spectroscopy
GDMS	:	Glow discharge mass spectrometry = SNMS : sputtered neutral mass spectrometry

Table 15 :

	AES	ESCA (XPS)	LEIS (ISS)	SIMS	BLE	HEIS
Element range	$\gg$ Li	$\gg$ He	$\gg$ H	$\gg$ H	$\gg$ H (except el. neg. el.)	$\gg$ C (high Z on low Z)
Quantit. analysis	Semiquantit. directly from peak height. Quantit. with standard or correction model			Standard or correction model necessary		Quantit. from peak height
Average detection limit (atomic)	0.1%	1%	0.1%	1 ppm	1 ppm	1% (surface) 10^{-4} (bulk)
Information depth	3 – 25Å	10 – 30Å	3 – 10Å	3 – 20Å	3 – 20Å	100Å – 2 μm
Lateral resolution	10^3Å (500Å)	1 mm	100 μm	1 μm (0.1 μm)	100 μm (1 μm)	100 μm (1 μm)
Compound inform.	(Restricted)	Yes	No	Yes	Yes	No
Structural inform.	No	No	Yes	No	No	Yes
Organic samples	No	Yes	No	Yes	No	No
Destructiveness	Small	No	Small	Destruction (sputtering) essential for method		Small
Insul. analysis	Charge compens.	Easy	Possible with charge compensation			Easy

1) <u>The elemental range</u> (line 1 of table 15) which is important
for a survey analysis (qualitative analysis and a rough estimate
of the concentration) is limited with AES and XPS by the excitation
mechanism. SIMS and to a less degree, BLE are the only methods
which can detect also hydrogen. HEIS allows elementary analysis
of C and heavier elements, with the restriction that the heavier
elements should preferably be situated at the surface to avoid
interference between high and low set element peaks. By means of
channelling techniques in single crystals the latter effect can be
partially overcome.

2) <u>Quantitative analysis</u> is the determination of the elementary
concentration from the measured signal, with accuracies $\leq$ 20 %. In
AES and ESCA semi-quantitative analysis (accuracies factor 2-3)
direct from the peak height is possible, by the use of a correction
model or of working curves (obtained from a calibration sample)
quantitative analysis is possible.

Quantitative LEIS analysis of the topmost layer from the
measured signal is possible without the use of standards or
correction models[144].

SIMS and BLE show wide variation of the sensitivity (up to a
factor 1000) from element to element and from matrix to matrix.
By the use of correction models or calibration curves semiquantitative and even quantitative analysis becomes possible. In contrast
to the other methods, quantitative analysis directly from the peak
heights is not possible . HEIS allows quantitative analysis
directly from the measured spectrum without the use of standards,
because values of scattering cross-section and stopping power are
available in tables, which allows one to calculate respectively
the concentration of the elements and the depth scale by means of
the measured data.

3) <u>Average limit of detection</u>. Here, SIMS and BLE are the favorites.
The values given in line 3 are average values. With SIMS, for
example, a detection limit of 1 ppb is possible in many cases
(elements with low ionisation energy), while other elements
depending on the experimental conditions, may have detection limits
of up to 100 ppm. Comparison of the different methods with regard
to their detection limits is in principle not possible, because the
material consumption is strongly dependent on the method. The
values given in line 3 are typical for a practical analysis of
thin films with average material consumption.

4) <u>Information depth</u> is an important quantity because the depth
resolution achieved in depth profiling by means of sputtering is
composed of information depths and other factors. When <u>estimating</u>
the depth distribution without sputtering (in AES and ESCA from a
comparison of the intensities of lines with different energies) the
information depth is the determining factor. LEIS and static SIMS
are the only methods which allow of a real monolayer analysis.
Material consumption by sputtering is less than one monolayer/hour.
HEIS gives simultaneous information about element concentration

from the surface to the penetration depth   (about a few micrometres)
without the use of sputtering.

5) Lateral resolution. When the scanning mode (microprobe) is used,
lateral resolution is of importance in element mapping (determination
of the element distribution across the surface). Also in microspot
analysis, i.e. analysis of small zones e.g. inclusions, local
inhomogeneties on the surface good lateral resolution is needed.

With AES a resolution of 10^3 Å is obtained with the help of a
fine beam from a thermionic cathode. A lower limit is given by the
(too small) electron current density. Recent developments using
field emission cathodes seem to push the lateral resolution towards
500 Å and even lower; power dissipation may then give problems.

ESCA has a 1 mm lateral resolution, because X-ray beams cannot
be focused to a smaller value. The approach of Hovland[145] who has
reported on a transmission-scanning ESCA instrument, opens new
possibilities to obtain better lateral resolution in this mode with
ESCA. LEIS : element mapping with beam diameters smaller than 100 μm
gives difficulties due to intensity loss (sensitivity is proportional
to the product of ion current density and bombarded area). The ion
current density should not be chosen too high, because of sputtering.

SIMS :   1 μm has already been obtained. Attempts to use a
field emission cathode with high ion current density give beam
diameters of 0.1 μm. Difficulties were encountered in the finding
of a compromise between erosion rate (proportional to ion current
density) and sensitivity (proportional to ion current density and
bombarded area), as discussed under LEIS. The same holds for BLE.
HEIS : at the moment beams of 100 μm down to about 1 μm are possible
with a good ion optics.

Lateral resolution of 1 μm has been achieved with SIMS and BLE.
Resolutions < 1 μm will not easily be obtained, due to intensity
loss and also because image aberations (mainly chromatic aberations)
cannot be completely neglected.

6) Compound information, i.e. information about a chemical state
(compound) of an element, can be expected from methods which are
matrix sensitive : this holds to a limited degree for AES (peak
shift and peak distortion for carbides, for instance). ESCA is the
only method which can be used for compound information, because
the peak position is directly influenced by a change of the binding
energy. SIMS gives compound information via the fingerprint
spectrum. BLE shows different peaks for elements and compounds. HEIS
and LEIS, which are based on binary collisions, give no compound
information.

7) Structural information, i.e. information on the position of atoms
in the lattice or on the surface, are obtained with respectively
HEIS and LEIS. The other methods are not sensitive to those
structural differences.

8) Only ESCA allows the analysis of organic samples with respect
to their composition and structure. SIMS gives information about
the structure (from the fragment pattern) and about the molecular

weight (from the protonated parent ion).
9) <u>Destructiveness</u>. In principle every measurement of a physical
system is destructive. The degree of destructiveness, however, is
different for the different methods.
Bombardment with X-rays in ESCA gives the least destruction, also
the compositions of organic samples under intense X-ray beams are
known.
        Electron bombardment in AES yields adsorption and desorption
phenomena; at higher electron current densities (scanning AES)
severe thermal effects occur in addition.
        In LEIS, only small sputtering occurs, due to the small ion
energy and dose.
        SIMS and BLE are based on a sputtering process and are
essentially destructive. Moreover they give a change of the target
by knock-on effects, selective sputtering, and implantation.
        In HEIS no sputtering occurs; however, lattice damage and the
implantation of ions can disturb the original state of the sample.
10) <u>Analysis of insulators</u> is nowadays possible with all techniques
using designs for charge compensation. The latter results from accumulation of primary particles (if charged) and   secondary electron
emission. ESCA shows the least charging, as found in practice.
Charge compensation by positive ions has proved succesful in
AES and so has the use of auxiliary ion or electron beams in SIMS.
BLE is the only technique where the emitted signal is
not influenced by charging. LEIS shows little charging, because
the primary ion current is very low.
        HEIS is relatively insensitive to charging, as long as it is
small with respect to the energy of the ions.

Conclusion

        SIMS appears to be a method which can be used for the most
divers problems. This does not mean, however, that SIMS is the
only technique and that is has no disadvantages. On the contrary,
all methods, including SIMS, have advantages and disadvantages.
The different methods complement each other, however, and should
be used in a good analytical laboratory either sequentially or,
if possible, simultaneously to solve research or analytical
problems. Modern instrumentation is already available in the
modular form, i.e. combining several techniques in one instrument.
These instruments have advantages when, for example, it is
necessary to investigate samples which should not be exposed to
air in order to avoid contamination, or when microspot analysis is
required. For routine analysis in an analytical laboratory with a
relatively large sample throughput, modular techniques may have
the disadvantage that the individual methods are not used to the
full. It can be expected that the use of all these modern techniques will further extend our fundamental and practical knowledge of solids, so that we can cope the better with future
challenges to the material sciences.

Appendix I

Determination of depth profiles[75]

As an illustration for the conversion of measured ion current as a function of bombardment time, into concentration as a function of depth (quantitation of depth profile) profiles implanted in single crystals are best suited, because the measured profile can be compared with the theoretical profile known both from theory (range distribution) and from experiment (implanted particle dose, type and energy). Moreover, artifacts e.g. cones and ridges do not build up on single crystals.

In order to measure an implantation profile, the sample mounted in the target chamber of the mass spectrometer is sputter-etched by continuous bombardment with an ion beam. The mass spectrometer is tuned to the required element whose ion current I_S is then recorded as a function of the bombardment time t (Fig. I,1). This then provides a measure of the depth profile of the element under consideration, provided that the element is homogeneously distributed in planes parallel to the original surface. If this condition is not fulfilled, wrong interpretations may result as illustrated in the following.

Fig. I,2 shows the Na^+ ion current as a function of time obtained from an aluminium target. Assuming a homogeneous lateral distribution, this curve would be interpreted as the Na concentration profile in Al. SIMS has been used to check the validity of this assumption. From figure I,3a, showing the Na distribution across the surface, one can see that Na is present only on certain parts of the target surface. Fig. I,3b taken at $t=t_2$ explains the decrease in the Na^+ peak : the area covered by Na is diminished, but the intensity of the Na line on the remaining spots stays the same as in the beginning. At $t=t_3$, only a faint Na^+ image is still visible. The decrease of the Na^+ peak in Fig. I,2 must therefore be interpreted as a decrease in the Na-covered area - the remaining spots emitting the same secondary ion current densities as in the beginning - and not as a diffusion of Na in the sample. This example illustrates the support which the imaging mode can provide for the general SIMS method in analysis of thin layers.

Conversion of bombardment time t into depth z

Implanted layers in general have homogeneous distribution of the implants parallel to the surface, so the above mentioned effects does not occur. We can then easily convert bombardment time into depth z provided that the erosion rate $\dot{z}$ i.e. the layer thickness sputtered per unit time is constant :

$$dz/dt = \dot{z} = constant \qquad (1)$$

The layer thickness sputtered after a given bombardment time is then found by integration of (1) as :

$$\int_0^t dz = z(t) = \int_0^t \dot{z}\,dt = \dot{z}t \qquad (2)$$

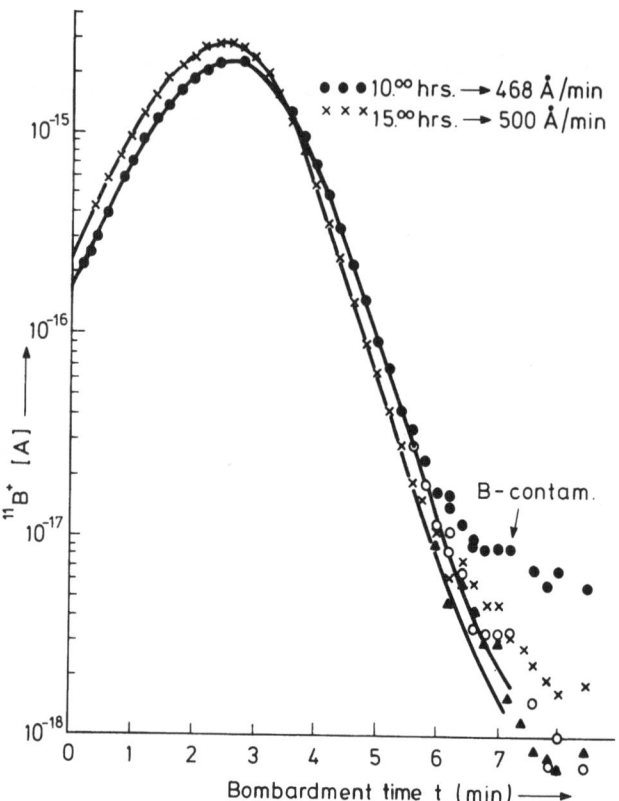

Fig. I,1 : B⁺ ion current (I_s) versus bombardment time[75].
Experiment started 30.9.1975 at 10.00 h.: ●●● as measured, ooo after subtraction of final level.
Experiment repeated at 15.00 h.: xxx as measured, ▲▲▲
after subtraction of final level.
Sample Si (414B), ¹¹B-implantation : 10^{15} cm⁻², 30 keV.
Primary ion current I_p = 1.5 μA (O_2^+, no oxygen gas ad-
mitted), bombardment area A_b = 1.2 x 1.2 mm², erosion
rate $\dot{z}$ = 468 Å.min⁻¹.
15.00 h.: the same sample as at 10.00 h., same conditions
of mass spectrometer, erosion rate $\dot{z}$ = 500 Å.min⁻¹.

The erosion rate $\dot{z}$ can be determined from equation (2) by bombarding a sample for a given time t = T and measuring the total
depth z(T) by means of e.g. a Talysurf, an interferometer or by
the variation in colour fringes. Knowing the erosion rate and the
bombardment time, then this time can be converted into a depth by
means of eq. (2). Another method for estimating the depth scale is
to use the implanted profile, or the profile of other implanted

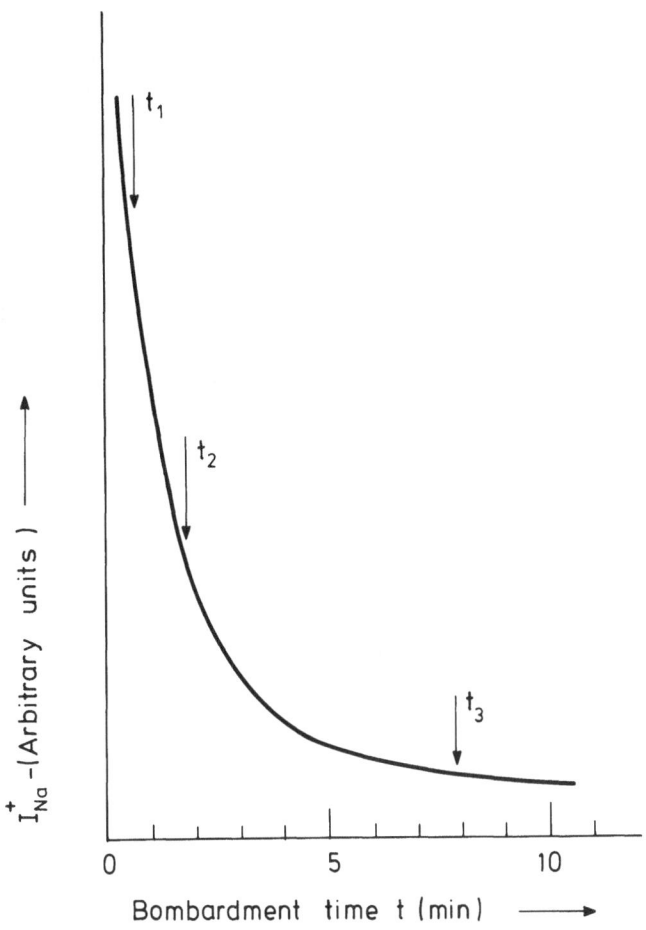

Fig. I,2 : Na$^+$-ion current as a function of time obtained by SIMS[75].
Na$^+$ distribution due to pressing a finger on an Al target.

species, as depth markers via theoretical relations between the
energy of the implanted ion and its range in a given medium,
provided diffusion and/or channeling does not occur.
The erosion rate can also be determined by sputtering samples of
known thickness. If the sputter yield S is known, the erosion rate
$\dot{z}$ (m.min^{-1}) can be estimated by means of the following formula :

$$\dot{z} = 6 \times 10^{-6} M j_p S / \rho \qquad (3)$$

where : M = mass number of the target atoms,
        ρ = density of the target material (g.cm^{-3}),
        j_p = primary ion current density (A.cm^{-2}) = primary ion

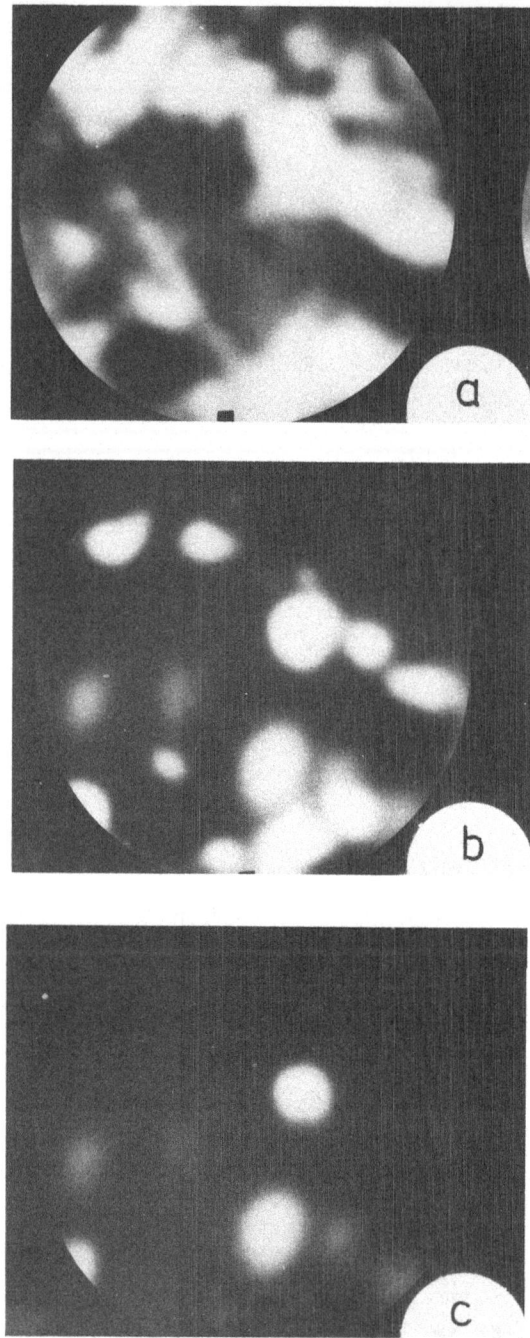

Fig. I,3 : Na$^+$-ion image from the same sample as in figure I,2
obtained : a: at time t = t$_1$, b: at time t = t$_2$, c : at
time t = t$_3$[75)].

current I_p/bombarded area (A_b),

S = sputter yield, depending on the target material and on mass, energy and angle of incidence of the projectile ion.

Equation (3) shows that, within given limits, the erosion rate can be selected by proper choice of the primary ion current density. The value of the erosion rate is dictated by the total depth to be analysed and the desire to keep the total time for a measurement below about half an hour. For a given vacuum the erosion rate must be large enough, however, to ensure that more particles per time unit are sputtered away than are adsorbed from the residual gas atmosphere.

The experimental conditions which led to figure 1 were as follows : primary ion current I_p = 1.5 μA (5.5 keV O_2^+ ions, no oxygen bleed-in) bombarded area A_b = 1.2x1.2 mm^2, i.e. :

$$I_p/A_b = j_p = 10^{-4} \text{ A.cm}^{-2}$$

With values for a silicon target of :

$$M = 28, \ \rho = 2.42 \text{ g.cm}^{-3} \text{ and } S = 6,$$

we find from eq. (3)

$$\dot{z} \approx 420 \times 10^{-10} \text{m/min} = 420 \text{ Å/min}.$$

This calculated value is of the same order of magnitude as the one found experimentally, using a Talysurf :

$$\dot{z} \approx 500 \times 10^{-10} \text{ m/min}.$$

Conversion of ion currents into concentration

The secondary ion current I_s is given by the product of absolute ion yield S^+ and concentration c_M of the detected element, and instrumental parameters. The dependence of S^+ on element and matrix and procedures for quantitative survey analysis have been discussed earlier in this paper.

For the quantitative analysis of implanted layers one is usually interested to determine the concentration of only a few elements (e.g. B, As, P) implanted in e.g. Si. A quantitative analysis (accuracy a few %) is then carried out by calibration of the SIMS signal with samples of known B content. This method we have found the most useful and most accurate during our experience in the last few years in which we have measured about 1000 implanted profiles.

The SIMS signal can be calibrated against signals obtained from the same sample with Rutherford back scattering, by classical chemical methods or by collecting a constant fraction of the sputtered material on an inert electrode and analysing the deposited material with classical chemical methods.

Determination of the absolute concentration profile using standard samples with known homogeneous concentration

For the conversion of ion current into concentration it is necessary to modify the standard formula of SIMS :

$$I^+ = j_p A_b S \beta^+ f c,$$ (4)

where I^+ = secondary ion current,
   j_p = primary ion current density,
   A_b = bombarded area,
   S = sputter yield,
   β^+ = degree of ionization,
   f = transmission of the instrument,
   c = fractional concentration = η/η_o,
   η = absolute concentration = number of atoms of an element
        which are actually present per cm^3 and η_o = maximum
        possible number of atoms of an element per cm^3.

The ion current measured in a time interval dt is then given by

$$I^+(t)dt = j_p A_b S \beta^+ f(1/\eta_o)\eta(t)dt = c'\eta(t)dt$$ (5)

With the transformation t = kz and the relation dt = kdz,
k = dt/dz = $1/\dot{z}$ we obtain from (5) :

$$I^+(t)dt = c' k \eta z(t)dz = (c'/\dot{z})\eta(z)dz$$ (6)

For the instantaneous value I(t) of the ion current one obtains
from (6) :

$$I^+(t) = c'\eta(z)$$ (7)

The integral of the ion current measured during the total bombardment
time (T) is obtained from (6) - keeping in mind that $\dot{z}$ = const.
and Z = the total layer thickness sputtered away in the time T-,
to be :

$$\int_o^T I(t)dt = (c'/\dot{z}) \int_o^Z \eta(z)dt$$ (8)

From (8) one finds :

$$\frac{\int I(t)dt}{\int \eta(z)dz} \dot{z} = c'$$ (9)

In order to determine c' one measures the ion current I_c from a
calibration sample with known, homogeneous distribution
$(\eta(z) = \eta_c$ = const.$)$ of the element under consideration. Obviously
then :

$$\int \eta_c dz = \eta_c Z,$$ (10)

and

$$\int I_c(t)dt = I_c T.$$ (11)

Having determined $\dot{z}$ as described above, eqs. (9)-(11) give,
keeping in mind that $Z/T = \dot{z}$, :

$$c' = \frac{I_c T}{\eta c z} \quad z = \frac{I_c(A)}{\eta_c(cm^{-3})} .$$

(12)

Conversion of I^+ into absolute concentration η can then be carried out by means of eq. (7).

Example :
A sample had been doped homogeneously with ^{11}B, the concentration of which had been determined by wet chemical analysis as $\eta_c = 1.3 \times 10^{19}$ cm^{-3}. The following calibration currents were measured :

a) at 10 h : $I_c = 2. \times 10^{-16}$ A, giving $c' = 1.54 \times 10^{-35}$ A.cm^3. Keeping the adjustment of the mass spectrometer the same, a boron profile was measured (see fig.I,1). Inserting the above found c' value in formula (7) a measured ion current of $I = 10^{-16}$ A was found to correspond to a concentration of $\eta = 6.5 \times 10^{18}$ cm^{-3}. This value, just as the value determined at 15 h, shows a systematic deviation by about a factor 1.5 from the one using the method as described in the next section.

b) At 15 h we have measured a ^{11}B current of the calibration sample : $I_c = 2.4 \times 10^{-16}$A, giving $C = 1.85 \times 10^{-35}$ A.cm^3, i.e. (see eq. (7)) a current of 10^{-16} A corresponds to a concentration of $\eta = 5.4 \times 10^{18}$ cm^{-3}.

Determination of the absolute concentration profile from a given total implantation dose
If for a sample the total implanted dose n(cm^{-2}) is known, conversion from ion current into concentration can be obtained in the following way : if one considers that in (9), the integral

$$\int \eta(z) \, dz$$

must be equal to the total number of particles (n) which have penetrated through 1 cm^2 of the surface, one obtains :

$$c' = \frac{\int I(t)dt(A.s)}{n(cm^{-2})} \quad \dot{z}(cm.s^{-1})$$

(13)

For the profile as shown in fig.I,1, we have determined :
a) at 10 h : $\dot{z} = 460$ Å.min$^{-1} = 7.67 \times 10^{-8}$ cm.s^{-1}.
From the measured curve we have determined the integral
$$\int_0^T I(t) \, dt, \text{ to be } 3.1 \times 10^{-19} A.s.$$

The total implanted dose has been determined during the implantation as n = 10^{15} cm^{-2}. From (13) one finds $c' = 2.38 \times 10^{-35}$A.cm^3. From (9) one finds with this value that a measured boron current of $I = 10^{-16}$A corresponds to a concentration of $\eta = 4.2 \times 10^{18}$ cm^{-3}.

b) For the measurement carried out at 15 h (see fig. I,1) we have determined : $\dot{z} = 500$Å min$^{-1} = 8.33 \times 10^{-8}$ cm s^{-1} and hence

$C'=3.1 \times 10^{-35}$ A.cm^3, giving the relation that a measured ion
current $I = 10^{-16}$A corresponds to a concentration of $\eta = 3.2 \times 10^{18}$cm^{-3}.
The systematic deviation of the values obtained with the 2 methods
we ascribe to a systematic effect in the wet chemical determination of η_c and in the determination of n in the implantation
machine. The values for η_c used in fig. I,4 are the ones determined
from the second method.

Determination of the total implanted dose from SIMS measurements

For a given implantation profile the total implantation dose can be
calculated, once C' and z have been determined from calibration
samples and Talysurf-crater measurement as described above, and
if $\int I(t)dt$ has been determined from the measured curve.
Obviously, rearrangement of (13) gives :

$$n = C' / \dot{z} \int I(t)dt \qquad\qquad (14)$$

The ion current versus bombardment time of figure I,1, converted
into a concentration versus depth profile is shown in figure I,4.
By comparison of this with figure I,1 one can see that the
deviations in the measured profiles $I_s(t)$, due to slight
variations of the primary ion current, are completely eliminated
by calibrating the scales of ion current and bombardment time,
resulting in coincident curves.

Another interesting effect is apparent from the tail in figure I,1:
the measured B current ends in horizontal lines of different
levels (5×10^{-18} and 1.7×10^{-18}A respectively). Two effects can be
made responsible for this result :

1) In the days preceding this measurement, we had measured glass
with relatively high B concentration. The boron, together with other
elements had been sputter-deposited onto the extraction electrode
situated approximately 5 mm opposite the sample. During the bombardment these deposited layers are hit by high energetic particles,
sputtered from the sample. As a consequence boron is resputtered
from the extraction electrode to the target and is measured
(memory effect). In the course of the measurement (between 10.00 h
and 15.00 h) other measurements on silicon samples had been carried
out; silicon was sputter-deposited on the extraction electrode and
the boron containing deposit from the last days was burried.
Therefore, the boron contamination in the signal decreased to a
value of 1.7×10^{-18}A. This shows how sputter deposition can reduce
memory effects.

2) The remaining final and constant value of the boron signal can
be ascribed to redeposition. In spite of all efforts to reduce
this disturbing effect a small but constant fraction of the target
is sputtered from the edge of the crater (see fig. 25) into that
central part of the crater. Since the boron concentration at the
edge of the crater may be some orders of magnitude larger than in
the center, this contribution will be noticeable the more, the
smaller the boron concentration becomes with increasing depth.
Subtraction of this constant contribution of the measured values

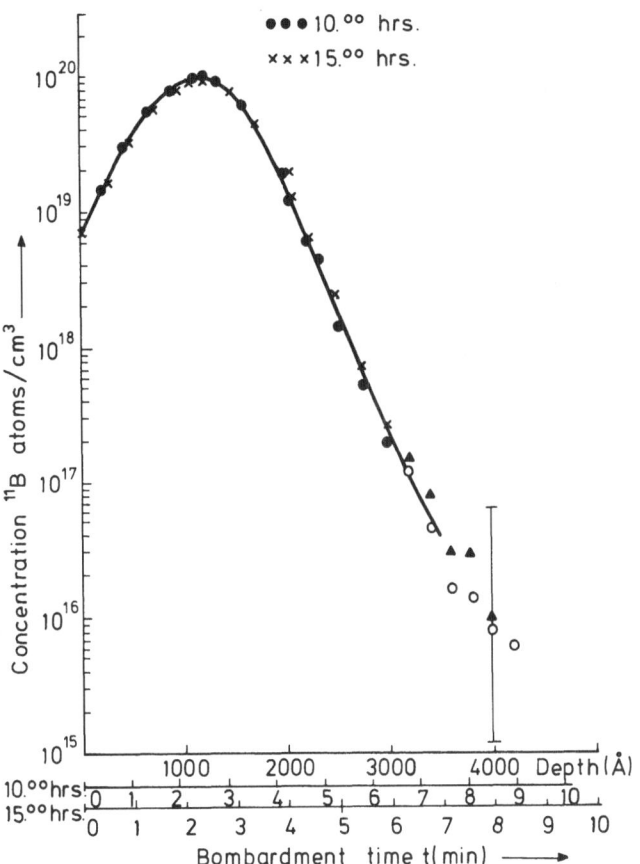

Fig. I,4 : Absolute B concentration versus depth obtained from
figure I,1 by conversion of I into c and t into z[75].
Absolute concentration found at 10.00 h.: ●●● as measured, ooo after subtraction of final level; 15.00 h.:
xxx as measured, ▲▲▲ after subtraction of final level.

gives the points indicated with circles and triangles respectively.
This final value of the boron signals can be attributed to redeposition and mass interference effects.
Studies of As and BF_2 implantations have been published elsewhere[75].
Automatic data handling for depth profiling
Some years ago we have started to handle our SIMS data by means of
a computer. This was found to be necessary due to the growing
number of in-depth analyses to be carried out with SIMS. Fig. I,5
is a plot of the profile obtained with one of the possible modes
which we have now at our disposal : the measured ion current as a
function of time is fed to three channels of different sensitivity.

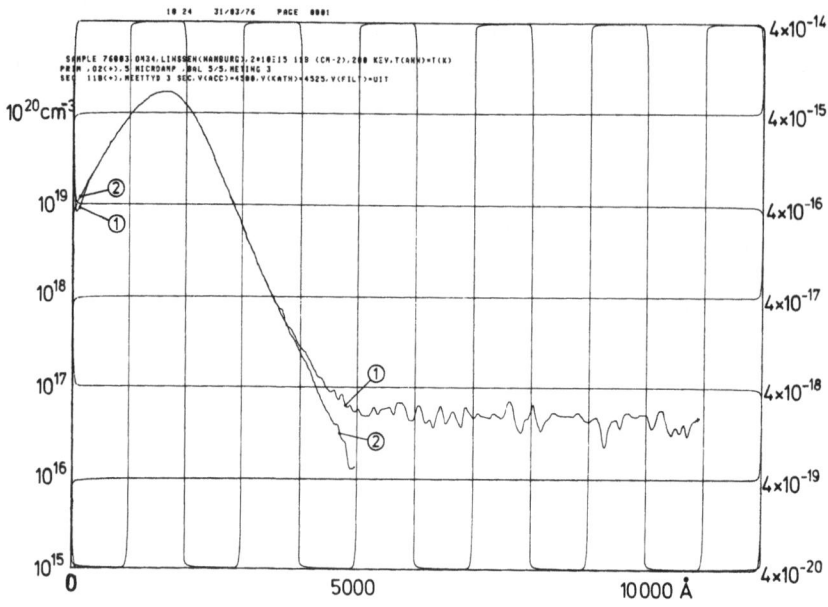

Fig. I,5 : Curves of computer output for a measured B profile in
Si[75].
Curve 1 : Raw data (use right ordinate = secondary ion
current in A). The constant B background (approximately
5×10^{16} cm^{-3}) is a memory effect from several B profiles
measured previously.
Curve 2 : Concentration as found after conversion, and
background plus front correction (use left ordinate
= concentration of ^{11}B atoms.cm^{-3}).

The computer automatically chooses that channel with the appropriate sensitivity so that a total dynamic range of up to 10^5 is
obtained.

Appendix II.
Nomogram for the determination of the detection limit c_{min}.

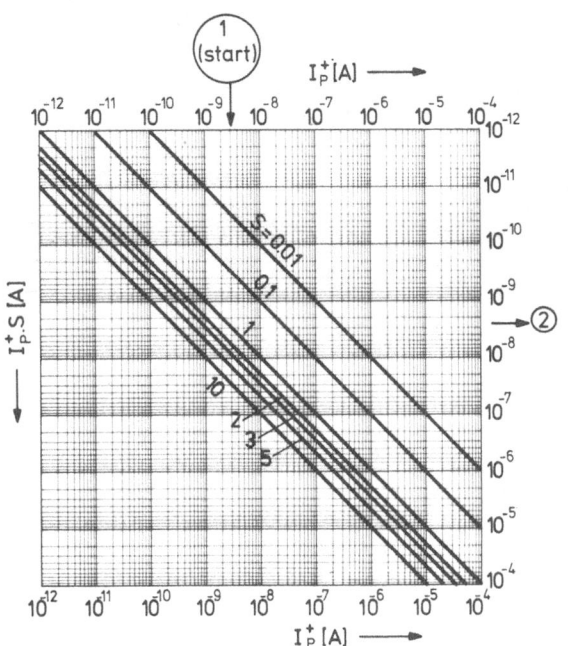

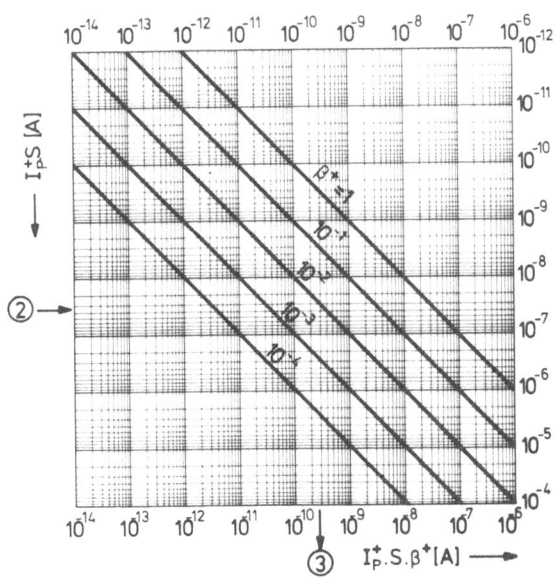

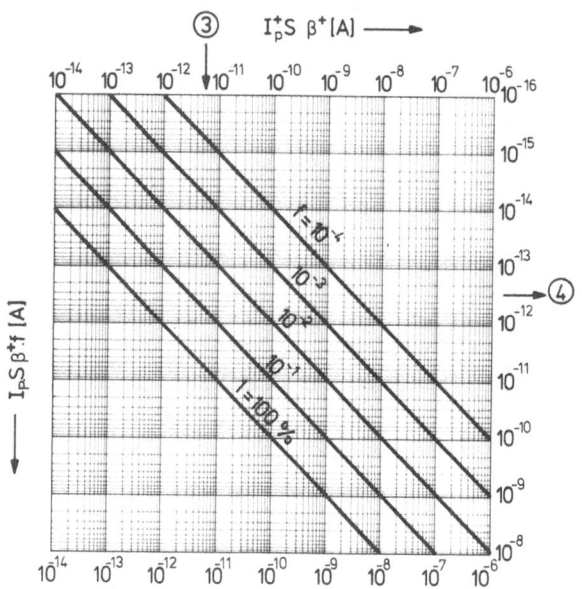

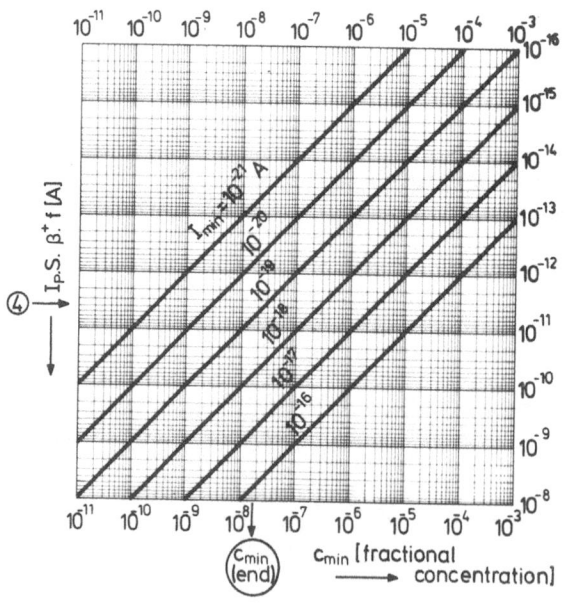

Acknowledgement

The author wishes to thank Dr. N. WARMOLTZ and Mr. H. WIELINK,
as well as the members of the typing and drawing offices, in hel-
ping with great enthousiasm to compose this manuscript in a ready-
for-print version.

References

1) F.L. Arnot and J.C. Milligan, Proc. Roy. Soc. Ser. A, 156,(1936) 538.
2) R.F.K. Herzog and F.P. Viehböck, Phys. Rev., 76 (1949) 855.
3) V.I. Veksler and M.B. Ben'iaminovich, Sov. Phys. Tech. Phys., 1 (1957) 1626.
4) V.E. Krohn, Jr., J.Appl. Phys., 33 (1962) 3523,A.Smith et al. J.Appl.Phys.,34(1963)2489. H.E.Beske, Z.Angew.Phys.,14(1962)30. V. Walther and H. Hintenberger, Z. Naturforsch. A, 18 (1963) 843. A. Benninghoven, Ann. Phys., 15 (1965) 113. J.A. McHugh and J.C. Sheffield, J. Appl. Phys., 35 (1964) 512.
5) R. Castaing, B. Jouffrey and G. Slodzian, C.R. Acad. Sci., 251 (1960) 1010. H.J. Liebl and R.F.K. Herzog, J. Appl. Phys., 34 (1963) 2893. H.J. Liebl, J. Appl. Phys., 38 (1967) 5277.
6) H.W. Werner, Dev. Appl. Spectroscopy, 7A Eds. E.L. Grove and A.J. Perkins, Plenum, New York (1969) 239.
7) H.W. Werner und H.A.M. de Grefte, Vakuum-Technik, 17 (1967) 37.
8) G. Carter and J.S. Colligon, Ion Bombardment of Solids, Heinemann Educational Books Ltd, London (1968).
9) H.W. Werner and H.A.M. de Grefte, Surface Sci., 35 (1973) 458.
10) A. Benninghoven, Surface Sci., 28 (1971) 541. A. Benninghoven, Z. Physik, 230 (1970) 403.
11) A. Benninghoven, D. Jaspers and W. Sichtermann, Appl. Phys., 11 (1976) 35.
12) H.W. Werner, Paper to be presented at 7th Intern. Vac. Congr. & 3rd Intern Conf. Solid Surfaces (Vienna 1977), to be published in Proc.
13) W. v.d. Weg, in Proceedings NATO Summer School on Material Characterization, Corsica, 1976 (Plenum Press).
14) F.G. Rüdenauer, Round table discussion on SIMS, 7th Intern. Mass Spectrom. Conf., (Florence 1976), in press.
15) P. Joyes, J. Physique, 29 (1968) 774. P. Joyes, J. Physique, 30 (1969) 243. P. Joyes, J. Physique, 30 (1969) 365.
16) G. Blaise and G. Slodzian, J. Physique, 31 (1970) 93. G. Blaise and G. Slodzian, J. Physique, 35 (1974) 237. G. Blaise and G. Slodzian, J. Physique, 35 (1974) 243.
17) Z. Sroubek, Surface Sci., 44 (1974) 47.
18) J.M. Schroeer, T.N. Rhodin and R.C. Bradley, Surface Sci., 34 (1973) 571.
19) W.H. Gries and F.G. Rüdenauer, Int. J. Mass Spectrom. Ion Phys., 18 (1975) 111.
20) P. Sigmund, Phys. Rev., 184 (1969) 383.
21) M. Cini, Surface Sci., 54 (1976) 71.
22) J. Antal, Phys. Lett., 55A (1976) 281.
23) G. Slodzian and J.F. Hennequin, CR Acad. Sci., 263 (1966)B1246.
24) A. Benninghoven, Z. Naturforsch., 22A (1967) 841.
25) R. Castaing and G. Slodzian, CR Acad. Sci., 255 (1962) 1893.

26) A. Benninghoven, Z. Physik, 220 (1969) 159.
27) C.A. Andersen and J.R. Hinthorne, Anal. Chem., 45 (1973) 1421.
28) H.W. Werner, Paper presented at the joint Japan-US Seminar on "Quantitative SIMS", Honolulu, Oct. 13-17, 1975.
29) F.G. Rüdenauer, W. Steiger and H.W. Werner, Surface Sci., 54 (1976) 553.
30) A.E. Morgan and H.W. Werner, Anal. Chem., 49 (1977) 927.
31) D.S. Simons, J.E. Baker and C.A. Evans, Jr.,Analyt. Chem., 48 (1976) 1341.
32) D.E. Newbury, Paper Nr. 234, Pittsburgh Meeting on Anal. Chem., Cleveland 1977.
33) R. Shimuzu, T. Ishitani and Y. Ueshima, Japan J. Appl. Phys., 13 (1974) 249.
34) F.G. Rüdenauer, W. Steiger, Vacuum 26 (1976) 537.
35) A.E. Morgan and H.W. Werner, Spectrochimica Acta B, (1977) in press. See also ref. 30.
36) A. Lodding, J.M. Gourgout, L.G. Petersson and G. Frostell, Z. Naturforsch., 29A (1974) 897.
37) G.R. Sparrow, Paper Nr. 348, Pittsburgh Meeting on Anal. Chem., 1977.
38) A.E. Morgan and H.W. Werner, Analytical Chem., 48 (1976) 699.
39) C. Plog, Thesis Univ. Münster, 1974.
40) H.W. Werner, H.A.M. de Grefte and J. van den Berg, Advances in Mass Spectrometry (Ed A.R. West) Vol. 6, p. 673. Applied Science Publishers, Barking, Essex (1974).
41) Z. Jurela, Int. J. Mass Spectr. Ion Phys., 12 (1973) 33.
42) H. Oechsner, Appl. Phys., 8 (1975) 185.
43) P. Sigmund, Rev. Roum. Phys., 17 (1972) 823.
44) R. Kelly in: Ion Surface Interactions,Gordon & Breach, R. Behrisch, W. Heiland Eds. London 1973.
45) J.A. McHugh, Methods of Surface Analysis (Ed A.W. Czanderna) Vol. 1 of Methods and Phenomena, p. 223. Elsevier Scientific Publishing Company, Amsterdam,Oxford, New York (1975).
46) A. Benninghoven and A. Müller, Surf. Sci. 39 (1973), 416.
47) P. Vallerand, Thesis, University Quebec, 1976.
48) H.W. Werner, Vacuum, 22 (1972) 613.
49) F.G. Rüdenauer, Int. J. Mass Spectrom. Ion Phys., 6 (1971) 309.
    H. Liebl, Int. J. Mass Spectrom. Ion Phys., 6 (1971) 401.
    G. Slodzian, Workshop on SIMS and IMMA, NBS Spec. Publ. 427, Gaithersburg, Md., (Eds K.F.J. Heinrich and D.E. Newbury) (1975) 33.
50) J.M. Morabito and R.K. Lewis, Anal. Chem., 45 (1973) 869.
51) H. Liebl, J. Phys. E: Sci. Instrum., 8 (1975) 797 .
52) L. Radermacher and H.E. Beske, Int. J. Mass Spectr· Ion Physics, 20 (1976) 333.
53) G. Slodzian, Rev. Phys. Appl. 3(4), (1968), 360.
54) H. Liebl, J. Vac. Sci. Technol., 12 (1975) 385.
55) W.O. Hofer and H. Liebl, Appl. Phys., 8 (1975) 359.

56) H.W. Werner and A.E. Morgan, J. Appl. Phys., 47 (1976) 1232.
    C.A. Andersen, H.J. Roden and C.F. Robinson, J. Appl. Phys.,
    40 (1969) 3419.
57) H.A. Storms, K.F. Brown and J.D. Stein, paper presented at the
    Japan-US joint seminar on "Quantitative SIMS", Honolulu,
    Hawaii, USA, Oct. 1975.
58) Unpublished work.
59) H.H. Brongersma, Rev. Sci. Instr., (1977) in press.
60) F. Schulz, K. Wittmaack and J. Maul, Radiat. Eff., 18 (1973) 211.
    J.A. McHugh, Workshop on SIMS and IMMA, NBS Spec. Publ. 427 (Eds
    K.F.J. Heinrich and D.E. Newbury) Gaithersburg, (1975) 179.
61) W.K. Hofker, H.W. Werner, D.P. Oosthoek and H.A.M. de Grefte,
    Proc. 3rd Intern. Conf. Ion Implantation in Semiconductors,
    Yorktown Heights, (1972), (Ed B.L. Crowder) New York,
    Plenum Press (1973) 133.
62) H. Liebl, paper given at the "Pittsburgh Conference", Cleveland,
    Ohio, March 1973.
    H. Liebl, Round table discussion on SIMS, 7th Intern. Mass
    Spectrom. Conf., Florence 1976.
63) R.F.K. Herzog, W.P. Poschenrieder and F.G. Satkiewicz, Proc.
    Intern. Conf. Ion Surface Interaction (Eds R. Behrisch and
    W. Heiland), Gordon & Breach, London (1972) 173.
64) H.W. Werner, A.E. Morgan, Paper given at the 7th Intern. Mass
    Spectrom. Conf., Florence 1976.
65) M. Bernheim, G. Blaise and G. Slodzian, Int. J. Mass Spectrom.
    Ion Phys., 10 (1972/73) 293.
66) D.K. Bakale, B.N. Colby and C.A. Evans, Jr., Anal. Chem., 47
    (1975) 1532.
    R. Hernandez, P. Lanusse, G. Slodzian and G. Vidal, Rech.
    Aerospatiale, 6 (1972) 313.
67) H.W. Werner and H.A.M. de Grefte, Surface Sci., 35 (1973) 458.
68) H.W. Werner and H.A.M. de Grefte, Rad. Eff., 18 (1973) 269.
69) H.W. Werner, paper presented at the 8th Colloquium on
    Metallurgical Analysis with Special Emphasis on Electron and
    Ion Probe Microanalysis, Vienna, Oct. 1976. Published in
    Mikrochimica Acta, Suppl. 7 (1977) 63.
70) N. Winograd, W. Baitinger, A. Shepard, R. Hewitt, G. Slusser,
    Paper Nr. 289, presented at the "Pittsburgh Conference"
    Cleveland, Ohio, March 1977.
71) A.E. Morgan and H.W. Werner, Surface Sci., 65 (1977) 687.
72) A. Benninghoven, E. Loebach; Surf. Sci., 39 (1973) 397.
73) Ya. M. Fogel, Intern. J. Mass Spectr. Ion Phys. 9 (1972) 109.
74) W.K. Hofker, Thesis, University Amsterdam, 1975.
75) H.W. Werner, Acta Electronica, 19 (1976) 53.
76) H.W. Werner, Vacuum, 24 (1974) 493.
77) See e.g. H.J. Mathieu, D.E. McClure and D. Landolt, Thin
    Solid Films, 38 (1976) 281.
78) S. Hofmann, Appl. Physics, 9 (1976) 59.
79) H. Doi, I. Kanomata and N. Sakudo, 7th Conf. Solid State Devices,
    Tokyo 1975; Suppl. Jap. J. Appl. Phys., 15 (1976) 71.

80) I.W. Drummond and J.P.V. Long, Nature, $\underline{215}$ (1967) 5277.
    H.J. Liebl, R.F.K. Herzog, J. Appl. Phys. $\underline{34}$ (1963) 2893.
81) J. Rouberol, J. Radioanal. Chem. $\underline{12}$ (1972) 59.
82) H.W. Werner, Vacuum, $\underline{22}$ (1972) 613.
83) J.P. Servais and V. Leroy, to be published.
84) G.H. Morrison and G. Slodzian, Anal. Chem., $\underline{47}$ (1975) 932A.
85) M. Prager, A. Wolf and K.H. Gaukler, Beitrage Elektr.
    Direktabbildg Oberfl. $\underline{7}$ (1974) 509.
86) F.G. Rüdenauer and H.W. Werner, unpublished.
87) H. Hickam and G. Sweeney, private communication 1977.
88) R. Buhl and A. Preisinger, Surf. Sci., $\underline{47}$ (1975) 344.
89) M.G. Dowsett, R. King and E.H. Parker, Surf. Sci. 1977, in press.
90) H.W. Werner, H.A.M. de Grefte, J. v.d. Berg, Radiation
    Effects, $\underline{18}$ (1973) 269.
91) H.W. Werner, H.A.M. de Grefte, J. v.d. Berg, Adv. Mass Spectr.
    $\underline{6}$ (1974) 673, Proc. Edinb. Conf., A.R. West Ed; Appl. Sci.
    Publ., Barking, Essex 1974.
92) D.V. McCaughan, R.A. Kushner and V.T. Murphy, Phys. Rev.
    Letters, $\underline{30}$ (1973) 614 and G. Thomas, private communication.
93) H.W. Werner and A.E. Morgan, J. Appl. Phys., $\underline{47}$ (1976) 1232.
94) A.J. Smith, D.J. Marshall et al. Vacuum, $\underline{14}$ (1964) 263.
95) H. Oechsner, Phys. Lett. $\underline{40A}$ (1972) 211.
96) E. Kay and J. Coburn, VI Int. Vac. Congr. Kyoto, 1974.
97) H. Oechsner and W. Gerhard, Surf. Sci., $\underline{44}$ (1974) 480.
98) J.W. Coburn, E. Taglauer and E. Kay, Japan J. Appl. Phys. Suppl.
    2, Pt 1, 1974.
99) G.E. Thomas and E.E. de Kluizenaar, Acta Electr., $\underline{18}$ (1975) 63.
100) C.W. White, D.L. Simms and N.H. Tolk, Science $\underline{177}$ (1972) 482.
101) H. Bach, Vacuum, $\underline{24}$ (1974) 469.
102) H.W. Werner and A.E. Morgan, Anal. Chem., $\underline{49}$ (1977) 927.
103) R.J. MacDonald and P.J. Martin, Surf. Sci., (1977), in press.
104) G. Blaise, Surf. Sci., $\underline{60}$ (1976) 65.
105) F. Hillenkamp and E. Unsöld, Appl. Phys., $\underline{8}$ (1975) 341.
106) G. Pittaway, Paper D12, 1ere Conf. Int. sur les source d'ions,
    Saclay 1969, INSTN, Saclay, France.
107) H. Heil, Zs.f.Physik, $\underline{120}$ (1942/43) 212.
108) M. von Ardenne, Tabellen der Elektronenphysik, VEB, Deutscher
    Verlag der Wissenschaften, Berlin 1956, Vol.I, p. 544.
109) H. Liebl and R.F.K. Herzog, J. Appl. Phys., $\underline{34}$ (1963) 2893.
110) H. Hintenberger and L. König in Adv. Mass. Spectr., Pergamon
    Press 1959, Ed. J.D. Waldron, p. 16.
111) H. Liebl, Int. J. Mass Spectr. Ion Physics, $\underline{22}$ (1976) 203.
    Cf. Also C.A. Evans Jr., Anal. Chem., $\underline{44}$ (1972) 67A.
    L. Bolduc and M. Baril, J. Appl. Phys., $\underline{43}$ (1972) 1655.
112) H.J. Roden and R.D. Fralick, Pittsb. conf. 1975, Paper No. 70.
    H. Liebl, J. Appl. Phys., $\underline{38}$ (1967) 5277.
113) L. Bolduc and M. Baril, J. Appl. Phys., $\underline{44}$ (1973) 657.
    H. Ewald and H. Liebl, Zs. Naturforsch. $\underline{A,10}$ (1955) 872.
114) M. Baril and P. Vallerand, Can. J. Phys., $\underline{52}$ (1974) 482.

115)R.Castaing and G. Slodzian, J. Microscopie (Paris)1(1962),395.

116)H. Ewald and H. Liebl, Zs. f. Naturforsch., A12 (1957) 28.

117)H. Wollnik, Nucl. Instr. Methods 59 (1968) 277.

118)T. Matsuo, H. Matsuda and H. Wollnik, Nucl. Instr. Meth.
     103(1972) 515, North—Holland Publ. Co.

119)Cf. H. Kienitz, Massenspektrometrie, Verlag Chemie, (1968)
     p. 74-98.

120)J. Vastel and J. Rouberol, unpublished.

121)W. Paul und U. von Zahn, Zs. f. Physik, 152 (1958) 143.

122)F.G. Rüdenauer, Vacuum, 22 (1972) 609.

123)P.H. Dawson and N.R. Whetten in : Advances in Electronics and
     Electron Physics, Acad. Press, N.York 1969, L. Marton Ed,
     p. 60.

124)P.H. Dawson and N.H. Whetten in : Dynamic Mass Spectrometry,
     Vol. 1, Heyden and Son, London 1970, p. 1.

125) U.v. Zahn, Diplomarbeit, Univ. of Bonn 1956.

126)W. Austin, A. Holm and J.H. Leck in Quadrupole Mass Spectrometry
     and its applications, P.H. Dawson Ed., Elsevier, Amsterdam 1976,
     p. 121.

127)K. Wittmaack, Rev. Sci. Instr. 47 (1976) 157, Int. J. Mass
     Spectr. Ion Phys. 11 (1973) 23.
     Z. Sroubek, Rev. Sci., Instr. 44 (1973) 1403.

128)M. Baril and P. Vallerand, Can. J. Phys., 48 (1970) 2487.

129)R. Schubert and J.C. Tracy, Rev. Sci. Instr., 44 (1973) 487.

130)H. Liebl, J. Appl. Phys., 38 (1967) 5277.
     A. Hurrle and G.Sixt, Appl. Phys., 8 (1975) 293.

131)W.L. Fite, Paper 204, Pittsburg Conference 1977. M.G. Dowsett,
     R.M. King and E.H.C. Parker, J. Phys. E: Sci. Instr.,8 (1975)704.

132)W.K. Huber, H. Selhofer and A. Benninghoven, J.Vac. Sci. Techn.
     9 (1972) 482.

133)I. Pelchowitch and J.J. Zaalberg van Zelst, Rev. Sci. Instr.
     23 (1952) 73.

134)H.W. Werner, H.A.M, de Grefte and J. v.d. Berg, Int. J. Mass
     Spectr. Ion Phys., 8 (1972) 459.

135)P. Richards and E.E. Hays, Rev. Sci. Instr., 21 (1950) 99.

136)N.R. Daly, Rev. Sci. Instr., 31 (1960) 264.

137)W. Schütze and F. Bernhard, Z. Phys. 145 (1956) 44.
     F. Bernhard and K.H. Krebs, Z. Phys., 161 (1961) 103.

138)H.H. Tuithof and A.J.H. Boerboom, Int. J. Mass Spectr. Ion
     Phys., 15 (1974) 105.

139)L. Radermacher and H.E. Beske, Int. J. Mass Spectr. Ion Phys.,
     20 (1976) 333.

140)H. Mai and H. Wagner, J. Sci. Instr., 44 (1967) 883.

141)M. Maurice, Paper given at ANRT-SIMS meeting, Paris 1977.

142)A. Savitzky and M. Golay, Anal. Chem.,36 (1964) 1627.

143)H.W. Werner, Paper to be presented at 7th Inter. Vac. Congress
     Vienna, 1977; C.A. Evans Jr., Anal. Chem., 47 (1975) 818A;
     A. Benninghoven, Appl. Phys., 1 (1973) 3; R.E. Honig, Thin
     Solid Films,31 (1975) 89.

144)H.H. Brongersma, private communication 1977.
145)C.T. Hovland, Appl. Phys. Lett. 30 (1977) 274.
146)T. Ishitani and R. Shimizu, Appl. Phys., 6 (1975) 241.
147)H.D. Hagstrum, Phys. Rev., 123 (1961) 758.
148)C.B.W. Kerkdijk, Thesis, Leiden, the Netherlands, 1975.
149)C.B.W. Kerkdijk, K.K. Schartner, R. Kelly and F.W. Saris,
     Nucl. Instr. Meth. 132 (1976) 427.
150)P. Sigmund, Phys. Rev. 184 (1969) 383.
151)H. Oechsner, Z. Physik, 261 (1973) 37.
152)H. Oechsner, Appl. Phys., 8 (1975) 185.
153.N. Laegreid and G.K. Wehner, Journ. Appl. Phys., 32 (1961) 365.
154)D. Rosenberg and G.K. Wehner, Journ. Appl. Phys., 33 (1962) 1842.
155)B.M. Gurmin, Yu. A. Ryzhov, J.J. Skarban, Bull. Acad. Sci.
     USSR, Phys. Ser., (USA) 33 (1969) 752.
156)H. Oechsner, Z. Physik, 238 (1970) 433.
157)A. Benninghoven and A. Mueller, Physics Letters, 40A (1972) 169;
     J.F. Hennequin, Journ. de Physique, 29 (1968) 957; H.W. Werner
     (see ref. 6); H.E. Beske, Z. Naturforsch. 22a (1967) 459.
158)H.A. Storms, K.F. Brown and J.D. Stein, Joint-US-Japanese Semi-
     nar, Hawaii 1975; C.A. Andersen and J.R. Hinthorne, Science
     175 (1972) 853.
159)H.A.M. de Grefte and H.W. Werner, unpublished.
160)H.W. Werner, Mikrochimica Acta, Suppl. 7 (1977) 63.
161)H.H. Tuithof, Thesis Univ. Amsterdam, 1977.
162)P.H. Dawson, 7th Int. Mass Spectr. Conf. Florence 1976.

AUTHOR INDEX

****

Satkiewich F.G.   See Herzog R.F.K.
Satoko C.212
Savitzky A., Golay M.   411
Sawatzky G.A. 256,258   See Antonides E.
Schaich W.L. 26,27;-,N.W. Ashcroft 2,3,94,98,131
Scheffler M. See Jacobi K.
Schiebner E.J. See Amelio G.F.
Schober O. See Christmann K.
Schotte K.D. 220
Schrieffer J.R. See Einstein T.E.
Schroeer J.M., Rhodin T.N., Bradley R.C.   333
Schubert R., Tracy J.C.   395
Schulz F., Wittmaack K., Maul J.   354
Schutze W., Bernhard F.   404
Schwentner N., Skibowski M., Steinmann W.   65
Schwarz J.A. See Jones R.H.
Schweitzer J.K. 215,219
Schwartz M.E. 198
Scmitz W. See Hillis H.
Scofield J.H. 194
Seah M.P. 263
Selhofer H.   See Huber W.K.
Sen S.K. 212
Servais J.P.,Leroy V.   362
Sevier K.D. 232
Sexton B. See Ibach H.
Shaw R.W. 198
Shay J., W.E. Spicer   82
Shay J.L.   83
Sheffield J.C.   See Mchugh J.A.
Shekhter S.S. 278,290
Shepard A.   See Winograd N.
Sheperd J.P.G. 196
Shevchik N.J. 202
Shigeishi R.A.And King D.A. 157
Shimizu R., Ishitani T., Ueshima Y.   334,336
Shimizu R.   See Ishitani T.
Shirley D.A. see Apai G.
Shirley D.A. see Banna M.S.
Shirley D.A.   55
Shirley D.A. 192,198,199,212,213,215,219
Shirley D.A. 232,235,237   See Kowalczyk S.P.
Shirley D.A.   See Pardee W.J.,See Williams R.S.
Siegbahn K. see Gelius U.
Siegbahn K. 55
Siegbahn K. 192,241,249   See Werme L.O.
Siekhaus W.J. See Jones R.H.
Sigmund P.   341,342,333

Werner H.W., Grefte H.A.M. De, Bers J.Van De   338,
   369,399,403,404
Werner H.W., Morgan A.E.   354,356,373,374
Werner H.W.   See Hofker W.K.
Werner H.W. See Rudenauer F.G.
Werner H.W. See Grefte H.A.M. De
Wernick J.H. 202,221
Wertheim G.K. 199,202,203,204,205,207,209,211,212,
   213,215,219,224
Wertheim G.K.   See Citrin P.H.
West K.W,207
Whetten N.H. See Dawson W.H.
White J.M. See Masushima T.
White J.M. See Close J.S.
White C.W., Simms D.L., Tolk N.H.   374,375
White C.W.,Tolk N.H. 291
Wille R.A. See Netzer F.P.
Williams D.R. ,204
Willis R.F. See Feuerbacher B.
Williams R.S.,Kowalczyk S.P.,Wehner P.S.,Apai G.
   ,Stohr J.
      Shirley D.A. 3
Wilson J.A. 204
Winograd N., Baitinger W., Shepard A., Hewitt R.
   , Slusser G.   357
Winograd N. 212
Winograd N. See Kim K.S.
Winsor H.V. See Wooten F.
Wittmaack K. 390,394
Wittmaack K.   See Schulz F.
Wohlleben D.K,207
Wolf A. See Prager M.
Wollnik H.   383
Wollnik H. See Matsuo T.
Woolsey I.J. ,215,219
Wooten F.,Huen T.,Winsor H.V. 12,26,28
Worthington C.R. & Tomlin S.G. 260,261
Wray L. See Gallon T.E.
Wright G.R. & Brion C.E. 105 & Brion C.E.
   , Van Der Wiel M.J. 105
Yafet Y. 214,223
Yang C.N. 103,104
Yates J. T., Madey T. E. And Erickson N. E. 148
Yellin E. See Yin L.
Yin L. & Tsang T., Adler I., Yellin E.
   250 & Tsang T., Adler I. 259
Yu K.Y.,Ling D.T.,Spicer W.E. 162
Zabolitskii E.I. ,205

Subject index

****

a-vector, alignment of -   76
absorption coefficient 107
absorption coefficient,differential optical -   59
absorption   drude 12,13;interband 12,13;rate 17
adsorbates 110
adsorption rate 179
adsorption isotherms 175
adsorption site 167
adsorption energies of H2 on W 163
adsorption energies of CO on Ni,Cu,Pd 163
adsorption, CO - on metals   72
adsorption, - of oxygen   65
advantage of wide energy range   64
AES 230   quantitative 260,264
As-Pd(XPS)   200,201,202
alloys(XPS) 198,199
angle-integreted cross section 104,105
angular dependence 5,12,18,19,32
antimonides(XPS)   207
ARPES   76
assymmetry parameter 104,107
Auger line shape analysis 147
Auger band structure 247,258
        current 260
        energy (atoms) 231,138,141   (solids) 225,
  238,241
        line 242,245,251
        process 231,237,243
        shifts 240
Auger techniques, comparison with -   72
Auger decay 30,45
Auger neutralization 274,276,278,281,289,291,294
  296,297,298,301
Auger deexcitation 276,278,281,287,291,294,295
  296,297,298
autoionization 298,301,300,333
backscattering 260,261
background pressure   353
bandstructure 125
beta-brass(XPS)   199,200
binding energy   198,212
                core 198
                valence 198
bloch wave 125,126,128

467